Soft and Biological Matter

"Soft and Biological Matter" is a series of authoritative books covering established and emergent areas in the realm of soft matter science, including biological systems spanning all relevant length scales from the molecular to the mesoscale. It aims to serve a broad interdisciplinary community of students and researchers in physics, chemistry, biophysics and materials science.

Pure research monographs in the series, as well as those of more pedagogical nature, will emphasize topics in fundamental physics, synthesis and design, characterization and new prospective applications of soft and biological matter systems. The series will encompass experimental, theoretical and computational approaches. Topics in the scope of this series include but are not limited to: polymers, biopolymers, polyelectrolytes, liquids, glasses, water, solutions, emulsions, foams, gels, ionic liquids, liquid crystals, colloids, granular matter, complex fluids, microfluidics, nanofluidics, membranes and interfaces, active matter, cell mechanics and biophysics.

Both authored and edited volumes will be considered.

More information about this series at http://www.springer.com/series/10783

Xiaogong Wang

Azo Polymers

Synthesis, Functions and Applications

Springer

Xiaogong Wang
Department of Chemical Engineering
Tsinghua University
Beijing, China

ISSN 2213-1736 ISSN 2213-1744 (electronic)
Soft and Biological Matter
ISBN 978-3-662-53422-9 ISBN 978-3-662-53424-3 (eBook)
DOI 10.1007/978-3-662-53424-3

Library of Congress Control Number: 2016955536

Printed on acid-free paper

This Springer imprint is published by Springer Nature
The registered company is Springer-Verlag GmbH Germany
The registered company address is: Heidelberger Platz 3, 14197 Berlin, Germany

Preface

This book is about azo polymers including their syntheses, properties, functions, and applications. Generally speaking, azo polymers are macromolecules that contain a significant amount of azo groups in their molecular structures, which should also include polymers containing aliphatic azo moieties. However, as their scope and importance are rather limited, the polymers containing aliphatic azo groups are not considered in this book. In this book, the term "azo polymer" is exclusively used to refer to the polymers containing aromatic azo chromophores, such as azobenzene derivatives in most cases. Various photoresponsive properties are the main characteristics of the azo polymers developed in past decades, and much more new azo polymers with such properties are emerging in recent years. Like any literature trying to record a rapidly developing area, this book faces the dilemma to balance between the introduction of basic concepts and the report of important progresses in this area. However, to answer the question what is more important is not trivial by any means. The selection of materials can only reflect our perspective as researchers interested in this fascinating area. To fulfill the object with abovementioned limitations, this book is organized as follows:

Chapter 1 is the introduction of this book, which discusses the background and terminology related to azo compounds and azo polymers. As azo groups play a critically important role in determining the functions of azo polymers, their structures and spectroscopic features are discussed first. Light absorption and electronic excitation of the aromatic azo cores are then discussed to understand the initial steps causing the photoresponsive variations of azo polymers. Photoisomerization of azo groups, which is the most important concept to understand the contents of this book, is briefly introduced here. Based on the understanding, azo functional structures are classified according to their spectral features and isomerization behavior. Finally, a brief retrospect of azo polymers is given to end this chapter.

In Chap. 2, following the brief discussion in Chap. 1, the *trans–cis* isomerization of azo chromophores is discussed in detail. A brief overview is given first to guide the reading. The discussion starts from the *trans–cis* isomers of azobenzene and its thermal isomerization. More detailed discussions are given to understand the excited states, intermediates, and pathways during the isomerization. Some pioneering models are discussed here as the background knowledge. Then, a thorough discussion of the photoisomerization mechanism of azobenzene is presented, starting from the famous rotation vs inversion controversy. More experimental results and theoretical models are discussed together with the understanding from computer simulation. Based on the knowledge of the azobenzene isomerization, the substituted effect is further discussed in order to understand the isomerization behavior of azobenzene derivatives and other azo chromophores, which are actually used in most azo polymers.

Chapter 3 is devoted to the azo polymer synthesis, which presents the different ways to obtain azo polymers. Like other functional polymers, azo polymers can be synthesized by polymerization and copolymerization of monomers that contain azo functional groups or by chemical modification of a suitable precursor polymer to introduce the azo functional groups after polymerization. The chain polymerization and step polymerization are discussed as the two typical ways to obtain azo homopolymers and random copolymers. Controlled radical polymerization is introduced as a new way to synthesize azo polymers with well-controlled molecular weight and distribution. Then, the ways to obtain azo block and graft copolymers, azo dendritic and star polymers, and azo hyperbranched polymers are discussed separately in detail. Finally, syntheses of azo polymers with π-conjugated backbone and azobenzene-containing polypeptides are presented as separate sections according to their unique characteristics.

In Chaps. 4, 5, and 6, three most widely investigated functions of azo polymers are discussed. Chapter 4 is devoted to the photoinduced orientation and anisotropy, which are intensively investigated for holographic recording and data storage based on the polarization gratings. The orientation induced by polarized light, structure–property relationship, mechanism and models, related materials, and possible applications in holography are presented in this chapter. In Chap. 5, another fascinating function of azo polymers, photoinduced mass transport, is discussed in detail. This photoinduced effect is unprecedented in polymers and other materials, which has been intensively studied to understand the mechanism and also for possible applications in many areas such as surface modification and patterning. This chapter discusses the original discovery and many subsequent investigations from materials to functions. In Chap. 6, one very interesting function of azo polymers is discussed, which is unique for the azobenzene-containing liquid crystal polymers with light cross-linking. These azo polymer networks, which are classified as azo liquid crystal elastomers, show many interesting properties and potential applications in actuators, shape memory, and others. In this chapter and also two chapters mentioned above, the discussion starts from the basic concepts and background as the introduction. Then, material design and preparation, structure–property relationship, and mechanism and models are sequentially discussed. Meanwhile, several selected cases are presented to show the characteristics and potential applications of the functions. Abundant references related to these rapidly developing areas are given in each chapter to facilitate further reading.

Beijing, China Xiaogong Wang

Acknowledgment

I am truly grateful to Prof. S. K. Tripathy and Prof. J. Kumar for triggering my interest in the fascinating field of azo polymers when the photoinduced surface-relief-grating formation was first observed by their eminent research group at the University of Massachusetts Lowell (UMASS Lowell). I express my gratitude to my colleagues and coworkers Dr. L. Li, Dr. J. I. Chen, Dr. S. Marturunkakul, Dr. S. Balasubramanian, Dr. K. G. Chittibabu, Dr. D. Y. Kim, Dr. K. Yang, and Dr. M. Sukwattanasinitt in UMASS Lowell for collaboration and academic advice during the period 1994–1997 and after that. I am indebted to Prof. P. Keller, Prof. T. Ikeda, and Prof. A. Natansohn for academic discussions. I offer my special thanks to all my colleagues and students at Tsinghua University for their creative thinking and consistent inspiration in the past two decades. Without their contributions to this area, it is almost impossible to complete this book in the current form. Finally, I am extremely grateful to the NSFC for sanctioning a series of research grants, among which a recent one is under project 51233002.

Contents

Chapter 1
Introduction

Abstract This chapter is the introduction of this book to discuss the background and terminology related to azo compounds and azo polymers. As azo groups play a critically important role in determining the functions of azo polymers, their structures and spectroscopic features are discussed first. Light absorption and electronic excitation of the aromatic azo cores are then discussed to understand the initial steps to cause the photoresponsive variations of azo polymers. Photoisomerization of azo groups, which is the most important concept for understanding the contents of this book, is briefly introduced here. Based on the understanding, azo functional structures are classified according to their spectral features and isomerization behavior. Finally, a brief retrospect of azo polymers is given to end this chapter.

Keywords Azo compound • Azo polymer • Light absorption • Electronic excitation • Photoisomerization • Spectral feature

1.1 Background and Terminology

In the past decades, many types of functional polymers have been synthesized for different purposes. Among them, one very unique category of functional polymers has attracted great attention for their interesting physical, chemical, and biological properties. This fascinating category with fast-growing species is characterized by –N=N– linkage as functional groups in the polymeric architecture. This functional structure is termed "azo group" as its common name and "diazene" by the systematic nomenclature. Generally speaking, this classification can be used for a variety of polymeric substances to involve organic, inorganic, and organometallic components. In this monograph, the name *azo polymer* is more specifically used to describe the organic polymers containing azo functional groups. Azo polymer is a nomenclature used to name polymers that contains azo groups as the functional groups to introduce a variety of functionalities [1, 2]. Azo polymer can be considered as macromolecules built up by the covalent linkage of a large numbers of much smaller units containing azo groups. There is a historical connection and a close correlation between azo polymers and their low-molecular-weight counterparts. Therefore, it is suitable to start the introductory remarks from *azo compounds* with low molecular weights.

X. Wang, *Azo Polymers*, Soft and Biological Matter,
DOI 10.1007/978-3-662-53424-3_1

Azo compounds are substances whose molecular structures contain two or more moieties linked by one or several azo bridging groups. The early synthesis of azo compounds can be traced back to the time when organic chemistry was in its infancy. Since P. Griess reported his discovery of diazotization and diazo compounds in 1858 [3], chemists have known a very simple and efficient way to prepare azo compounds. This classic reaction, known as azo coupling, can form an azo linkage by a reaction between an aromatic or a heteroaromatic diazo compound with a coupling component. A simple azo compound can be represented by a core molecular structure formed by two moieties bonded to the azo bridging group. Properties of azo compounds are not only dependent on the azo linkage but also controlled by the groups on both sides of the bridge [3, 4]. When they are aryl and alkyl groups, the substances are termed aromatic and aliphatic azo compounds, whose properties are completely different due to the structural difference.

Aromatic azo compounds have the core structures with two or more aromatic rings linked by one or several azo linkages. Many aromatic azo compounds are brightly colored owing to their ability to absorb visible light (400–700 nm). By proper molecular design, such as extending the conjugation, incorporating different aromatic rings, and introducing electron donors and acceptors, almost a full spectral range of colors can be obtained [5]. These colorful compounds are known as azo colorants and have been widely used as excellent dyestuffs and pigments, i.e., azo dyes and pigments. The term "chromophore" has been coined to name an atom or a group of atoms that serves as a unit in light absorption for these colorful organic substances [6]. Therefore, light absorption groups and structures containing azo groups are also referred to as azo chromophores.

Aliphatic azo compounds are generally referred to as substances in which at least one side of the azo group is bonded to an alkyl group. There exists significant distinction between the properties of aliphatic azo compounds and their aromatic counterparts. The aliphatic azo compounds are usually unstable without the aromatic conjugated effect. One well-known reaction of the aliphatic compounds is the thermolytic fission into nitrogen and free radicals [2, 3]. This type of compounds has been widely used as initiators for radical polymerizations of vinyl monomers, such as azobisisobutyronitrile (AIBN) as a typical example.

Since the middle of last century, azo polymers have emerged as new frontiers of scientific research and industrial practice. A huge number of azo polymers have been synthesized and intensively investigated for different purposes. People realized that almost no single type of functional groups in organic chemistry can exhibit such diversified functionalities as the azo group regarding the physical, chemical, and biological properties [2]. More importantly, the functionalities not only come from the functional groups but also originate from the macromolecular nature of polymers. Azo polymers are macromolecular substances that possess both azo functionalities and characteristic properties of extraordinary large molecules. Researchers in this area have developed many different ways to incorporate azo functional groups into various polymeric structures. This approach has created fascinating ways to combine the unique properties of the azo group with those of high-molecular-weight polymers. Nowadays, the azo polymer study is so widely ranged and opens up a field with great growth potential.

In the early stage of azo polymer development, the thermal dissociation property of aliphatic azo moieties was used to prepare polymeric azo initiators; the biodegradability of the azo groups was explored to prepare biodegradable polymers; and colorful nature of the structures was used to develop polymeric azo dyes and pigments [1, 2]. Since then, fundamental researches and application explorations of azo polymers have gone far beyond the scopes of this early stage. Recent research interest mainly focuses on polymers containing aromatic azo moieties. The term azo polymer is almost exclusively used to identify the polymers containing aromatic azo chromophores. These azo polymers show a variety of *photoresponsive* properties, which will be presented in this book in detail. In many cases, azo polymers are synonymous with photoresponsive polymers. Polymers containing aliphatic azo moieties are usually unstable, and one of the major applications is a macroinitiator used for preparing block and graft copolymers through radical polymerization. As the scope and importance are rather limited, polymers containing aliphatic azo moieties will not be considered in this book. Here the term "azo polymer" is exclusively used to refer to the polymers containing aromatic azo moieties.

The photoresponsive functions of azo polymers are directly related to the *photoisomerization* of azo groups. The azo group in aromatic azo compounds can exit in two configurations, the "*trans*" or "*E*" and the "*cis*" or "*Z*" forms [7, 8]. The conversion between the isomers is termed the *trans–cis* (*E–Z*) isomerization. The *trans–cis* isomerization can be triggered by the light irradiation and heat. Photoisomerization is a special type of photochemical reactions, which undergoes the configuration change through the excited states. The photoisomerization of azo functional groups in polymers can cause hierarchical structure and property variations from molecule to aggregated states. These photoresponsive properties and their mechanisms have been widely explored for scientific curiosity and also for real applications.

The investigations of azo polymers not only greatly extend the fundamental understanding but also lead to the development of diversified photoresponsive materials and prototype devices for future applications. In the past decades, wisely utilizing light has become the core of modern technologies and frontier of scientific explorations, which greatly motivates azo polymer researches. Azo polymers have been actively investigated to address some critically important issues in this rapidly developing area, which is often inspired by the creative sparks emerging there. The researches have accumulated a tremendous amount of knowledge and become a very active field for functional polymer study and new material developments. Azo polymers have become bright new members in this prestigious old family of molecules with –N=N– double-bonded linkage as their signified emblems. All these contents related to azo polymers will be discussed in this book in the following chapters.

1.2 Azo Functional Structure

Azo group (–N=N–) is critically important to identify azo polymers from other functional polymers. However, properties of azo polymers are not only dependent on the azo linkage itself but also controlled by the groups on both sides of the bridge. Therefore, the azo cores including both –N=N– and aromatic rings on both sides should be considered as the azo functional structures to define the features of azo polymers. Due to the light absorption nature, these rigid core structures are often referred to as azo chromophores. The aromatic azo cores usually contain two or more aromatic rings linked by azo groups (–N=N–) formed from sp^2-hybridized nitrogen atoms. The azo groups can be covalently bonded to benzene rings, naphthalene rings, aromatic heterocycles, and other aromatic rings. Depending on the number of the azo linkages, it can be referred to as mono-, dis-, tris-, and tetrakisazo structures, which are similar to the nomenclature of corresponding azo compounds. The aromatic or heteroaromatic rings can contain different substituents, which are often used to introduce some specific properties. By adjusting the numbers of azo linkages and the types of rings and substituents, it is possible to design and synthesize a vast number of aromatic azo core structures with different properties for various applications.

Figure 1.1 gives some typical azo functional structures that can be considered as azobenzene derivatives. For the azobenzene moieties and other aromatic azo cores, the nitrogen atoms of the azo groups and the adjoining carbon atoms are sp^2 hybridized. The sp^2-hybridized orbitals overlap to form σ-frameworks of the structures, and the unhybridized *p* orbitals of nitrogen and carbon atoms overlap to form π-bonding molecular orbitals. For the nitrogen atoms, there are nonbonding valence shell electrons (designated *n* electrons), which are localized mainly on their atomic nuclei. The aromatic azo core structures are usually quite stable owing to the resonance stabilization. The molecular structure parameters of the functional structures can be approximately estimated from the parameters of the corresponding azo compounds, which are obtained by X-ray diffraction and electron diffraction [9, 10]. For *trans*-azobenzene, the chemical bond length between the phenyl ring and azo group is 1.41 Å; the N=N bond length is 1.26 Å; the N–N–C bond angles are about 113°, as obtained from the gas electron diffraction [10].

One important characteristic of the azo functional structures is the ability to absorb light in specific wavelength ranges. Owing to the existence of the aromatic azo cores, azo polymers usually show strong light absorption in the ultraviolet (UV) and/or visible spectral range. The energy acquired from the absorbed light will trigger various photoresponsive variations in hierarchic levels from the molecular to aggregated states. In most cases, the energy of the IR light (wavelength = 0.7–500 μm) is not high enough to excite electrons, which is mainly related to the transition between molecular vibration states. On the other hand, the absorption of UV (wavelength = 200–400 nm) and visible light (wavelength = 400–700 nm) can cause the excitation of electrons from occupied low-energy orbital to previously unoccupied high-energy orbitals of the aromatic

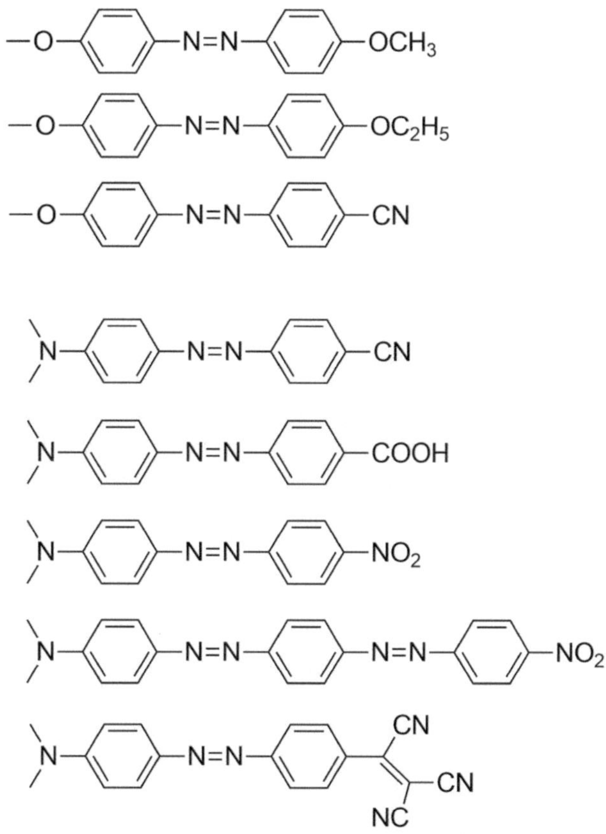

Fig. 1.1 Some examples of typical azo functional structures

cores. For azo polymers without other conjugated structures, the strong absorptions appearing in above wavelength range are mainly related to the electronic transitions of the aromatic azo core structures. The electronic excitation can cause energetic, structural, and dynamic variations owing to the light absorption. One of the most important variations at the molecular level is the *trans–cis* photoisomerization of the azo cores as discussed below in detail.

Besides the optical and photonic characteristics, there are other functions that can be attributed to the aromatic azo cores. Because of the π-conjugation between the azo groups and aromatic rings, the azo functional structures containing many π-electrons can show properties of large π-conjugated systems. Aromatic azo cores belong to the rigid molecular architecture possessing the anisotropic geometry, such as elongated or disklike shapes. The polymers containing these rigid rodlike and disklike structural units will form liquid crystal phases under proper external conditions such as temperature and solvent. The aromatic azo cores with flexible spacers and tails are often used as the building blocks to construct liquid crystal azo polymers.

1.3 Electronic Excitation and Spectroscopy

The aromatic azo core structures can absorb light very efficiently to cause electronic excitation, which is a key step to initiate the photoresponsive variations of azo polymers. The light absorption is directly related to the electronic transitions in a molecule upon light absorption. Electronic excitation spectroscopy can supply important information of the *electronic excitation* and excited states. Quantum mechanics provides the solid basis to understand the light absorption and electronic excitation and spectra [11, 12]. According to the molecular orbital theory, for a given nuclear geometry, molecular electronic configurations are generated by filling the molecular orbitals (MOs) with the available electrons. For most light absorption processes, only the lowest energy configuration (ground configuration) and the few low-energy excited configurations need to be considered. The excited electronic configurations are generated by adding electrons to the antibonding MOs. An electronic configuration considering spin correlation by the Pauli principle is termed a molecular electronic state, where a state with paired spins or unpaired spins is termed a singlet or triplet state (Fig. 1.2). A state energy diagram can generally be obtained by experiments and theoretical calculations to display the relative energies of the ground state and the excited states. A transition with a zero or nonzero probability of occurrence is termed forbidden or allowed transition.

The electronic absorption spectrum of an azo polymer usually exhibits several bands with different intensities and partially superimposed in visible and ultraviolet spectral range (Fig. 1.3). Without other chromophoric structures, the light absorption can be exclusively attributed to the electronic transitions between the ground and excited states of the aromatic azo cores in polymeric structures. According to the Einstein–Bohn relationship, the frequency and wavelength of the absorbed light are correlated with the energy difference between these states by the following equation:

$$\Delta E = h\nu = \frac{hc}{\lambda} \tag{1.1}$$

In simple molecular orbital terms, the transitions are treated as the change in the occupation pattern of a set of orbitals, which is supposed to be the same in the ground and excited states. Each band in the absorption spectrum can be attributed to one particular electronic transition (the one-electron excitation approximation).

For azo polymers, the σ-frameworks of the polymers are not directly involved in the light absorption process and need not to be considered at the initial stage. If the polymers do not contain other π-conjugated structures, the light absorption mainly causes the electron excitation of the azo chromophores, which usually show spectroscopic features similar to the corresponding aromatic azo compounds. The characteristic absorption bands include the transition of an n electron from the nonbonding orbital (n orbital) of nitrogen to the antibonding π^* orbital and the transition of a π electron from the bonding π orbital to the antibonding π^* orbital [14, 15]. The transitions are referred to as the $n \rightarrow \pi^*$ transition and $\pi \rightarrow \pi^*$ transition, and the corresponding excited states as $^1(n, \pi^*)$, $^1(\pi, \pi^*)$ for singlet

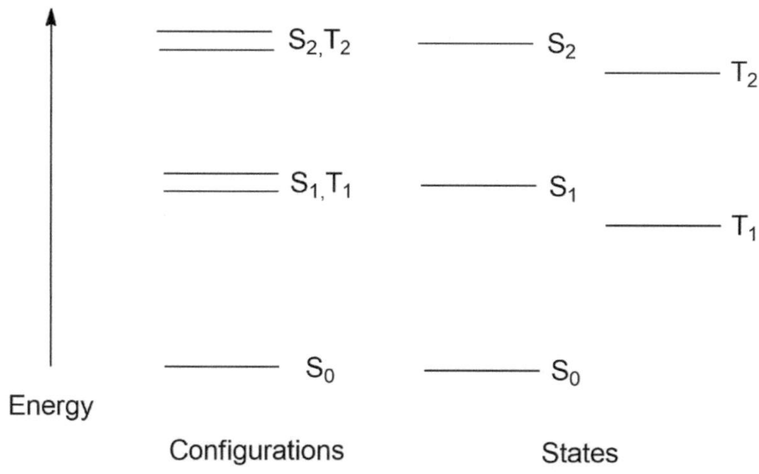

Fig. 1.2 A simple state diagram, S_0, S_1, and S_2; and T_1 and T_2 are singlet and triplet states

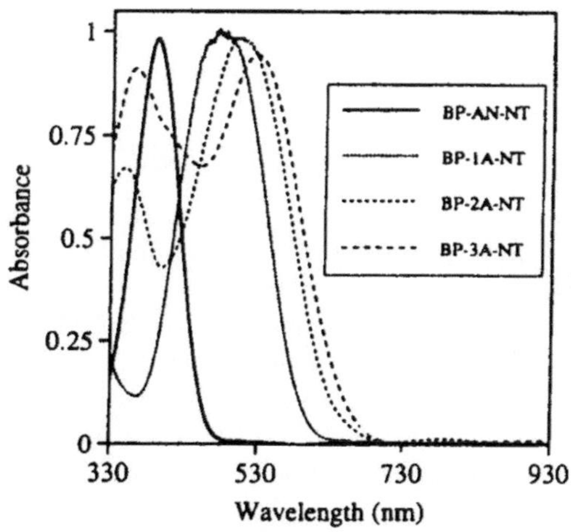

Fig. 1.3 UV–vis absorption spectra of azo polymers containing mono-azo chromophores (BP-1A-NT), dis-azo chromophores (BP-2A-NT), and tris-azo chromophores (BP-3A-NT) with amino and nitro as electron-donating/electron-accepting groups, compared with that for a polymer containing 4-nitroaniline moieties (BP-AN-NT). Structures of the azo polymers can be seen in Fig. 3.9 (Chap. 3) (Reprinted with permission from Ref. [13] Copyright 1997 American Chemical Society)

states and $^3(n, \pi^*)$, $^3(\pi, \pi^*)$ for triplet states. An aromatic azo core can have many π electrons and at least two electron lone pairs for each azo group. Theoretically, there are different transitions related to these n and π electrons. For the simplest aromatic azo core, azobenzene, there are 12 π electrons and 4 n electrons. Due to the different probability of occurrence and different absorption wavelengths, many of the absorption bands cannot be seen from the absorption spectra. For most cases, only absorption bands corresponding to the two lowest excited states can be seen, which are simply referred to as $S_1(n \rightarrow \pi^*)$ and $S_2(\pi \rightarrow \pi^*)$ transition bands.

Under normal circumstances, an absorption spectrum of azo polymers is characterized by the wavelength of the maximum absorbance of an absorption band (λ_{max}) and the absorption intensity of the band. The absorption intensity is characterized by the absorbance (A), extinction coefficient (ε), and oscillator strength (f). The empirical Beer–Lambert law is often used to correlate the absorbance with the extinction coefficient at a particular wavelength as

$$A = \log\left(\frac{I_0}{I}\right) = \varepsilon c l \tag{1.2}$$

where I_0 and I are the intensities of the incident monochromatic radiation and transmitted radiation, c is the concentration (or density), and l is the path length of the radiation through the sample. If c is in molarity units, l is in cm, and logarithm is to be base 10, ε is specified as the molar extinction coefficient. In some cases, the light-absorbing process can be characterized by the oscillator strength of an absorption band. The oscillator strength (f) is given by the equation:

$$f = 4.315 \times 10^{-9} \int \varepsilon dv \tag{1.3}$$

where v is the energy in wave number of the absorption band. It is a measure of the integrated intensity of absorption over a whole band, which is correlated with the probability of a given electronic transition and can be calculated quantum mechanically. However, due to the superposition of the absorption bands, it is not always feasible to experimentally obtain these values for azo polymers.

For quantitative quantum mechanics calculation, the absorption of light is regarded as a time-dependent perturbation process, and the Schrödinger equation has the following form:

$$(H_0 + H')\Psi(x,t) = E\Psi(x,t) \tag{1.4}$$

where H_0 is the static Hamiltonian operator, H' is the perturbation operator, and E is the energy of the system [12, 16]. The eigenfunction $\Psi(x, t)$, which is a function of space and time coordinates, can be expanded in terms of wave functions of the unperturbed system:

$$\Psi(x,t) = \sum a_k(t)\Psi_k \tag{1.5}$$

Therefore, the effect of the light absorption can be treated as a time-dependent mixing of the initial wave functions of the molecule with all the other possible wave functions. This probability is given by the square of the corresponding coefficient

$\alpha_k(t)$ in Equation (1.5). Both $[\alpha_k(t)]^2$ and the oscillator strength f are proportional to the square of the transition moment $< \Psi_i | \boldsymbol{\mu} | \Psi_f >$, where $\boldsymbol{\mu}$ is the dipole moment operator and Ψ_i and Ψ_f are the wave functions for the initial and final states [16]. Theoretically, the light absorption can be calculated by the above expressions. However, for a macromolecule such as azo polymer, it is hardly possible and also unnecessary to directly calculate these wave functions. As a reasonable simplification, a model compound with structure similar to the aromatic azo cores in the polymers can be used to carry out the calculation.

Even with sophisticated computer technology today, it is still a mathematically formidable task to solve the Schrödinger equation to obtain the exact molecular wave functions for a polyatomic system. The Schrödinger equation can be solved only by introducing a series of approximations. Born–Oppenheimer approximation factorizes the total wave function Ψ into electronic wave function φ, nuclear wave function χ, and spin wave function ξ [11]. The approximation is proposed on the grounds that the motions of electrons in orbitals are much more rapid than nuclear motions and the electron spin motions due to magnetic interaction are weakly interacted with the motion of electrons in orbitals and the motion of nuclei in space. On the basis of the Born–Oppenheimer approximation, the electronic configuration and nuclear geometry of a molecule can be treated separately. By using the Born–Oppenheimer approximation, a series of selection rules can be obtained to predict the transition probability from an initial state to a final state. The selection rules are used to establish the correlation between the observed spectroscopic features and the excited states. We refer the reader to the textbooks for these selection rules [such as 16, 17]. For more precise calculation, the zero-order approximation only considering the "pure" states needs to be improved by considering state mixing as the first-order or higher-order corrections. More about calculations of the electronic spectra of azobenzene and its derivatives can be seen in references [such as 18–20].

As the excited states have higher energy than the ground state, the excited states will revert to the ground state with emission of a photon. Electronic emission spectroscopy can provide valuable information concerning excited states. If both electronic absorption and emission spectra are obtained, a static state energy diagram can often be deduced [17]. As shown in Fig. 1.2, a state energy level diagram is usually used to represent the electronic ground and excited states. However, as the dissipation of the excess energy of the excited species can also undergo radiationless transitions and energy transfer quenching, the light emission cannot be observed for most cases.

For azo functional groups, the *trans–cis* isomerization is a main pathway for the dissipation of the excess energy of the excited states. Therefore, the radiative transitions (fluorescence and phosphorescence) are usually not observed for azo polymers by ordinary spectroscopic methods. This is similar to their low-molecular-weight counterparts. Most aromatic azo compounds do not emit at room temperature but can show fluorescence under specific conditions, such as in strong acids at 77 K. Aromatic azo compounds with *push–pull* substituents can show weak fluorescence in nonacidic solution at low temperature or in rigid matrices [4, 14, 15]. The low temperature and rigidity of the surrounding environment decrease the possibility to deactivate the emitting state through radiationless transition [21]. Owing to this reason, light emitting has been seen for azobenzene-containing block copolymer micelles [22]. In recent years, the femtosecond time-resolved fluorescence spectroscopy has been used to study the photoisomerization mechanism and dynamics of aromatic compounds, which will be discussed in Chap. 2. However, such a study on azo polymers has rarely been seen in literature yet.

1.4 Classification of Azo Functional Structures

Azo functional structures can be categorized according to their structures and properties. Classification according to their spectroscopic characteristics is a convenient and straightforward way. Due to their wide applications as dyestuffs and pigments, the electronic absorption spectra of aromatic azo compounds in ultraviolet and visible spectral range have been intensively investigated for many years. According to their spectroscopic features, aromatic azo compounds have been classified into three types: *azobenzene type*, *aminoazobenzene type*, and *pseudo-stilbene type* [14, 15]. In most cases, the aromatic azo cores in azo polymers show similar spectral features to the corresponding aromatic azo compounds and can be categorized by a similar way. Azo polymers containing different types of azo functional structures usually show distinctive photoresponsive properties. Figure 1.4 shows the chemical structure of three azo polymers containing these types of the azobenzene moieties and UV–vis spectra of the polymers. Their spectral features and corresponding electron transitions are discussed as follows.

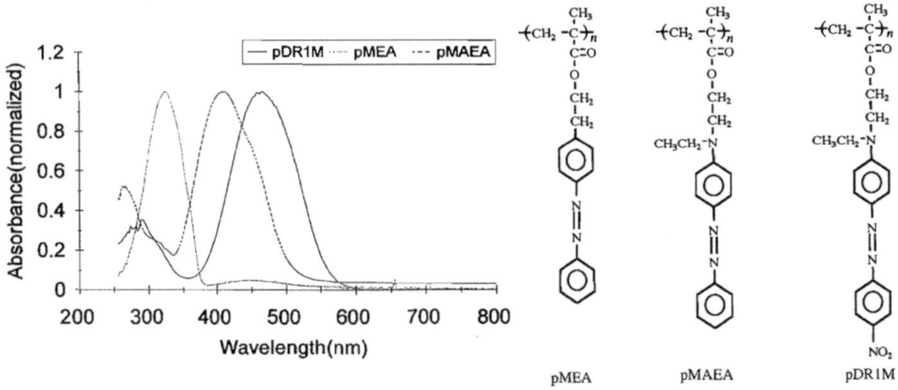

Fig. 1.4 Chemical structure and UV–vis spectra of pMEA, pMAEA, and pDR1M polymer films, which contain azobenzene, aminoazobenzene, and pseudo-stilbene types of functional groups (Reprinted with permission from Ref. [23]. Copyright 1995 NRC Research Press)

Azo functional structures are classified as the azobenzene type if the absorption spectra show a low-intensity $n \rightarrow \pi^*$ transition band in the visible region and a high-intensity $\pi \rightarrow \pi^*$ band in the UV region [14, 15]. It has been well recorded in the literature; typical azobenzene-type molecules with the *trans* configuration show three distinct absorption regions in the UV–vis spectra. The two absorption bands with high intensities ($\varepsilon = 10{,}000$–$20{,}000$) appear at 220–240 and 320–360 nm, and a weak band appears at 430–460 nm [24]. Historically, the bands are assigned as *E*-band, *K*-band, and *R*-band, which arise from the electronic excitations of the benzene rings along, π-conjugated system ($\pi \rightarrow \pi^*$) and azo linkage ($n \rightarrow \pi^*$) [6, 24]. The low intensity of the $n \rightarrow \pi^*$ transition is attributed to the forbiddance of the transition for the molecules with C_{2h} symmetry. The spectral variations of $\pi \rightarrow \pi^*$ and $n \rightarrow \pi^*$ transitions are often used to characterize the *trans–cis* isomerization occurring for the molecules. The excitation energy and absorption bands of azobenzene-type molecules have been calculated by semiempirical LCAO-SCF-CI [18] and ab initio methods [19, 20]. These spectroscopic features of azobenzene-type molecules can be used to classify similar azo functional structures.

Azo functional structures are assigned to the aminoazobenzene type if their spectra are identified by superimposed $n \rightarrow \pi^*$ and $\pi \rightarrow \pi^*$ bands, where the low-intensity $n \rightarrow \pi^*$ band is often buried beneath the intense $\pi \rightarrow \pi^*$ band [14, 15]. This spectral characteristic is normally related to the substituted effect on the azobenzenes. The amino and hydroxyl groups on the rings of the azobenzenes displace the $\pi \rightarrow \pi^*$ band considerably to longer wavelengths (bathochromic shift), but they only show slight influence on the $n \rightarrow \pi^*$ band position (in most cases to shorter wavelength) [6]. The bathochromic shift of the *K*-band can be observed for azobenzene derivatives with single *p*-substituent irrespective of its chemical nature and for many with two substituents [25]. However, only a strong electron-donating group, such as amino, can cause the superimposition of the $n \rightarrow \pi^*$ and $\pi \rightarrow \pi^*$ bands. The absorption intensity of the *K*-band is also affected by the substitution [24]. For azobenzene derivatives bearing the electron-donating groups with lone electron pairs, e.g., a dimethylamino group in the *p*-position, a significant redshift band ($\lambda_{max} = 410$ nm) and the high extinction coefficient ($\varepsilon = 28{,}300$) are observed for its benzene solution [26]. The excitation energy and oscillator strength of some substituted azobenzenes have been theoretically calculated by ab initio methods [19, 27, 28].

Azo functional structures are classified as pseudo-stilbene type if their spectra show a long wavelength $\pi \rightarrow \pi^*$ band in visible range [14, 15]. Owing to the intensive absorption in visible light ranges, the pseudo-stilbene-type azo compounds are typically used as dyestuffs. The spectra of such chromophoric structures can exhibit the reversed sequence of (n, π^*) and (π, π^*) states in the energy level diagrams. This reversal of energy levels is caused by the decrease of the energy of the $^1(\pi, \pi^*)$ state, which can be caused by *push–pull* substitution on azobenzenes. For example, when introducing electron-withdrawing-NO_2 group in the p'-position of dimethylaminoazobenzene, an obvious bathochromic shift ($\lambda_{max} = 445$ nm) with $\varepsilon = 33{,}500$ is observed for its solution in cyclohexane [29]. The $\pi \rightarrow \pi^*$ absorption band is further redshifted in the polar solvents [30]. Due to the strong overlap of the absorption bands in visible range, the UV–vis absorption spectra are poorly resolved, and the exact positions of the bands often cannot be determined. The UV–vis spectra given in Figs. 1.3 and 1.4 exhibit the characteristics of polymers containing pseudo-stilbene-type azo chromophores. Moreover, the charge-transfer band relative to the ground state also appears in the visible range as indicated by resonance Raman scattering [31–33]. The excited states in visible light range include the mixed charge-transfer and $n \rightarrow \pi^*$ excitations as well as excited $\pi \rightarrow \pi^*$ states. The interaction between them is sensitively influenced by the electron-donating/electron-withdrawing substituents [33]. The excited states and excitation energy of some *push–pull* disubstituted azobenzene derivatives have been theoretically studied by ab initio calculations [34]. The *push–pull* type of azo functional structures is widely used to prepare azo polymers, but their electronic structures are not fully elucidated yet.

1.5 Photoisomerization

Aromatic azo compounds can change their molecular structure from one stereochemical configuration to another when photoexcited or thermally triggered. Accordingly, the isomerization is classified into photoisomerization and thermal isomerization. The thermal isomerization without the light excitation is a process where a molecule changes its configuration in its ground state over energy barriers to reach a new configuration. A photoisomerization means a process involving the excited states through different pathways, which could include various radiative and radiationless transitions. The *trans–cis* photoisomerization is one of the most important properties of aromatic azo compounds, which has been widely explored to develop azo polymers with various photoresponsive functions. Many properties of azo polymers discussed in this book are directly related to the isomerization caused by light absorption and relaxation of the excited states.

Hartley first observed the *trans–cis* photoisomerization of azobenzene (Fig. 1.5), which has been intensively studied by many researchers after his pioneering work. Hartley reported that both forward and backward isomerization reactions of *azobenzene* were activated by light and the thermal reaction was very slow [7]. The *cis*-form in the crystalline state can be stably kept in the dark, but is slowly converted into *trans*-form in solution [8]. Instead of repeated fractional

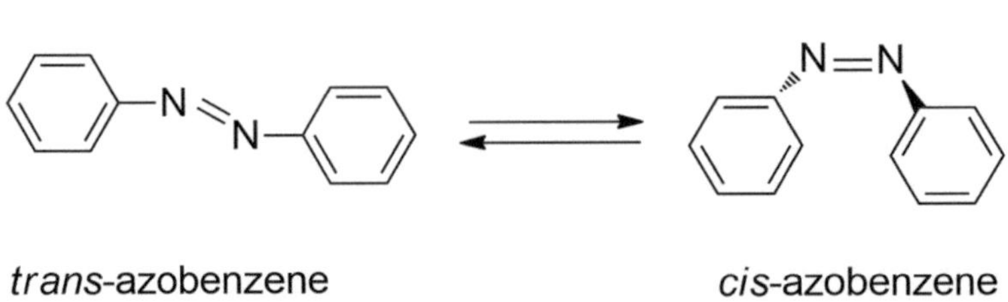

Fig. 1.5 *Trans–cis* isomerization of azobenzene

crystallizations used by Hartley, a relatively simpler isolation procedure was proposed, which used column chromatography to separate the *cis* isomer from the irradiated solutions [25, 35]. This method can be used for separation of isomers of azobenzene and other substituted azobenzenes. However, in many cases, the backward isomerization of substituted azobenzenes is too fast to permit isolation, and the identification is mainly relied on spectroscopic techniques. The *trans*- and *cis*-azobenzene show $n \rightarrow \pi^*$ transition bands at 440 nm ($\varepsilon \sim 500$) and 430 nm ($\varepsilon \sim 1500$) and $\pi \rightarrow \pi^*$ transition bands at 314 nm ($\varepsilon \sim 17{,}000$) and 280 nm ($\varepsilon \sim 5100$), respectively [36].

For aromatic azo compounds with a stable *cis*-form, the spectral variation corresponding to the *trans–cis* isomerization can be recorded with ordinary UV–vis spectroscopy. This method has also been widely used to investigate the isomerization of azo polymers. Figure 1.6 shows a typical UV–vis spectrum of an azo polymer after irradiation with UV light for different time periods. Specific spectroscopic techniques, such as flash spectroscopic technique, are required to detect the processes involving fast backward isomerization [37]. As discussed in Chap. 2, the ultrafast time-resolved spectroscopic techniques and theoretical calculations have supplied more powerful methods to study the *trans–cis* isomerization in recent years.

For azobenzene and many of its derivatives, the *trans–cis* isomerization is completely reversible, and fairly drastic tests fail to show the presence of any side reactions [39]. Possible side reactions, such as photocyclization and photoreduction, are rather inefficient processes with very low quantum yield compared to the isomerization [36]. Because of its extremely important role in the fundamental understanding and applications, the *trans–cis* photoisomerization has been thoroughly investigated to elucidate its mechanism. A very significant amount of research efforts has been devoted to this topic, which will be discussed in Chap. 2 in detail.

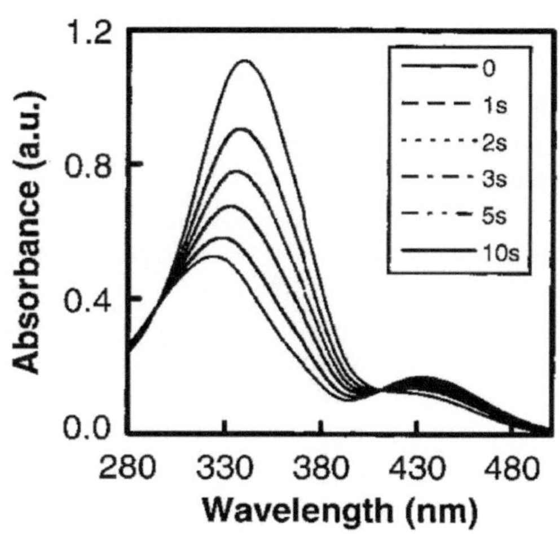

Fig. 1.6 Typical UV–vis spectrum of an azo polymer after irradiated with UV light for different time periods (Reprinted with permission from Ref. [38]. Copyright 2001 American Chemical Society)

1.6 Azo Polymers in Retrospect

Over the past decades, many azo polymers have been synthesized, investigated, and applied in diversified areas. According to the available literature, people started to introduce azo functional groups into polymers in the 1960s. Many types of azo polymers have been synthesized and investigated since then. One might not be able to figure out who was the first to open the door, from where the wonderful journey started, but some pioneering works and significant developments can be readily traced from the published literature. Some of the milestone developments and available references are briefly introduced below.

The syntheses of azo polymers can be traced back to the 1960s [40–42]. The term "photoresponsive" started to be used in some early reports [40]. Nuyken has reviewed these early stage researches on azo polymers [1]. The azo polymers are divided into labile main-chain, labile side-chain, stable main-chain, and stable side-chain categories accordingly. The stable categories are azo polymers containing aromatic azo moieties in main chain and side chain, which have been much more intensively studied in the following decades. One of the motivational factors at that stage was the use of azobenzene moieties as photochemical probes to understand the details of light–polymer interactions at microscopic and molecular levels [43]. Inspired by these early studies, researches on azo polymers as photoresponsive and functional polymers have attracted more and more attention. The major developments before the 1990s have been summarized by several review articles and one monograph [2, 44, 45].

The later surge in azo polymer researches has gone far beyond these pioneering studies, which was ignited by many exciting discoveries in different research areas. The researches have explored many previously unknown fields of azo polymers, such as liquid crystallinity (LC), nonlinear optical (NLO) properties, photorefractive properties, monolayer and multilayer assemblies, self-assembly in solutions, and many others. Several unique properties of azo polymers have been discovered, which are photoresponsive variations classified into phase transition, birefringence and reorientation, surface pattern formation, and deformations in various forms, among others. The current research area of azo polymers has been shaped by these important discoveries and developments. The principles and knowledge obtained from these research areas will be discussed in the following chapters in detail.

The progresses and developments in the past decades have been chronicled in a series of review articles. We refer the interested reader to a few of the references for some selected topics: the early developments related to liquid crystallinity, optical nonlinearity, Langmuir–Blodgett technique, and light-induced birefringence of azo polymers [46]; photochromic LC azo polymers and photoalignment modulation of LC systems [47–50]; photoinduced surface-relief structures and all-optical patterning of azo polymer films [51, 52]; photoinduced motions in azo polymers [53]; NLO azo polymers [54, 55]; photoisomerization-driven processes and optical storage in ultrathin films [56]; photoswitching from molecule conformation to biomaterials [57–59]; liquid crystalline elastomers and actuators [60–62]; photomechanical effects [63]; photoresponsive microphase-separated azo block copolymers [64–66]; light-responsive block copolymer micelles [22]; photoresponsive amphiphilic azo polymers [67], holography, micropatterning, and directional photofluidization lithography; and other applications [68–70].

References

1. Nuyken, O.: Azo polymers. In: Kroschwitz, J.I. (ed.) Encyclopedia of Polymer Science and Engineering, vol. 2, pp. 158–175. Wiley, New York (1985)
2. Kumar, G.S.: Azo Functional Polymers: Functional Group Approach in Macromolecular Design. Technomic Publishing Co. Inc., Lancaster (1992)
3. Zollinger, H.: Azo and Diazo Chemistry: Aliphatic and Aromatic Compounds. Interscience Publishers Inc., New York (1961)
4. Zollinger, H.: Color Chemistry: Syntheses, Properties and Applications of Organic Dyes and Pigments, 2nd edn. VCH publishers Inc., New York (1991)
5. Hunger, K., Mischke, P., Rieper, W., Raue, R.: Azo dyes. In: Gerhartz, W. (ed.) Ullmann's Encyclopedia of Industrial Chemistry, vol. A3, 5th edn, pp. 245–324. VCH, Weihheim (1985)
6. Burawoy, A.: Studies in the light absorption of organic compounds. Part VIII. azo-compounds. J. Chem. Soc. 1865–1869 (1937)
7. Hartley, G.S.: The *cis*-form of azobenzene. Nature **140**, 281–281 (1937)
8. Hartley, G.S.: The *cis*-form of azobenzene and the velocity of the thermal *cis* → *trans* conversion of azobenzene and some derivatives. J. Chem. Soc. 633–642 (1938)
9. Harada, J., Ogawa, K., Tomoda, S.: Molecular motion and conformational interconversion of azobenzenes in crystals as studied by X-ray diffraction. Acta. Cryst. B **53**, 662–672 (1997)
10. Tsuji, T., Takashima, H., Takeuchi, H., Egawa, T., Konaka, S.: Molecular structure and torsional potential of trans-azobenzene. A gas electron diffraction study. J. Phys. Chem. A **105**, 9347–9353 (2001)
11. Levine, I.N.: Quantum Chemistry, 3rd edn. Allyn and Bacon Inc., Boston (1983)
12. Hameka, H.F.: Advanced Quantum Chemistry: Theory of Interactions Between Molecules and Electromagnetic Fields. Addison-Wesley Publishing Company, Massachusetts (1965)
13. Wang, X.G., Kumar, J., Tripathy, S.K., Li, L., Chen, J.I., Marturunkakul, S.: Epoxy-based nonlinear optical polymers from post azo coupling reaction. Macromolecules **30**, 219–225 (1997)
14. Rau, H.: Photoisomerization of azobenzenes. In: Rabek, J.F. (ed.) Photochemistry and Photophysics, vol. 2, pp. 119–141. CRC Press, Boca Raton (1990)
15. Rau, H.: Azo compounds. In: Dürr, H., Bouas-Laurent, H. (eds.) Photochromism: Molecules and Systems, pp. 165–192. Elsevier, Amsterdam (1990)
16. Barltrop, J.A., Coyle, J.D.: Principles of Photochemistry. John Wiley & Sons Ltd, New York (1978)
17. Turro, N.J.: Modern Molecular Photochemistry. The Benjamin/Cummings Publishing Co. Inc., California (1978)
18. Beveridge, D.L., Jaffé, H.H.: The electronic structure and spectra of *cis*- and *trans*-azobenzene. J. Am. Chem. Soc. **88**, 1948–1953 (1966)
19. Åstrand, P.O., Ramanujam, P.S., Hvilsted, S., Bak, K.L., Sauer, S.P.A.: *Ab initio* calculation of the electronic spectrum of azobenzene dyes and its impact on the design of optical data storage materials. J. Am. Chem. Soc. **122**, 3483–3487 (2000)
20. Fliegl, H., Köhn, A., Hättig, C., Ahlrichs, R.: Ab initio calculation of the vibrational and electronic spectra of *trans*- and *cis*-azobenzene. J. Am. Chem. Soc. **125**, 9821–9827 (2003)

21. Shimomura, M., Kunitake, T.: Fluorescence and photoisomerization of azobenzene-containing bilayer membranes. J. Am. Chem. Soc. **109**, 5175–5183 (1987)
22. Zhao, Y.: Light-responsive block copolymer micelles. Macromolecules **45**, 3647–3657 (2012)
23. Ho, M.S., Natansohn, A., Barrett, C., Rochon, P.: Azo polymers for reversible optical storage. 8. The effect of polarity of the azobenzene groups. Can. J. Chem. **73**, 1773–1778 (1995)
24. Gore, P.H., Wheeler, O.H.: Absorption spectra of aromatic azo and related compounds. III. Substituted azobenzenes. J. Org. Chem. **26**, 3295–3298 (1961)
25. Cook, A.H., Jones, D.G.: *cis*-Azo-compounds. J. Chem. Soc. 1309–1315 (1939)
26. Brode, W.R., Gould, J.H., Wyman, G.M.: The relation between the absorption spectra and the chemical constitution of dyes. XXVI. Effect of solvent and of temperature on the cis-trans isomerization of azo dyes. J. Am. Chem. Soc. **75**, 1856–1859 (1953)
27. Chen, P.C., Chieh, Y.C., Wu, J.C.: Theoretical study of the electronic spectra of azobenzene dyes. J. Mol. Struct.: THEOCHEM **715**, 183–189 (2005)
28. Briquet, L., Vercauteren, D.P., Perpète, E.A., Jacquemin, D.: Is solvent trans-azobenzene twisted of planer. Chem. Phys. Lett. **417**, 190–195 (2006)
29. Nishimura, N., Sueyoshi, T., Yamanaka, H., Imai, E., Yamamoto, S., Hasegawa, S.: Thermal cis-to-trans isomerization of substituted azobenzenes II. substituent and solvent effects. Bull. Chem. Soc. Jpn. **49**, 1381–1387 (1976)
30. Gabor, G., Fischer, E.: Spectra and *cis-trans* isomerization in highly bipolar derivatives of azobenzene. J. Phys. Chem. **75**, 581–583 (1971)
31. Cataliotti, R.S., Murgia, S.M., Paliani, G., Poletti, A., Zgierski, M.Z.: Resonances Raman scattering in Dyes derived from azobenzene. J. Raman Spectrosc. **16**, 251–257 (1985)
32. Rava, R.P.: Deconvoluting the visible absorption spectrum of methyl orange using inverse Raman transform techniques. J. Chem. Phys. **87**, 3758–3765 (1987)
33. Cataliotti, R.S., Morresi, A., Paliani, G., Zgierski, M.Z.: Resonances Raman scattering in azo dyes. J. Raman Spectrosc. **20**, 601–604 (1989)
34. Crecca, C.R., Roitberg, A.E.: Theoretical study of the isomerization mechanism of azobenzene and disubstituted azobenzene derivatives. J. Phys. Chem. A **110**, 8188–8203 (2006)
35. Cook, A.H.: The preparation of some *cis*-azo-compounds. J. Chem. Soc. 876–881 (1938)
36. Griffiths, J.: Photochemistry of azobenzene and its derivatives. Chem. Soc. Rev. **1**, 481–493 (1972)
37. Wildes, P.D., Pacifici, J.G., Irick Jr., G., Whitten, D.G.: Solvent and substituent effects on the thermal isomerization of substituted azobenzenes. A flash spectroscopic study. J. Am. Chem. Soc. **93**, 2004–2008 (1971)
38. Wu, L.F., Tuo, X.L., Cheng, H., Chen, Z., Wang, X.G.: Synthesis, photoresponsive behavior, and self-assembly of poly(acrylic acid)-based azo polyelectrolytes. Macromolecules **34**, 8005–8013 (2001)
39. Birnbaum, P.P., Style, D.W.G.: The photo-isomerization of some azobenzene derivatives. Trans. Faraday Soc. **50**, 1192–1196 (1954)
40. Lovrien, R., Waddington, J.C.B.: Photoresponsive systems. I. Photochromic macromolecules. J. Am. Chem. Soc. **86**, 2315–2322 (1964)
41. Ravve, A., Fitko, C.: Polymer formation through diazonium coupling. J. Polym. Sci. A **2**, 1925–1940 (1964)
42. Kamogawa, H., Kato, M., Sugiyama, H.: Syntheses and properties of photochromic polymers of the azobenzene and thiazine series. J. Polym. Sci. A **6**, 2967–2991 (1968)
43. Williams, J.L.R., Daly, R.C.: Photochemical probes in polymers. Prog. Polym. Sci. **5**, 61–93 (1977)
44. Kumar, G.S., Neckers, D.C.: Photochemistry of azobenzene-containing polymers. Chem. Rev. **89**, 1915–1925 (1989)
45. Irie, M.: Properties and applications of photoresponsive polymers. Pure Appl. Chem. **62**, 1495–1502 (1990)
46. Xie, S., Natansohn, A., Rochon, P.: Recent developments in aromatic azo polymers research. Chem. Mater. **5**, 403–411 (1993)

47. Ichimura, K.: Photoalignment of liquid-crystal systems. Chem. Rev. **100**, 1847–1873 (2000)
48. Ikeda, T.: Photomodulation of liquid crystal orientations for photonic applications. J. Mater. Chem. **13**, 2037–2057 (2003)
49. Shibaev, V., Bobrovsky, A., Boiko, N.: Photoactive liquid crystal polymer systems with light-controlled structure and optical properties. Prog. Polym. Sci. **28**, 729–836 (2003)
50. Yu, Y.L., Ikeda, T.: Alignment modulation of azobenzene-containing liquid crystal systems by photochemical reactions. J. Photochem. Photobio. C **5**, 247–265 (2004)
51. Viswanathan, N.K., Kim, D.Y., Bian, S.P., Williams, J., Liu, W., Li, L., Samuelson, L., Kumar, J., Tripathy, S.K.: Surface relief structures on azo polymer films. J. Mater. Chem. **9**, 1941–1955 (1999)
52. Yager, K.G., Barrett, C.J.: All-optical patterning of azo polymer films. Curr. Opin. Solid State Mater. Sci. **5**, 487–494 (2001)
53. Natansohn, A., Rochon, P.: Photoinduced motions in azo-containing polymers. Chem. Rev. **102**, 4139–4175 (2002)
54. Delaire, J.A., Nakatani, K.: Linear and nonlinear optical properties of photochromic molecules and materials. Chem. Rev. **100**, 1817–1845 (2000)
55. Yesodha, S.K., Pillai, C.K.S., Tsutsumi, N.: Stable polymeric materials for nonlinear optics: a review based on azobenzene systems. Prog. Polym. Sci. **29**, 45–74 (2004)
56. Oliveira Jr., O.N., dos Santos Jr., D.S., Balogh, D.T., Zucolotto, V., Mendonca, C.R.: Optical storage and surface-relief gratings in azobenzene-containing nanostructured films. Adv. Colloid Interface Sci. **116**, 179–192 (2005)
57. Renner, C., Moroder, L.: Azobenzene as conformational switch in model peptides. ChemBioChem **7**, 868–878 (2005)
58. Yager, K.G., Barrett, C.J.: Novel photo-switching using azobenzene functional materials. J. Photochem. Photobio. A **182**, 250–261 (2006)
59. Ercole, F., Davis, T.P., Evans, R.A.: Photo-responsive systems and biomaterials: photochromic polymers, light-triggered self-assembly, surface modification, fluorescence modulation and beyond. Polym. Chem. **1**, 37–54 (2010)
60. Yu, Y.L., Ikeda, T.: Photodeformable polymers: a new kind of promising smart material for micro- and nano-applications. Macromol. Chem. Phys. **206**, 1705–1708 (2005)
61. Jiang, H.Y., Kelch, S., Lendlein, A.: Polymers move in response to light. Adv. Mater. **18**, 1471–1475 (2006)
62. Yu, H.F., Ikeda, T.: Photocontrollable liquid-crystalline actuators. Adv. Mater. **23**, 2149–2180 (2011)
63. Barrett, C.J., Mamiya, J.I., Yager, K.G., Ikeda, T.: Photo-mechanical effects in azobenzene-containing soft materials. Soft Matter. **3**, 1249–1261 (2007)
64. Seki, T.: Photoresponsive self-assembly motions in polymer thin films. Curr. Opin. Solid State Mater. Sci. **10**, 241–248 (2006)
65. Seki, T.: Smart photoresponsive polymer systems organized in two dimensions. Bull. Chem. Soc. Jpn. **80**, 2084–2109 (2007)
66. Seki, T., Nagano, S.: Light-directed dynamic structure formation and alignment in photoresponsive thin films. Chem. Lett. **37**, 484–489 (2008)
67. Wang, D.R., Wang, X.G.: Amphiphilic azo polymers: molecular engineering, self-assembly and photoresponsive properties. Prog. Polym. Sci. **38**, 271–301 (2013)
68. Shishido, A.: Rewritable holograms based on azobenzene-containing liquid-crystalline polymers. Polym. J. **42**, 525–533 (2010)
69. Lee, S.W., Kang, H.S., Park, J.K.: Directional photofluidization lithography: micro/nanostructural evolution by photofluidic motions of azobenzene materials. Adv. Mater. **24**, 2069–2103 (2012)
70. Priimage, A., Shevchenko, A.: Azopolymer-based micro- and nanopatterning for photonic applications. J. Polym. Sci. B. **52**, 163–182 (2014)

Chapter 2
Trans–Cis Isomerization

Abstract Following the brief discussion in Chap. 1, the *trans–cis* isomerization of azo chromophores is discussed in detail in this chapter. A brief overview is given first to guide the reading, and the discussion starts from the *trans/cis* isomers of azobenzene and its thermal isomerization. More detailed discussions are given to understand the excited states, intermediates, and pathways during the isomerization. Some pioneering models are discussed here as the background knowledge. Then, a thorough discussion of the photoisomerization mechanism of azobenzene is presented, starting from the famous rotation vs inversion controversy. More experimental results and theoretical models are discussed together with the understanding from computer simulation. Based on the knowledge of the azobenzene isomerization, the substituted effect is further discussed in order to understand the isomerization behavior of azobenzene derivatives and other azo chromophores, which are actually used as the functional groups in most azo polymers.

Keywords *Trans–cis* isomerization • Photoisomerization • Excited state • Intermediate • Pathway • Rotation and inversion • Substituted effect

In the past decades, *trans–cis* isomerization of aromatic azo compounds has been intensively investigated because it is of great importance for fundamental understanding and practical applications. The knowledge obtained from these studies has provided valuable insights into the photoresponsive properties of azo polymers. This chapter focuses on the discussion of the *trans–cis* isomerization of azobenzene and its derivatives. The understanding will be used to rationalize the photoresponsive behavior of azo polymers discussed in the following chapters. Instead of a comprehensive survey of aromatic azo compounds, azobenzene and its derivatives are selected as representative materials for the deep discussion. It is not only because they have been most intensively studied as the typical aromatic azo compounds, but also because they are the versatile functional groups used in almost every type of azo polymers. The *trans−cis* isomerization is one of the typical photochemical processes, which obeys some general rules of this discipline. Some background knowledge about photochemistry will be helpful to comprehend the discussions of the isomerization behavior, which can be seen in the previous textbooks [such as 1, 2].

X. Wang, *Azo Polymers*, Soft and Biological Matter,
DOI 10.1007/978-3-662-53424-3_2

2.1 An Overview

The *trans–cis* isomerization of azobenzene and its derivatives is a process for molecules to change stereochemical configurations (*trans* and *cis* isomers) in different electronic states. The thermal isomerization without the light excitation is a process occurring in the ground state overcoming the energy barriers. On the other hand, the photoisomerization involves the electronic excited states in the pathways of the configuration changes. Most of the photoisomerization processes only involve the ground state S_0 and few lowest singlet and triplet excited states such as S_1, S_2, T_1, and T_2. Generally speaking, the processes can include different radiative and radiationless transitions between the states, such as absorption, emission, internal conversion (transitions between states of the same spin), and intersystem crossing (transitions between excited states of different spin). A diagram used to depict the states and their interconversion is known as a Jablonski diagram (Fig. 2.1). Photoisomerization is a special process to dissipate the excess energy of the excited species by changing the nuclear configurations, which are usually accompanied by other processes such as radiative transitions, radiationless transitions, vibration relaxation, and energy transfer quenching. The transition from one isomer to the other can go through different pathways and intermediate states. The study of the *trans–cis* isomerization is to understand the nuclear geometry, state energy, pathway, and dynamics of the time-dependent evolution of the isomers and excited species.

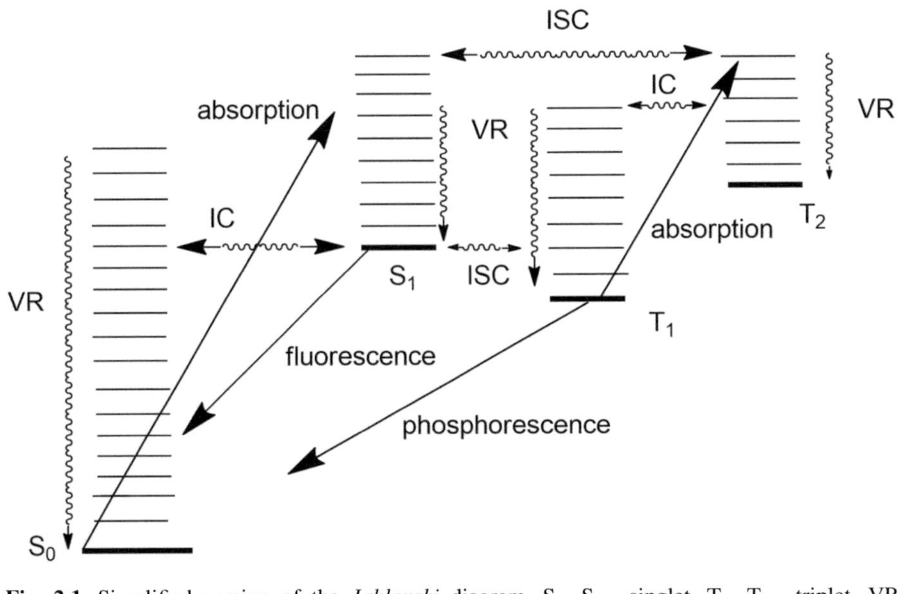

Fig. 2.1 Simplified version of the *Jablonski* diagram. S_0, S_1…singlet, T_1, T_2…triplet, VR vibrational relaxation, IC internal conversion, ISC intersystem crossing

Spectroscopic analyses can supply critically important pieces of information to understand the processes. Through solving the Schrödinger equation and other approaches, theoretical calculations can provide quantitative data of the electronic potential energy for the nuclear configurations in the electronic ground and exited states. Based on the results, *potential energy surfaces* (*PESs*) can be created to map the potential energy versus the nuclear configurations for the states. In most cases, *a potential energy curve* (*PEC*) is drawn to represent the PES as it is more readily visualized. Each point on a potential energy curve represents a specific nuclear geometry (horizontal axis) with a specific energy (vertical axis). PESs can supply a unified way to consider the nuclear geometry, state energy, pathway, and dynamics of the *trans–cis* isomerization. PESs calculated on the basis of the Born–Oppenheimer approximation can be reliably used for many cases. However, the approximation is invalid near degeneracies where potential energy surface cross. In this case, a small change in nuclear geometry will cause a large change in the electronic wave function [1].

In a PES representation, *trans* and *cis* isomers are shown as some minimum energy points on the ground energy surface separated by energy barriers. The *trans–cis* isomerization through intermediate structures is shown by a representative point moving on the PESs. The isomerization processes involve the PESs of the ground state S_0 and few singlet and triplet excited states such as S_1, S_2, T_1, and T_2. The thermal isomerization without the light excitation is visualized as a process where the representative point moves on the ground surface over energy barriers to reach a new minimum. A photoisomerization means a process involving the excited surfaces through different pathways, where the radiative and radiationless transitions are also shown by the representative point.

According to the Franck–Condon principle, the electronic transitions occur in an exceedingly short interval of time so that the initial and final nuclear geometries must be very similar [1]. Therefore, the light absorption and emission are usually visualized by "vertical" jumps between the energy surfaces for a representative point. After the electronic transitions, the structure change corresponds to a process where the representative point moves away from the Franck–Condon region through a specific pathway or different pathways on the PESs to finally reach a new minimum on the ground surface. A photochemical process can be classified into adiabatic and nonadiabatic, depending on whether the process occurs completely on an excited surface or involves crossing of different electronic states [2]. For a nonadiabatic process, the Born–Oppenheimer approximation (adiabatic approximation) is no longer valid. Several semiclassical approaches to deal with its dynamics have been developed, such as the Ehrenfest method and the surface hopping approach [3].

For years, various experimental and theoretical methods have been used to study the *trans–cis* isomerization of azobenzene and its derivatives. The stable structures, excited states, and details of different radiative and radiationless transitions, such as absorption, emission, internal conversion, and intersystem crossing, have been gradually unfolded through persevering investigations. However, due to the complexity of the issue, some mysteries related to the microscopic mechanism of the photoisomerization have not been fully disentangled until the current stage. The nuclear geometry, excited states, isomerization pathways, and dynamics are among the most important issues related to the *trans–cis* isomerization and will be discussed below.

2.2 *Trans-/Cis*-azobenzene and Thermal Isomerization

Azobenzene is the simplest aromatic azo compound and a parent structure of many derivatives. Due to its cornerstone role, it is of particular interest to understand its stable structures and isomerization. For a long time, *trans*-azobenzene was known as the only form of this compound. Hartley first reported the existence of the second form of azobenzene to distinguish it from the normal form (*trans*-form). He found that this unstable form, existing in solutions after the light irradiation, could be separated by repeated solvent extraction and crystallization [4]. By analyzing its properties, he concluded that this new form with the polar moment of 3.0 Debye was the *cis* isomer of azobenzene. Since then, the isomeric structures and isomerization of azobenzene have been intensively investigated.

The *trans* isomer of azobenzene corresponds to the energy minimum of the ground state, and the *cis* isomer with higher energy will go back to *trans*-form through thermal isomerization. In the crystalline state, the *cis*-form can be kept indefinitely in the dark, but is converted into the *trans*-form in solutions [5]. An equilibrium mixture containing 15–40 percentage of the *cis* isomer can be quickly obtained by photoisomerization [6]. The exact position of the photochemical equilibrium is influenced by the temperature and the nature of the solvent. The stable structure of *trans*-azobenzene (*t*-AB) is controlled by two competing factors, i.e., the conjugative stability and stereo repulsive interaction [7]. The π-conjugation between the azo and phenyl groups favors the planar conformation, while the repulsion between the hydrogen atoms and the lone pair electrons in the nitrogen atoms tend to distort it. A variety of experimental and theoretical studies has been carried out to determine the exact structures of *t*-AB in the crystals, gas phase, and solutions.

Indicated by the X-ray diffraction studies, there are two crystallographically independent sites in the crystal of *trans*-azobenzene (*t*-AB) [8, 9]. The molecule at site 1 (molecule 1) is planar and the molecule at site 2 (molecule 2) shows a rotation of the ring from the planar position around the C–N link of about 17° at room temperature [8]. Although both molecules have the C_i symmetry, the N=N bond length [1.189 Å] and N–Ph bond length [1.473 Å] of the molecule 2 are respectively shorter and longer than the corresponding lengths of molecule 1 [1.249 and 1.431 Å] [9]. The presence of dynamic disorder at the site 2 is attributed to the torsional vibration of the N–Ph bonds.

It is more difficult to reach unanimous agreement about the exact structure of *trans*-azobenzene in gas phase and solution. The molecular structure of *t*-AB in gas phase has been investigated by gas electron diffraction [7, 10]. The early study showed that the dihedral angle between each phenyl ring and the CNNC plane is about 30° [10]. However, it could not be determined whether the two phenyl rings rotate in the same direction (C_i symmetry) or in the opposite direction (C_2 symmetry). Later, the gas electron diffraction study was carried out by analyzing the data with the structural constraints obtained from the second-order Møller–Plesset (MP2/6-31 + G*) calculations [7]. This later study shows that the stable conformation is a planar structure with C_{2h} symmetry. Disagreement between the C_i symmetry and C_{2h} symmetry also existed for *t*-AB in solutions [11]. Vibrational spectroscopy has been used to characterize the structure with assistance of the normal mode calculation at different theoretical levels [12–14]. The vibrational analysis and ab initio calculations at MP2 level indicate that the optimized structure is planar with C_{2h} symmetry [11].

Most of the theoretical studies and ab initio calculations suggest that the planar structure with C_{2h} symmetry is the stable structure of the *trans*-azobenzene [14–16]. On the other hand, the theoretical study also shows that the calculations using different methods and basis sets will result in different stable conformations [13]. The Hartree–Fock and density functional theory (DFT) calculations lead to the almost planar conformation. Nevertheless, MP2 calculation utilizing the diffuse basis set (6−31 + G*) demonstrates that the two equivalent nonplanar structures correspond to the minimum energy, where the phenyl rings are rotated by 18.5° from the N=N–C plane. In this case, the planar conformation is a transition state connecting the nonplanar conformations. However, as the calculated potential energy barrier of the transition state is only 0.35 kcal/mol, *t*-AB may appear to possess the planar structure if the thermal average is considered [17]. The density functional theory (DFT) in the local-density approximation (LDA) calculation also shows that the potential energy in the vicinity of the minima is weakly dependent on the dihedral angles [18]. Most of above calculations correspond to the case of azobenzene in vacuum. Briquet et al. considered the role of solvent in MP2 and DFT calculations; it also showed that in solutions, energy difference between the planar and twisted *t*-AB is negligible [19].

As *cis*-azobenzene (*c*-AB) has higher energy than *t*-AB, it is much more difficult to obtain a pure sample for the analysis. Although only few investigations have been carried out to determine its structure, there seems to be no controversy related to stable structure of *c*-AB. The crystal structure of *c*-AB was measured by the three-dimensional X-ray method [20]. It indicated that the molecule is severely distorted from the planar structure, where the phenyl ring is rotated by 53.3° about the C–N bond relative to the N=N–C plane. This experimental result about the *c*-AB structure has been supported by theoretical studies and ab initio calculations [13, 15, 16, 18, 21, 22].

From above investigations, we can establish the following picture for the stable structures of the isomers. *t*-AB has a planar structure with C_{2h} symmetry, and *c*-AB with the *cis* configuration is severely distorted from the planar structure by the rotation around C–N bonds. As an example, Fig. 2.2 shows the optimized geometries of the *trans*/*cis* isomers in the ground state, calculated at the 14 electrons in

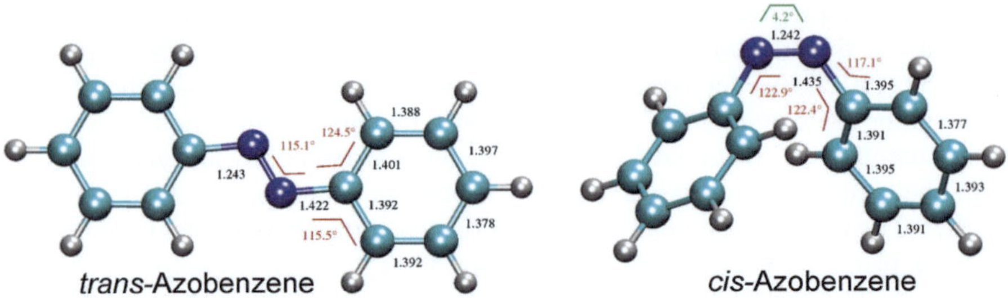

Fig. 2.2 Optimized geometries of the trans/cis isomers in the ground state, calculated at the 14 electrons in 12 orbitals (14/12) CASSCF level of the theory (Reprinted with permission from Ref. [15]. Copyright (2004) American Chemical Society)

12 orbitals (14/12) CASSCF level of the theory [15]. Thermal isomerization represents the interconversion of the *cis* and *trans* isomers of azobenzene on the electronic ground state. The *t*-AB corresponds to the energy minimum on the PES of the ground state. The enthalpy of *t*-AB → *c*-AB reaction is 11.7 ± 1.3 kcal/mol measured at 24.5–25.5 °C in *n*-heptane [23]. The downhill *cis*-to-*trans* conversion is a unimolecular reaction taking place slowly in solutions in the dark. The rate of the *cis* → *trans* thermal isomerization can be described by the Arrhenius equation [5, 24]:

$$k_t = A\exp\left\{-\frac{E_a}{RT}\right\} \tag{2.1}$$

The pre-exponential factor A and activation energy E_a are $4.5 \times 10^{11}\ s^{-1}$ and 23.7 kcal/mol in benzene, where the exact values depend on the solvents [24].

2.3 Excited States, Intermediates, and Pioneering Models

For azobenzene and its derivatives, also for most aromatic azo compounds, the isomerization in both the *trans* → *cis* and *cis* → *trans* directions can be induced by irradiating with light in proper wavelengths. In addition to the direct excitation, the photoisomerization is possible to be induced by triplet sensitization. Upon the light irradiation, *trans* and *cis* isomers of azobenzenes transfer from the ground states to the excited states. The isomerization goes through different intermediates upon the conditions. Determining the excited states and intermediates is a critically important task to understand the photoisomerization processes. In the early stage of the investigations, the understanding of the excited states and intermediates was mainly achieved by the studies on the influences of the excitation wavelength, temperature, and solvent on the quantum yields. Some important results obtained from the studies are introduced below.

The wavelength dependence of the photoisomerization has been intensively studied since the early reports by Fischer et al. [25] and Zimmerman et al. [26]. The results reveal some direct connections with the mechanism of the isomerization. The quantum yields of azobenzene are significantly dependent on which band is excited at the irradiation wavelength but are approximately constant throughout a single absorption band [26, 27]. The average *quantum yields* of azobenzene in isooctane are *cis* → *trans* 0.42 ± 0.04 and *trans* → *cis* 0.11 ± 0.01 for the UV band ($\pi \rightarrow \pi^*$) excitation and *cis* → *trans* 0.48 ± 0.05 and *trans* → *cis* 0.24 ± 0.02 for the visible band ($n \rightarrow \pi^*$) excitation [26]. There are more results in literature about the isomerization quantum yields for azobenzene on the $n \rightarrow \pi^*$ and $\pi \rightarrow \pi^*$ excitation, obtained under different conditions [28–31]. Although the quantitative values are not exactly the same, the results are agreed on two important points. (i) The sum of the quantum yields $\phi_{c \rightarrow t} + \phi_{t \rightarrow c}$ is significantly smaller than unity [26]. (ii) The *trans* → *cis* quantum yield is always larger for the low-energy ($n \rightarrow \pi^*$) excitation compared to that obtained from the high-energy ($\pi \rightarrow \pi^*$) excitation. This unexpected result obviously contradicts Kasha's rule (the rapidity of internal conversion) according to which higher energy states should be completely deactivated to the lowest excitation one [29, 30].

The quantum yields of photoisomerization of azobenzene show some correlation with the temperature variation in the range from room temperature to extremely low temperature [32]. The yield for the *trans* → *cis* transformation decreases sharply on cooling, while that for *cis* → *trans* reaction changes little in the same temperature range [33]. At a very low temperature, the *trans* → *cis* transformation yield approaches zero (<0.01) for $\pi \rightarrow \pi^*$ excitation, but it does not show this tendency for the $n \rightarrow \pi^*$ excitation. In the case of 2, 2′-azonaphthalene, the *trans* → *cis* quantum yields for both the $n \rightarrow \pi^*$ and $\pi \rightarrow \pi^*$ vanish when temperature is lower than -140 °C [33]. It indicates that energy barriers exist somewhere between the electronically excited singlet state and the ground state of the *cis* isomer [32, 33]. The *trans–cis* photoisomerization also shows a correlation with the solvent properties. The quantum yields for *trans* → *cis* ($\phi_{t \rightarrow c}$) and *cis* → *trans* ($\phi_{c \rightarrow t}$) photoisomerization of azobenzene were determined when irradiated at 317 and 439 nm in solutions with different polarity [28]. The results show that at both irradiation wavelengths, $\phi_{t \rightarrow c}$ increases and $\phi_{c \rightarrow t}$ decreases with increasing polarity of the solvent, but their sum ($\phi_{t \rightarrow c} + \phi_{c \rightarrow t}$) almost remains constant. The quantum yields of the photoisomerization show reliance on the solvent viscosity. The viscosity effect of the solvents is attributed to the change of molecular volume somewhere along the path of the photoisomerization [34].

Photosensitized isomerization is a way to study the mechanisms of photochemistry of azobenzene. However, the choice of suitable sensitizers for the experiments is not always an easy task, which could be the reason to cause significant disagreement on the results. The triplet energies varying from 45 to 66.6 kcal/mol were used by Jones and Hammond to achieve the photoisomerization of azobenzene. They reported that about 2 % *cis* isomer was formed in the photostationary states [35]. By using triphenylene as the sensitizer, 25 % *cis* isomer was detected in the photostationary state [36]. The disagreement between those results was later analyzed and attributed to the different triplet energies of the sensitizers [37]. The study of Ronayette et al. suggests that there exist two triplet levels with different isomerization yields in the *trans*-form. After energy transfer in the photosensitized isomerization, the lower triplet leads to isomerization with a quantum yield about 0.5, while the higher triplet rarely leads to the *trans* → *cis* isomerization.

At the same time, several models were proposed to rationalize the observations about the exited states and intermediates of the photoisomerization. Kearns summarized those important experimental results required to be considered by a reasonable model [38]. (i) The quantum yields, $\phi_{t\to c}$ and $\phi_{c\to t}$, decrease with decreasing temperature, (ii) the activation energy of the *cis* → *trans* transformation is smaller than that of the *trans* → *cis* transformation, (iii) the sum of $\phi_{t\to c}$ and $\phi_{c\to t}$ is less than unity at all temperatures studied, and (iv) excitation of an azo molecule to its lowest (n, π^*) singlet state results in a greater photoisomerization quantum yield than does excitation to the higher-lying (π, π^*) state. Typically, the early-stage models were represented by a Jablonski diagram including the two isomers and intermediate of azobenzene, where each of the energy levels is represented by the lowest energy point of the PES.

The above experimental results (i) and (ii) suggest that in an electronic excited state, there is a barrier separating the two configurations, which correspond to the *trans* and *cis* configurations in the ground state after Franck–Condon transition [26]. The isomerization undergoes a thermal rate process over the energy barrier. The barrier heights for the *trans* → *cis* transformation and *cis* → *trans* transformation is different, which is higher for the former. The experimental result (iii) infers that there is no common excited initial state of the two isomers [32]. Stegemeyer put forward a quantitative model supposing two intermediate states with different energies and another model involving only one common intermediate state for comparison [39]. The calculated results of the two-intermediate-state model (Fig. 2.3) are consistent with the results of Zimmerman et al. [26] and Fischer et al. [32, 33] about the quantum yields and their temperature dependence. The result (iv) can render a critically important clue to the isomerization mechanism, which will be discussed in Sect. 2.4.

It was highly debatable whether triplet states become involved in the photoisomerization and should be included in the model. By comparing the results of direct and sensitized isomerization, Jones and Hammond suggested that decay from electronic excited singlets does not involve a pathway through the lowest triplet state of the system [35]. Accordingly, mechanisms involving the crossing of the singlets to higher triplets are unattractive because of the expected rapid decay to the lowest triplet. Based on different sensitized result, Fischer argued that the result favors a crossing of the pathways in the triplet sensitized and the direct photoisomerization of azobenzene [36]. One possible pathway suggested is the intersystem crossing, following excitation of the first or second singlet levels. In the same article, Fischer also indicated that the obvious discrepancy between his and Jones' results might have a deeper reason. Ronayette et al. attributed the

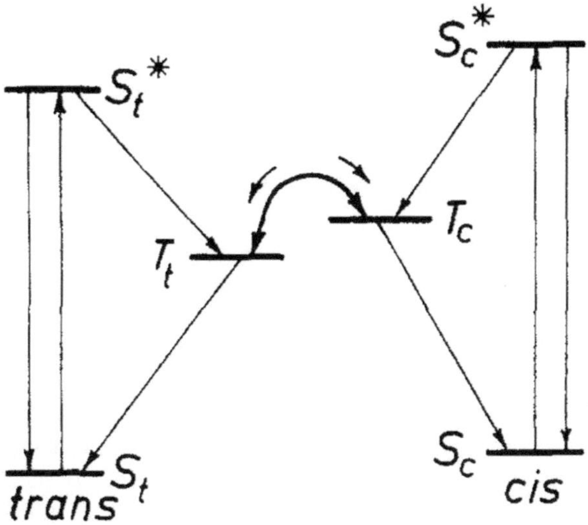

Fig. 2.3 The two-intermediate-state model proposed by Stegemeyer (Reprinted with permission from Ref. [39]. Copyright (1962) American Chemical Society)

discrepancy to the two lowest triplet levels with different isomerization yields and proposed a model including the two triplets [37, 40]. After studying the solvent and triplet donor effects on *cis–trans* photoisomerization of azobenzene, an alternative mechanism has been proposed by Bortolus and Monti [28]. They suggested that an excited state common to both isomers is involved and this state cannot be the triplet. The direct photoisomerization of azobenzene should occur in the singlet manifold as suggested by Jones and Hammond.

2.4 Photoisomerization Mechanism of Azobenzene

As discussed in above section, the early-stage investigations on azobenzene and its derivatives yielded a series of puzzling issues. Since then, more and more advanced instruments and calculation methods have been adopted to address the issues. However, the debatable issues related to the photoisomerization mechanism of aromatic azo compounds have not been fully clarified yet. Although most of the experimental results are not arguable, their interpretations related to the mechanisms are often subjects of heated debates. The most controversial issue is related to exact pathways of the configuration change through the electronic states involved in the photoisomerization. Due to the simplicity, azobenzene and its derivatives have been intensively studied as extremely valuable systems to understand the mechanism of the photoisomerization. In this section, azobenzene as the simplest compound in this series is used to discuss the mechanism. The photoisomerization of azobenzene derivatives and substituted effect will be discussed in Sect. 2.5.

2.4.1 Rotation vs Inversion Controversy

Two possible mechanisms have been commonly considered for the isomerization of double bonds [41]. One involves the rotation of substituent at one side of the double bond with respect to that at the other side. The torsion and twisting around an axil along the double bond are known as rotation mechanism. Instead of the term "*rotation*," torsion and twisting are also used by researchers to indicate this isomerization mechanism. The second mechanism, which is available to imines and azo compounds but not olefin, has been termed by Curtin et al. as the "lateral shift mechanism." The isomerization involves the shift of the substituent attached to nitrogen from one side of the molecule through a linear transition state. In the transition state, the nitrogen atom adopts a linear *sp*-hybridized orbital and the π bond almost remains intact. This mechanism is later referred by most authors as the *inversion* mechanism. Figure 2.4 shows the schematic illustration of the rotation and inversion mechanisms. The inversion has been widely accepted as a more reasonable mechanism for the thermal isomerization of azobenzene [34, 41–43].

However, for the photoisomerization, the situation is much more complicated and has aroused considerable controversy in the past decades. The controversy can trace its history back to the early stage of the photochemical study. It has been known for a long time that the photoisomerization of stilbene is through a rotation around an axis through the C=C double bond [34, 44]. At the same time, researchers doubt whether the rotation mechanism is correct for the isomerization of azo compounds. As early as 1941, Magee et al. suggested that there is additional possibility for azo compounds to change from *cis* to *trans* by retaining a plane structure throughout the process [44]). As the lowest (*n*, π*) state can participate in

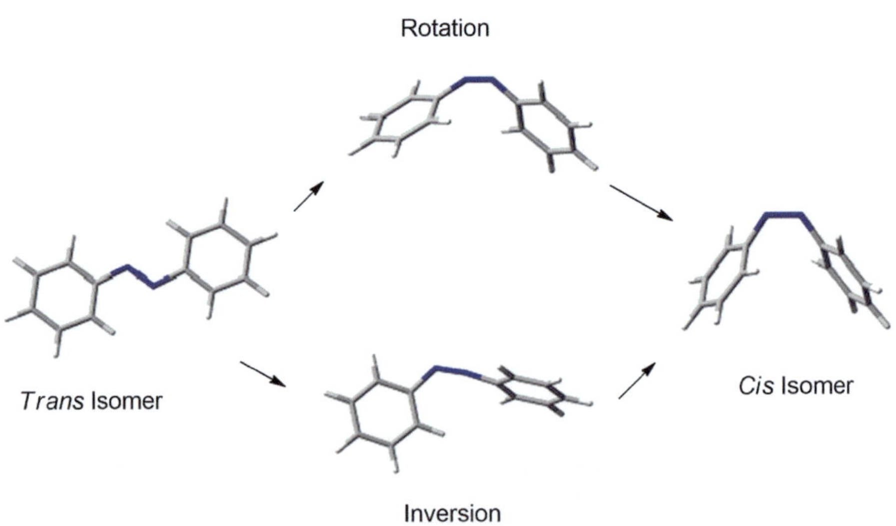

Fig. 2.4 The schematic illustration of the rotation and inversion mechanisms

the photochemical process, the photoisomerization behavior of azobenzene could be more involved than that of stilbene. Both rotation and inversion isomerization can be possible pathways, involving the ground state S_0 and few lowest singlet and triplet excited states such as the S_1, S_2, T_1, and T_2 states. These relaxation pathways are referred to as the nonadiabatic transition processes, which involve the transitions from one electronic PES to another. The relaxation pathways on the different states have been intensively investigated in order to understand the exact mechanisms of the photoisomerization processes.

Before 1980s, the tools and facilities used to investigate the photoisomerization mechanism are rather limited. The conventional spectroscopic instruments could supply little information related to the excited states and the details of the radiationless nonadiabatic processes. The computational facilities available at that time period can hardly afford the ab initio calculations for a polyatomic molecule such as azobenzene. The simplified quantum calculation methods and simple model compounds were used to provide insight into the controversial issues. For example, the extended Hückel calculation was carried out for isomerization of N_2H_2 [45]. Nonempirical molecular orbital calculations were performed on the S_0, S_1, T_1, and T_2 states of azomethane and selected states of diimine and azoethane [46]. However, there is no guarantee that these results are justifiably applicable to azobenzene. In the 1980s, significant progresses were made to establish models and theories from the available experimental evidences and theoretical calculations. The correlation diagram and qualitative PESs were proposed to understand the main features of PESs for possible isomerization pathways from *trans* to *cis* isomers. Two well-known models considering both rotation and inversion mechanism were proposed by Rau et al. [47] and Monti et al. [48].

The model of Rau and coworkers is based on a series of investigations on the photoisomerization behavior of a special group of azobenzenes, which include azobenzenophanes, azobenzene-capped crown ether, and *o,o,o′,o′*-substituted azobenzenes [47, 49, 50]. A common feature of this group of the azobenzenes is that the molecules are supposed to be unable to isomerize by rotation owing to the hindered structures. As mentioned above, a violation of Kasha's rule is typically observed for azobenzene and many of its derivatives. It means that the quantum yield of *trans*-to-*cis* isomerization is obviously higher for $n \rightarrow \pi^*$ excitation than that for $\pi \rightarrow \pi^*$ excitation. However, for these sterically hindered azobenzenes, Rau and coworkers found that $\phi_{t \rightarrow c}$ has the same value no matter for the $n \rightarrow \pi^*$ or $\pi \rightarrow \pi^*$ excitations, i.e., the quantum yields are actually wavelength independent in these cases. Based on the observations, Rau proposed the rotation-inversion dual-pathway model [47]. It states that in the S_1 (n, π^*) state, azobenzene isomerizes through inversion path with a high quantum yield; in the S_2 (π, π^*) state, it isomerizes by rotation with a low quantum yield [47, 49]. Bortolus et al. studied the *trans–cis* photoisomerization of azobenzene–cyclodextrin inclusion complexes [51]. The result indicates that for a partial or total block of the rotation about the –N=N– double bond, the inversion of the nitrogens could be the sole isomerization pathway.

According to Rau's model, azobenzene in its S_2 (π, π*) state can find a minimum at the 90° twisted configuration in the rotation pathway, which is dynamically coupled to the ground state [47]. The mechanism is based on a supposition that the internal conversion from S_2 (π, π*) to S_1 (*n*, π*) for the *trans* isomer is very slow. Therefore, for azobenzene and many aromatic azo compounds, the S_1 (*n*, π*) state is bypassed by the rotation pathway. If the rotation pathway is blocked, the isomerization from the S_2 (π, π*) of the *trans* isomer favors internal conversion (IC) to S_1 (*n*, π*). As a result, the isomerization takes the inversion pathway in the S_1 state. This model can explain the same $\phi_{t \to c}$ for both $n \to \pi^*$ and $\pi \to \pi^*$ excitations observed by Rau et al. for the special group of azobenzenes mentioned above.

Monti et al. performed the ab initio and configuration interaction (CI) computations for azobenzene over a minimal basis set of STO/3G atomic orbitals [48]. The energy sequence and the wave functions were computed for the excited states of the *trans* and *cis* isomers and two representative geometries along the rotation and inversion isomerization paths. The qualitative PESs proposed by Monti et al. give a unified interpretation of the photoisomerization mechanism. According to the model, upon excitation to S_1 (*n*, π*) state, the molecule will first relax through the inversion path to the shallow minimum and then relax either to *cis* or *trans* geometry of the ground state. Another possible way for the deactivation of S_1 state is through the S_1 (*n*, π*) to T_2 (π, π*) intersystem crossing (ISC). However, this slow process cannot compete with the rapid decay of the S_1 because the spin–orbit integral is zero. Following the excitation to the S_2 (π, π*) state, the rotation pathway is more favorable for the isomerization. The azobenzene twists from the *trans* or *cis* configuration to an energy minimum close to the $\varphi = 90°$ then from which decays very quickly to the S_1 state. In the vicinity of this geometry, the IC to the S_0 can also be efficient as the energy gap between the two states is narrow. This model supplies a quantitative explanation of the results obtained by Rau and Lüddecke, which show that the isomerization quantum yields are the same for molecules with blocked rotation pathway, no matter whether S_1 or S_2 state is initially excited.

The models proposed by Rau and Monti et al. can reasonably explain many experimental results. Both agree that rotation and inversion are the possible isomerization pathways after direct excitation to a singlet state. However, the rotation and inversion controversy has not been completely settled. Both mechanisms have their weaknesses. As indicated by Rau, if the temperature dependence of the quantum yields was measured, the results could be a way to judge how well the models reflect the physical reality. However, as indicated by Rau, neither of them could fully rationalize the observations at that time [29, 30].

2.4.2 Experimental Reexamination

In recent years, rapid developments in analytical instruments have supplied many new approaches to explore the photoisomerization mechanism, which are almost inconceivable for researchers several decades ago. The ultrafast time-resolved spectroscopic techniques are powerful methods and have provided many valuable pieces of information on this subject. The early-stage exploration of the time-resolved spectroscopic studies on the photoisomerization mechanism can be traced back to the reports of emission from excited states of azobenzene [52, 53]. In that stage, the results obtained from the time-resolved spectroscopy supported the dual-pathway mechanism proposed by Rau et al. [54–57].

Femtosecond time-resolved UV–vis absorption spectroscopy has been used to study the photoisomerization dynamics [55]. A short-lived transient absorption is observed at 370–450 nm, which is assigned to the absorption related to the S_2 (π, π^*) excited state of *trans*-azobenzene. Data fitting shows that the decay of this absorption can be divided into a shorter-lived decay component of lifetime (ca. 1 ps) and a longer-lived decay component (ca. 13 ps). The shorter-lived lifetime is attributed to the decay of the *trans*-azobenzene from the S_2 state after Franck–Condon transition, and the longer-lived one is attributed to the decay of an intermediate “bottleneck” state to S_0 state of *trans*-azobenzene. These results have been explained by the authors using the potential energy curves obtained by Monti et al. (Fig. 2.5). The femtosecond transient absorption measurements have also been carried out by Nägele et al. for the azobenzene excited at 435 nm [54]. Two components, 0.17 ps and 2 ps, are observed for the excited-state absorption of *cis*-azobenzene. Correspondingly, a slower initial photoreaction is observed for the *trans* isomer with the time constants of 0.32 and 2.1 ps. The results were used by the authors to support the inversion mechanism in the S_1 excited state.

Further investigation by Lednev et al. shows that the transient absorption at 400 nm decays with a dominant component of ca. 0.9 ps and a weaker component of ca. 15 ps [56]. The transient band (400 nm) rises synchronously as a transient band at 475 nm decays with a lifetime < 200 fs. On excitation at 503 nm (close to the S_1 origin) and 390 or 420 nm (above S_1 origin), the transient bands decay with lifetimes ca. 0.6 and 2.5 ps, respectively. The dual-pathway mechanism proposed by Rau et al. has been used by the authors to rationalize the results. The extremely fast component (<0.2 ps) is attributed to the motion along the rotation coordinate. The decay with lifetime ca. 2.5 is believed to be the inversion in the S_1 state. The same method has been used to study an azobenzene derivative, in which azobenzene is capped by an azocrown ether [57]. The result also supports the proposal that rotation is the dominant pathway for the S_2 excited state, and inversion is the dominant pathway for the S_1 state.

However, more recently, almost every mechanistic interpretation proposed before has been carefully reexamined by the sophisticated experimental investigations. Those widely accepted pioneering models have been challenged by these new observations. At first, the mechanism involving the direct isomerization on the S_2

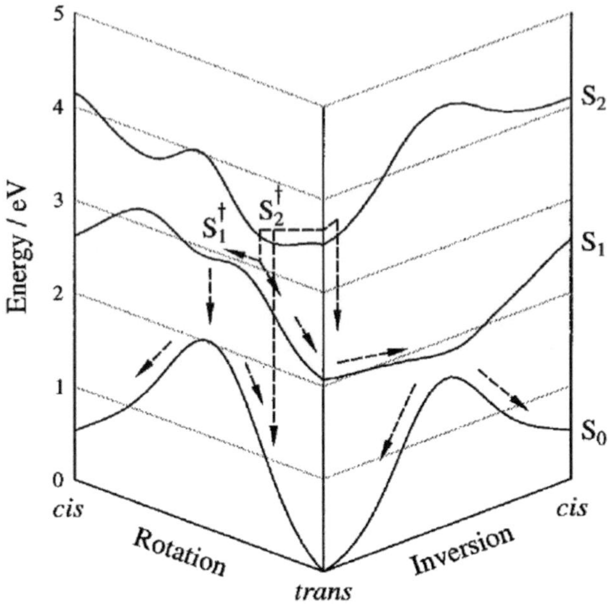

Fig. 2.5 Schematic representation of the S_0, S_1, and S_2 potential energy curves of azobenzene and possible decay routes following excitation to the t-S_2 state (Reprinted with permission from Ref. [55]. Copyright (1996) American Chemical Society)

potential energy surface has been challenged by new experimental results. The femtosecond time-resolved fluorescence spectroscopy has been used to study the electronic relaxation and isomerization mechanism after the $S_2 \leftarrow S_0$ photoexcitation [58]. The lifetimes of the S_2 and S_1 states are estimated to be ~0.11 ps (S_2) and ~0.5 ps (S_1) by the time-resolved measurements. The femtosecond time-resolved fluorescence study indicates that the quantum yield of the $S_2 \rightarrow S_1$ relaxation is almost equal to 1. On the basis of the above experimental results, the authors argued that the rotation pathway cannot exist for the S_2 excited state. The S_2 (π, π^*) state is relaxed very rapidly to the vibrational level of the excited S_1 (π, π^*) state (Fig. 2.6). The photoisomerization of azobenzene is through the inversion pathway of the S_1 state no matter the initial excited state is S_1 or S_2 [58]. According to this model, the violation of Kasha's rule is attributed to the presence of an additional relaxation pathway for the high vibrational levels in the S_1 state. Above result is supported by the study on the femtosecond fluorescence dynamics of rotation-restricted azobenzenephanes [59]. Upon excitation to the S_2 state, the ultrarapid $S_2 \rightarrow S_1$ relaxation (~0.1 ps) was observed for both the rotation-free and rotation-restricted molecules. This study also indicates that the *trans–cis* photoisomerization of azobenzene always occurs in a lower-lying electronic state (S_1 state) even if it is initially excited into the S_2 state.

Femtosecond time-resolved absorption experiments have been performed on both *cis*- and *trans*-azobenzenes after UV and visible light irradiations [60, 61]. For the *trans* $\rightarrow$ *cis* photoreaction, the transient decay times for 266 and 340 nm excitation ($\pi \rightarrow \pi^*$) are identical, which are 0.13, 0.42, 2.9, and 1.2 ps. Three slower decay times of 0.42, 2.9, and 1.2 ps and associated spectra well match

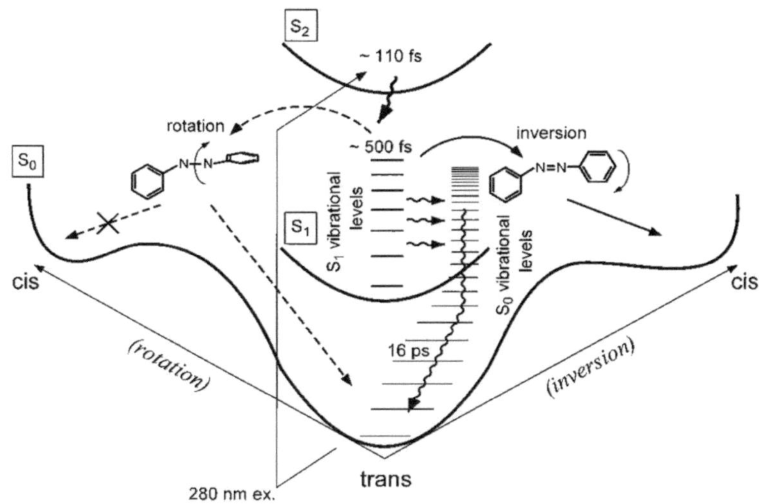

Fig. 2.6 Schematic diagram of the relaxation and isomerization pathway of trans-azobenzene after the $S_2(\pi, \pi^*) \leftarrow S_0$ photoexcitation in hexane (Reprinted with permission from Ref. [58]. Copyright (2001) American Chemical Society)

those obtained from the direct $n \rightarrow \pi^*$ excitation. The decay time of 0.13 ps is assigned to the radiationless relaxation from the S_2 state to the S_1 potential energy surface. The time components of 0.42 and 2.9 ps are ascribed to the movement out of the Franck–Condon region and internal conversion to the *cis* ground state. The slowest time of 12 ps is assigned to the cooling of the hot ground state by surrounding solvent. This time scale is consistent with that obtained from the study on the intermolecular energy transfer to solvent by femtosecond infrared spectroscopy [62]. As discussed for the *trans*-azobenzene, the dynamics initiated by $n \rightarrow \pi^*$ and $\pi \rightarrow \pi^*$ excitation are similar for *cis*-azobenzene. The spectroscopic findings are consistent with those observations on the two-step relaxation for the isomerization [55, 56, 58]. The authors suggested that the first step is a fast relaxation from the S_2 state to the S_1 PES. Then, both *trans* → *cis* and *cis* → *trans* isomerizations involve large-amplitude motion on the S_1 PES [61].

Besides those mechanisms considering pathways through the excited S_1 and S_2 states, the possibility of the isomerization through higher excited states has also been suggested. Photoisomerization mechanism and dynamics of azobenzene have been studied by femtosecond (fs) time-resolved photoelectron spectroscopy (TRPES) and ab initio molecular dynamics (AIMD) [63]. The TRPES spectra recorded in the wavelength range from 280 to 340 nm show two photoelectron bands α and β with distinct lifetimes of 0.17 ps and 0.42 ps. The two bands are assigned to the ionization of S_2 state and ionization of S_3 or S_4 ($S_{3,4}$) state. The S_2 excitation is delocalized over the molecule and $S_{3,4}$ involve excitation localized in the phenyl rings. By analyzing the results, Schultz et al. suggested the mechanistic model shown in Fig. 2.7. The existence of two different pathways is responsible for

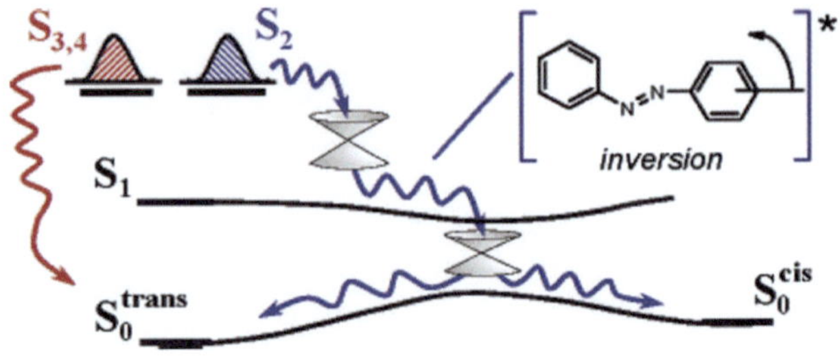

Fig. 2.7 Proposed electronic relaxation pathways for S_2 and $S_{3,4}$: $S_2 \rightarrow S_1$ internal conversion occurs with $\tau = 170$ fs at planar geometry (Reprinted with permission from Ref. [63]. Copyright (2003) American Chemical Society)

the different lifetimes of these two states. The S_2 state quickly decays to planar S_1 state and a conical intersection involving inversion in the S_1 state can be approached within 50 fs. On the other hand, the relaxation of the $S_{3,4}$ state involves a different pathway with the reduced quantum yield for the photoisomerization [63]. Through a comprehensive analysis of resonance Raman intensities, the role of a third state S_3 (n^2, π^{*2}) in the relaxation mechanism of S_2 has been proposed [64]. According to the mechanism, there are two competing pathways for the relaxation. One is the direct rotationless $S_2 \rightarrow S_1$ (C_{2h}) internal conversion, and the other is $S_2 \rightarrow S_3$ internal conversion along the rotational pathway followed by the decay to S_0 and S_1 states separately.

Above investigations provide crucial dynamical information on the excitation to the S_2 and higher energy excited states. The controversy mainly surrounds the pathways of the relaxation from the higher excited state to the S_1 state. Most previous studies agree that the photoisomerization on the S_1 PES is through the inversion pathway [55–57, 61, 63]. However, it does not mean that there are no debates about the isomerization mechanism in S_1 state. The study of the isomerization mechanism in S_1 state is also involved in the rotation vs inversion controversy as discussed below.

One important evidence to support the inversion mechanism among others has been acquired from the study on the electronic and vibrational relaxation of excited *trans*-azobenzene in solutions by picosecond time-resolved Raman spectroscopy [65]. The lifetime of the S_1 state, estimated from the transient Raman bands, shows a significant dependence on the solvents, which is ~12.5 ps in ethylene glycol and ~1 ps in hexane. The lifetime of vibrationally excited S_0 state of azobenzene is ~16 ps. Therefore, the time constant of ~1 ps is attributed to the electronic relaxation to the S_0 state. The longer lifetime (~12.5 ps) is caused by the dissipation of a significant amount of the energy to the surrounding solvent in the S_1 state. One important observation from the study is that the NN bond stretching frequency in the S_1 state is very close to that of the S_0 state and the one-to-one correspondence can also be made for the other Raman bands of the S_0 and S_1 states. It evidences that the NN bond in the S_1 state has the double bond nature and azobenzene possesses a planar structure in the excited state [65]. It is consistent with the planar transition state of the inversion pathway. The inversion mechanism is also supported by the femtosecond fluorescence up-conversion spectroscopy of a rotation-restricted azobenzene after the excitation to the S_1 state [66].

On the other hand, this inversion isomerization mechanism on the S_1 state has been challenged by measuring the anisotropy of femtosecond fluorescence after the excitation to the S_1 (n, π^*) state of azobenzene in hexane and ethylene glycol solutions [67]. The prominent depolarization is observed for the fluorescence in hexane (nonviscous solvent) in contrast to the indiscernible anisotropy change in ethylene glycol (viscous solvent). A new mechanistic model is proposed to rationalize the observation (Fig. 2.8). According to the model, the isomerization can occur through the rotation and concerted inversion pathways after the S_1 excitation. The rotation channel is operative for the photoisomerization of *trans*-azobenzene in a nonviscous solvent. The depolarization for the fluorescence is then attributed to the variation of orientation of the transition dipole moment along the –CNNC– rotation pathway in seeking an efficient conical intersection. When the rotation pathway is obstructed in a viscous solvent, the concerted inversion should be the pathway to cause the ultrafast electronic deactivation [67].

The concerted inversion with the in-plane symmetric motion is also evidenced by the observed indiscernible anisotropy change in the viscous solvent. Femtosecond fluorescence study indicates that the photoisomerization of rotation-restricted azobenzenephanes in the S_1 state might occur through a conical intersection of the concerted inversion coordinate [59]. More recently, azobenzene derivatives with hydrogen bonding to induce isomer locking have been synthesized to study the isomerization mechanism [68]. Inhibiting the concerted inversion pathway in an azobenzene shows the effect to prevent the isomerization after $S_2 \leftarrow S_0$ excitation.

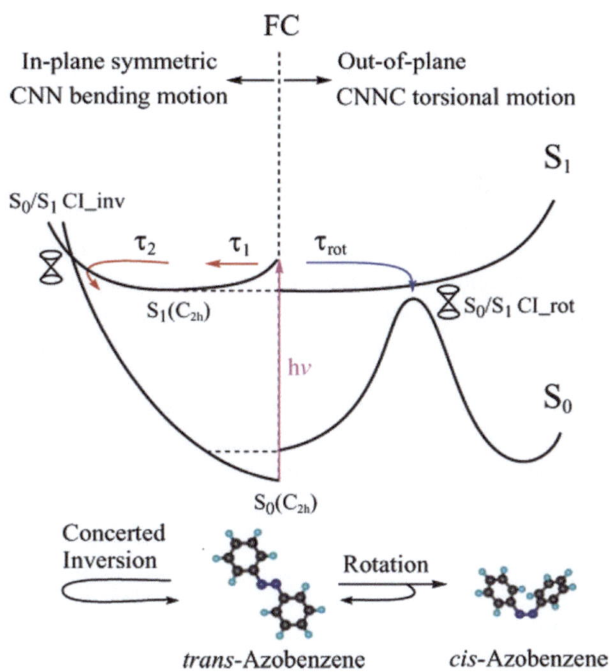

Fig. 2.8 A representation to illustrate two relevant reaction channels responsible for the observed dynamics of photoisomerization of *trans*-azobenzene on the S_1 potential energy surface in various solvents (Reprinted with permission from Ref. [67]. Copyright (2004) American Chemical Society)

It supplies a proof for the concerted inversion mechanism in the *trans* → *cis* isomerization.

For both *cis-* and *trans*-azobenzene, the initial excited-state isomerization dynamics and lifetimes following the S_1 excitation have been investigated by using of resonance Raman intensity analysis and fluorescence quantum yield measurement [69]. The extremely short lifetime (0.1 ps) is observed for the *cis*-azobenzene, which is consistent with the previous studies [54, 61]. The very weak fluorescence quantum yield of the *cis* isomer can be attributed to this short lifetime. The study indicates that the *trans* and *cis* isomers show the different initial nuclear motions on the PESs. The *cis*-azobenzene shows relatively larger excited-state displacement along the rotation coordinate, which provides the driving force for the ultrafast isomerization. In contrast, the *trans*-azobenzene mainly shows the CNN bend and CN and NN stretch modes after the S_1 excitation. The results indicate that following the S_1 excitation, the *cis*-azobenzene isomerizes rapidly through the rotation pathway, while isomerization of *trans*-azobenzene proceeds via inversion pathway.

2.4.3 Theoretical Studies

Paralleling the experimental investigations, theoretical explorations based on the quantum chemistry and other calculations have also been carried out in recent years. High accuracy quantum chemical calculation, such as ab initio calculation, is a powerful tool used to explore the photoisomerization *mechanism.* PESs obtained from the calculations can supply deep insights into the nuclear geometry, state energy, pathway, and dynamics of the *trans–cis* isomerization. The structure variations in response to the light irradiation can be investigated for light irradiation with different wavelengths, which can interpret the experimental results on basis of rational analysis. However, for a long time, due to the limited computation capacity, the quantum chemical calculation had been carried out by semiempirical methods, such as CNDO/2 [42]. The ab initio calculations were only performed for the ground state of azobenzene [70] and azomethane as the simplest azoalkane [21, 71]. The minimal basis set CI calculation performed by Monti et al. was among the few systematical investigations reported during that period [48]. The quantitative PESs of azobenzene were obtained by the study for the ground and excited states, which supported the dual-pathway photoisomerization mechanism.

In recent years, this situation has been completely changed owing to the rapid developments on the computational facility and softwares. The ab initio calculations have been widely used to explore the photoisomerization mechanism of azobenzene and its derivatives. In a pioneering study, the photoisomerization of azobenzene under $n \rightarrow \pi^*$ and $\pi \rightarrow \pi^*$ excitation was investigated by ab initio calculation [43]. Some most important nuclear geometries were obtained by optimizing with the complete active space self-consistent field (CASSCF). The calculations suggest that inversion is the preferred pathway for ground and probably also for $n \rightarrow \pi^*$ excitation, while rotation occurs after $\pi \rightarrow \pi^*$ excitation. This result obtained from the theoretical investigation is consistent with the dual-pathway mechanism [29, 30], which was generally accepted at that time. On the other hand, the authors also indicated that considering the shape of the S_1 energy curves, the rotation pathway cannot be definitely ruled out by a small modification of the PES.

Following the early stage of the ab initio calculations, the excitation energies and global potential energy surfaces have been calculated to obtain more reliable information. Some totally different models have been proposed to account for the photoisomerization mechanism. Some critically important issues are targeted by the theoretical investigations, which intended to: (1) determine whether a direct isomerization on the S_2 PES can occur, (2) find the minimum energy pathway on the S_1 PES, and (3) identify the possible relaxation pathways following the excitation to the S_2 and higher energy states. In accordance with the ultrafast time-resolved spectroscopic studies, most theoretical investigations agree that the direct isomerization on the S_2 PES is impossible. On the other hand, the quantum chemical calculations suggest that the rotation pathway is the minimum energy pathway on the S_1 PES, which is not completely in agreement with the most interpretations about the experimental results.

The PESs of the S_0, S_1, S_2, and S_3 (n^2, π^{*2}) states have been obtained by CASSCA and multi-reference single double configuration interaction (MRSDCI) calculations [22]. A conical intersection (CI) between the ground state S_0 and the S_1 state is found near the midpoint of the rotation pathway. On the other hand, the calculation only finds local minimum at the *cis* and *trans* configurations in the S_2 state. The calculation shows that the photoisomerization of azobenzene cannot occur on the S_2 PES, which is consistent with the experimental investigations [61, 63, 65]. The authors indicate that the isomerization after the excitation to S_2 state should proceed through the deexcitation to the S_1 or S_3 states. Isomerization involving S_3 state has also been suggested by the calculation of Cattaneo and Persico [43]. It means that the S_2 state can nonadiabatically relax to S_3 state through the CI, and the S_3 state strongly interacts with S_0 and S_1 states.

Based on the calculation, Ishikawa et al. proposed that the isomerization on the S_1 PES is through the rotation pathway because of the existence of the conical intersection (CI) between the S_0 and S_1 in the torsion coordinate [22]. This mechanism contradicts to some previously widely accepted models [47, 49, 65]. Fujino and Tahara have proved that the NN bond in the S_1 state has the double bond nature, and azobenzene possesses a planar structure in the excited state [65]. Regarding to this critically important evidence to support the inversion mechanism for isomerization in the S_1 state, Ishikawa et al. argue that the calculated bond order of the NN bond of the S_1 state is approximately 1.0, which is far from the double bond nature. The double nature only reflects the bond order increase in the vicinity of the conical intersection (CI) along the rotation pathway [22]. In recent years, this rotation mechanism on the S_1 state has been supported by the several theoretical investigations as discussed below.

The mechanism of photoisomerization in the lowest electronic states (S_0, S_1, T_1) has been investigated by use of the quantum chemical calculations [15]. The PESs for the isomerization were calculated using unconstrained optimization and minimum energy path (MEP) calculations at the CASSCF level. The CAS second-order perturbation theory (MS-CASPT2) calculations were performed on the optimized relevant structures to account for the correlation energy. Both the transition state (TS) on the S_1 PES and S_0–S_1 conical intersections (CIs) were searched to elucidate the mechanism. The calculations show that the only TS with activation energy of 2 kcal/mol appears at 60° of the CNNC torsion from the *trans* isomer (Fig. 2.9). The lowest energy conical intersections (CIs), lying 2 kcal/mol above S_1 minimum, exist on the CNNC torsion pathway at angles in the range 90–95°. The lowest conical intersection (CI) along the inversion pathway has an energy ca. 25 kcal/mol higher than the S_1 minimum, and the molecule possesses the highly asymmetric structure with one NNC angle of 174°. According to these results, the relaxation in the S_1 state mainly involves the rotation pathway, and the inversion may play a role only when molecule is excited with the light with the energy at least 25 kcal/mol higher than the S_1 minimum [15, 72]. The activation energy obtained from the

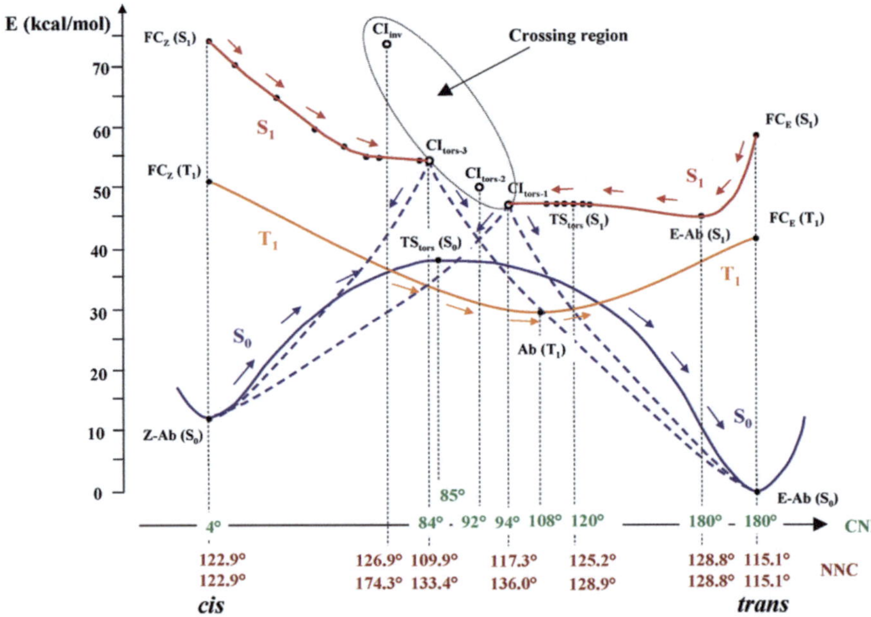

Fig. 2.9 Singlet (S_0 and S_1) and triplet (T_1) reaction paths for the *trans*(E) → *cis*(Z) isomerization in azobenzene. Open circles represent S_1–S_0 CI. The horizontal axis represents the CNNC torsion coordinate (*green* numbers). Values for NNC angles are also reported (*brown* numbers). The shaded region highlights the S_1–S_0 crossing space and embraces the low-energy (torsion) and high-energy (inversion) deactivation funnels. The paths of T_1 and S_1 radiationless decays are also shown (Reprinted with permission from Ref. [15]. Copyright (2004) American Chemical Society)

above calculation gives a reasonable way to explain the temperature dependence of the quantum yields of photoisomerization observed by Fischer [32].

The CASSCF calculation was carried out to scan the excited-state relaxation along the CNN bending (inversion), the CNNC torsion (rotation), and the concerted CNN bending (concerted inversion) pathways [73]. The result shows that after being excited to the S_1 state, the inversion is highly unfavorable due to a substantial energy barrier and a large S_0–S_1 energy gap. The rotation pathway is essentially barrierless with a conical intersection (CI) close to the midpoint of the torsion coordinate. On the other hand, when molecule is excited to S_2 state, the concerted inversion channel may be open and the S_0–S_1 CNN bending motion (concerted inversion) would become energetically favorable to produce more *trans* isomers. This calculation is supported by femtosecond fluorescence results discussed above [67].

The above theoretical investigations all indicate that the photoisomerization in the S_1 state mainly involves the rotation pathway, and inversion is a highly unfavorable pathway under this condition. Although CASSCF result is highly dependent on the active space selected for the calculation, those results are highly consistent with each other for the different selections. This rotation isomerization mechanism is also supported by calculations using constrained density functional theory (DFT) and time-dependent DFT [16, 18]. The theoretical investigations show that even for azobenzenophanes and azocrown ether-capped azobenzene derivative, the rotational pathway is the feasible and dominant photoisomerization mechanism [74, 75].

More recently, the influence of the condensed phase environment on the photoisomerization mechanism has been theoretically studied. The quantum classical two-scale techniques have been used for investigating isomerization in the liquid phase, where the photoactive chromophore and photo-inactive environment are treated with quantum mechanical (QM) method and molecular mechanics (MM), respectively [3, 76, 77]. The nonadiabatic QM/MM (na-QM/MM) simulation shows that the molecule undergoes a pedal motion-like translocation of the innermost nitrogen atoms in the S_1 state to cause the change in the central dihedral angle. The simulation rules out the inversion at the nitrogen atoms. The *trans* → *cis* photoisomerization is considerably slowed down in the liquid compared to the gas phase, while the *cis* → *trans* process is barely affected by the environment [77]. This difference has been attributed to the significant hindrance of the steric neighbor interactions to the reorientation of the phenyl rings.

In a separate study, the photoisomerization of azobenzene in vacuum and various solvents has been studied by a modified molecular dynamic simulation adopting an ab initio torsion–inversion force field [78]. Simulation shows that the photoisomerization in vacuum predominantly follows the torsional (rotation) pathway as found by others. On the other hand, the dominant isomerization mechanism is a mixed rotation-inversion one in solution, where the inversion contribution increases with the solvent viscosity increase. In the condensed phase, the probable pathways are more likely to be a mixed mechanism instead of a pure rotation or inversion pathway [76, 78]. These results provide a new and valuable perspective on solving the long-lasting rotation-inversion controversy. As most theoretical calculations were carried out for isomerization in vacuum, the disagreements with the experimental results could be attributed to the influences of medium and environment.

The isomerization mechanism for azobenzene to be excited to the S_2 state is more controversial compared with that of the photoisomerization in the S_1 state. Many experimental and theoretical results support that the photoisomerization of azobenzene cannot occur on the S_2 PES [22, 58, 65]. However, different pathways have been suggested for the isomerization after the excitation to S_2 state. According to mechanism proposed by Fujino and Tahara, photoisomerization of azobenzene occurs through the inversion pathway of the S_1 state after the high efficient $S_2 \rightarrow S_1$ relaxation [65]. Cattaneo and Persico suggested that the S_2 state can nonadiabatically relax to S_3 state through the conical intersection (CI) and isomerization occurs through the strongly interacted S_0 and S_1 states [43]. Schultz et al. indicated that the S_2 state quickly decays to planar S_1 state and isomerization takes place through the inversion conical intersection (CI) in the S_1 state [63]. Ishikawa et al. suggested that the molecule excited to S_2 state should proceed through the deexcitation to the S_1 or S_3 states [22]. The isomerization on the S_1 PES passes the rotational pathway via the conical intersection (CI) between the S_0 and S_1 states. Diau revivified the concept of the concerted inversion mechanism, which has been originally proposed for acyclic azoalkanes [46]. According to this mechanism, when molecule is excited to S_2 state, the concerted inversion channel may be open and cause the S_0–S_1 CNN bending motion [73]. All these models can provide a reasonable explanation to the non-Kasha behavior of photoisomerization of azobenzene observed by experimentalists. However, there seems to be no decisive factors to select superior to its competitors.

This issue has been further investigated through different theoretical approaches. The involvement of the doubly excited states related to the S_2 excitation has been suggested. The photoisomerization of azobenzene along the NN twisting (rotation) has been studied by using the CASSCF method, followed by the multistate CAS second-order perturbation theory (MS-CASPT2) [79]. The calculation shows that the singlet state S_3 (n^2, π^{*2}) has a deep minimum at about 90° of the twisting. According to the potential energy curves (PECs), the relaxation in the S_2 state undertakes the $S_2 \rightarrow S_3$ conical intersection (CI) radiationless transition, which is extremely fast, i.e., faster than the vertical $S_2 \rightarrow S_1$ decay at the *trans* state. Once on the S_3 PES, the molecule can jump on the S_1 PES through S_3–S_1 conical intersection (CI). After reaching the S_1 state, the isomerization occurs through the mechanism of the S_1 state. On the other hand, if molecule reaches the minimum of the S_3 PES, the decay to the ground state does not contribute to the *trans–cis* isomerization [79].

In a recent report, the role of the doubly excited-state S_{DE} (π^2, π^{*2}) has been thoroughly investigated by ab initio CASPT2/CASSCF calculation [80]. The critical structures, conical intersections (CIs), and minimum energy pathways related to the S_2 (π, π^*) and other excited states are determined through the calculation. The result shows that S_{DE} is immediately populated after the molecule is excited to the S_2 state. On this doubly excited state, azobenzene follows the torsion (rotation)

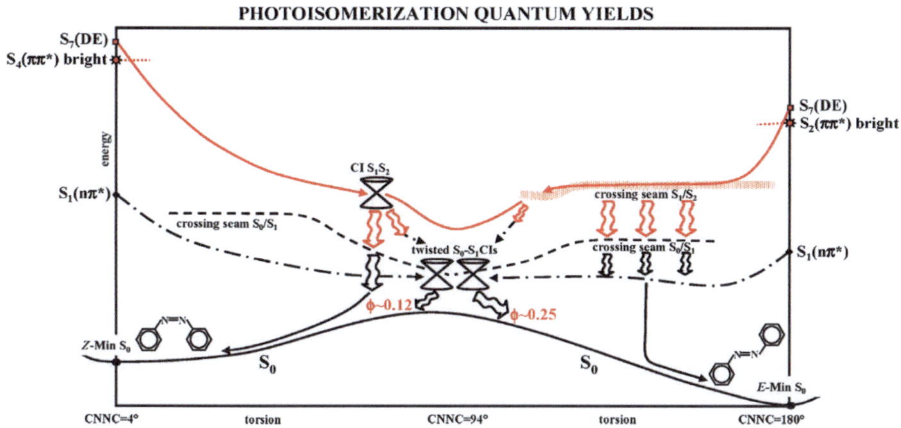

Fig. 2.10 Photoisomerization pathways and yields from the lowest (π, π*) state starting from the *trans* (E, *right side*) and *cis* (Z, *left side*) isomers (Reprinted with permission from Ref. [80]. Copyright (2008) American Chemical Society)

pathway and decays to the S_1 (*n*, π*) state at a variety of geometries through the fast internal conversion (Fig. 2.10). The existence of the S_{DE} state and seams of conical intersections (CIs) are found to be critically important in promoting the conversion from the S_2 to S_1 state. According to the model, the photoisomerization upon the S_2 excitation predominantly occurs along the rotational pathway, i.e., the same route for the S_1 state. However, during the process, the inversion coordinate is not frozen, and the NNC and CNN angles deviate from the equilibrium values at the S_1–S_0 CIs.

The above theoretical investigations indicate that after being excited to the S_2 state or other high-energy states, the photoisomerization mechanism could be more complicated than that suggested by previous simplified views. For a real system, the energy dissipation and relaxation pathways could be affected by surroundings and should be more complicated.

2.4.4 Simulation of Photoisomerization Dynamics

In recent years, computer simulation has provided a new approach to complement the theoretical and experimental investigations. The rapid-growing computational power makes it possible to simulate the dynamic process of the photoisomerization. As discussed above, a large amount of valuable information has been obtained by the femtosecond time-resolved techniques. The PESs and isomerization pathways obtained by the theoretical studies have also provided deep insights into the mechanism behind the photochemical process. On the other hand, the ab initio calculation cannot achieve the characteristic lifetimes to be directly compared with the experimental results. To solve the problem, the computational simulations have been used to provide the dynamical information on the photoisomerization process.

Based on this, a unified picture is gradually emerging to understand the photoisomerization mechanism and dynamics of azobenzene.

A semiclassical surface hopping approach has been used to simulate the photoisomerization dynamics of azobenzene [81]. The internal conversion and geometrical relaxation were taken into account to describe the *trans* → *cis* and *cis* → *trans* isomerization processes. The results suggest that for both S_1 and S_2 excitation, the torsion around the N=N double bond is the preferred mechanism. The inversion pathway in the S_1 PES is not energetically forbidden but costs much higher energy. The simulation correctly reproduces the different quantum yields for the S_1 and S_2 excitation observed by experiments. After the $\pi \rightarrow \pi^*$ excitation, the populated S_2 and S_3 states undergo a very fast decay to S_1 state (within 100 fs) through conical intersections (CIs) very close to the Franck–Condon regions for both *trans* and *cis* isomers. Because of the unfavorable slope of the S_2 and S_3 in this range, the torsional motion is slower than that with the $n \rightarrow \pi^*$ excitation. Therefore, the non-Kasha behavior of photoisomerization of azobenzene is attributed to the competition between the inertia of the torsional motion and the deactivation of the excited state. The motion of the wave packet on the PESs was simulated to compare with the lifetimes obtained with the transient absorption and emission spectra. The results are in good agreement with those obtained by Fujino and others.

The full multiple spawning (FMS) method for nonadiabatic wave packet dynamics was compared with the model using a semiclassical surface hopping (SH) method for simulation of the photoisomerization after $n \rightarrow \pi^*$ excitation [82]. Both methods show the agreement on the essential features of the mechanism. Later, a quantum decoherence correction was introduced in the SH algorithm, which improved the agreement with quantum wave packet calculations [83]. The improved SH simulation indicates that both photoisomerization and $S_1 \rightarrow S_0$ internal inversion go through the N=N torsional pathway, but the N–C bond rotation and NNC bending vibration also occur in the process [84]. Increasing the solvent viscosity shows effect to delay the torsional motion in *trans* → *cis* photoisomerization and the excited-state decay, while the *cis* → *trans* conversion is less affected. The simulation well reproduces the measured quantum yields for the *trans* → *cis* and *cis* → *trans* conversions. The result reveals the reason for the $\phi_{t \rightarrow c}$ increase with the increasing solvent viscosity, which is attributed to the intra/intermolecular vibrational energy redistribution occurring in the S_1 state.

Dynamic simulations can provide detailed understanding of the photoisomerization processes in femtosecond scale. The simulation with semiclassical electron–radiation–ion dynamics (SERID) shows that the *cis* → *trans* photoisomerization of azobenzene follows the rotation pathway after irradiation with a 100 fs laser pulse [85]. The excitation and deexcitation involve multiple steps, where the molecule always shows a superposition of the electronic states. The simulation reveals the influences of laser pulse parameters on the dynamical process through intricate sequence of connected steps [86]. For S_1 excitation, only a single avoided crossing is required along rotational pathway, while for S_2 excitation, another avoided crossing is required, and it occurs in either of two geometries near *cis* or *trans* configurations. The isomerization following both S_1

($n \rightarrow \pi^*$) and S_2 ($\pi \rightarrow \pi^*$) excitation has been studied by the dynamic simulation [87]. The result shows that the S_2 excitation causes the relaxation through the doubly excited state (π^2, π^{*2}), which preferably decays to the S_0 state of the *trans* configuration through partially twisted structure. The non-Kasha behavior of photoisomerization is attributed to the opening of this decay channel to the *trans* configuration after the $\pi \rightarrow \pi^*$ excitation. The SERID simulation shows that the *cis* → *trans* photoisomerization occurs predominantly through the rotation pathway even under a nontrivial resisting force [88].

As discussed above, similar to the ab initio theoretical calculations, majority of the dynamic simulations support the rotation mechanism for the photoisomerization. In addition to the on-the-fly dynamic simulation and dynamic simulation at the DFT level mentioned above, ab initial dynamic simulation (AIMD) including nonadiabatic transitions has been carried out in the recent years [89]. The result shows that the *trans* → *cis* photoisomerization undergoes a conventional rotation pathway. However, the *cis* → *trans* isomerization occurs through a two-step rotation process including the rotation of the central NN part and two phenyl rings (Fig. 2.11). The quantum yields obtained from the simulation is in good agreement with the corresponding experimental results.

2.5 Substituted Effect

Although azobenzene is an extremely important molecule used to study the photoisomerization, this compound itself is rarely incorporated in the azo polymers even as a dopant for a guest–host system. On the other hand, a variety of azobenzene derivatives have been used to develop azo polymers with different functions. Therefore, it is necessary to understand the substituted effects on the isomerization behavior and properties, which are briefly discussed in this section. The substituted effects have been studied, almost since the initial investigation on the isomerization of azobenzene, and have produced many valuable results reported in the literature. This section mainly concentrates on the azobenzene derivatives containing electronic donors and/or acceptors, which are most widely used as functional structures in azo polymers. Especially, *p*-donor–*p*′-acceptor azobenzenes, also known as *push–pull* substituted azobenzenes, are important for many applications. By the classification given in Chap. 1, many of these azo compounds can be categorized as aminoazobenzene-type and pseudo-stilbene-type azobenzenes [29, 30]. The azobenzene derivatives show unique spectral characteristics and solvent effects, which will be presented first before discussing their isomerization behavior.

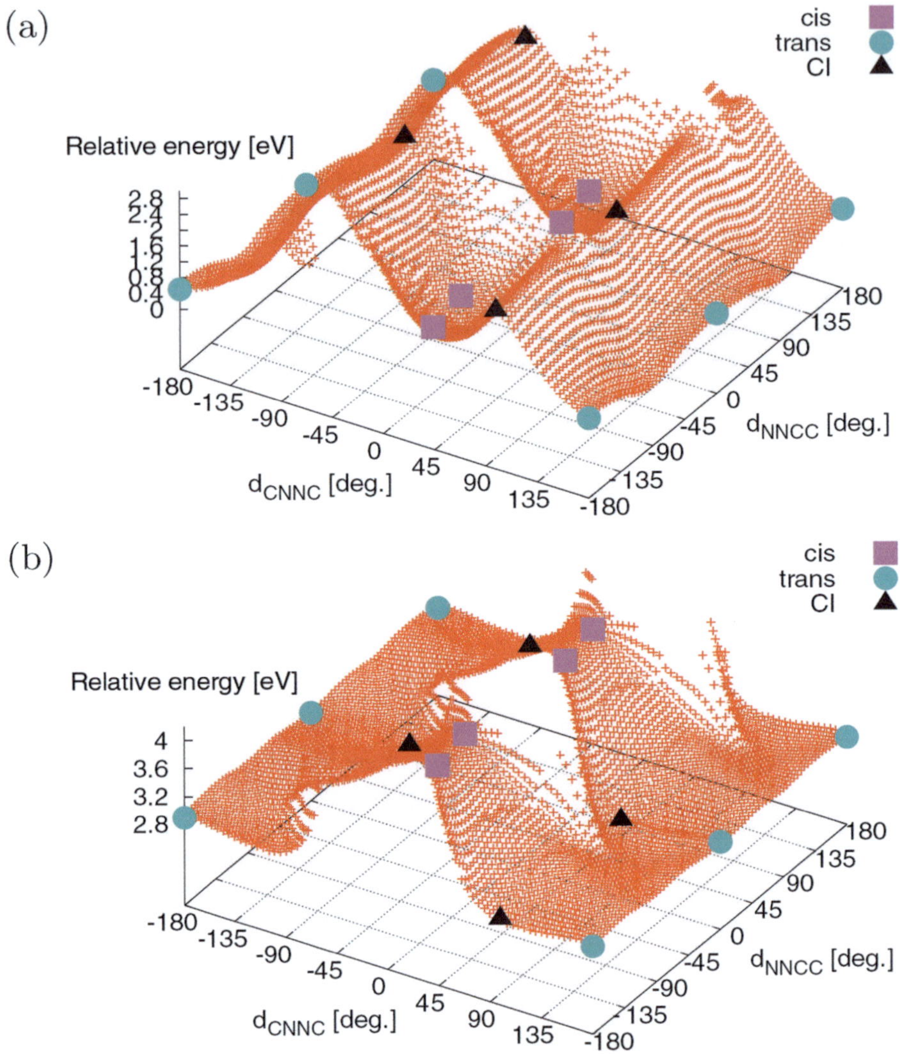

Fig. 2.11 PES as a function of CNNC and NNCC dihedral angles for (**a**) S_0 and (**b**) S_1 states (Reprinted from Ref. [89] with the permission of AIP Publishing)

2.5.1 Spectroscopic Characteristics

Electronic donors (such as $-NR_2$, $-OR$) can significantly displace the absorption band of azobenzene to a longer wavelength and enhance its intensity if they are on the *p*-position of the conjugated system [90–93]. The order of the increasing bathochromic (red-shifted) effect is $H < OH < NH_2 < N(CH_3)_2$ [94]. This effect will be remarkably amplified if there is an electronic acceptor on the p' position

[95]. Such an absorption band from the whole conjugated systems is referred to as a K-band, and the light-absorbing structures are known as K-chromophores. The high-intensity K-band can overlap with the very low-intensity R-band from the azo group for many substituted azobenzenes, which show strong absorption in the visible light range. In nowadays terminology, it means that their $n \rightarrow \pi^*$ and $\pi \rightarrow \pi^*$ transition bands overlap in this wavelength range. The absorption spectra of azo chromophores and dyes can be significantly affected by the solvents [95, 96].

For 4-donor–4′-acceptor-substituted azobenzenes, the strong absorption bands appear in the visible range, and their positions depend on the identity of the substituents and solvents. The dye characteristic for these compounds has been attributed to the shift of the $\pi \rightarrow \pi^*$ transition band to the visible region and/or the enhanced $n \rightarrow \pi^*$ absorption band because of the substituent perturbation [97]. Excitation profiles (EPs) of resonance Raman scattering have been measured to yield information on the transition responses [98–101]. As indicated by the results, there are two electronic states with the large electronic transition dipole moments, where the excitation energy is in the range 2.3–3.0 eV relative to the ground state. These states have been assigned to the mixed charge-transfer and $n \rightarrow \pi^*$ excitations. The interaction between them and higher-lying $\pi - \pi^*$ state is sensitive to the nature of the electronic donor and acceptor groups [101].

2.5.2 *Thermal Isomerization of Substituted Azobenzenes*

For azobenzene, there is almost no serious disputation over the thermal isomerization mechanism. On the other hand, thermal isomerization of substituent azobenzenes is a more involved issue for understanding its mechanism compared with that of azobenzene. Hartley reported that the thermal *cis* → *trans* isomerization of 4-amino, 4-dimethylamino, and 4-hydroxy-azobenzenes is more rapid than that of azobenzene [5]. The absorption spectra of a series of *cis*-azobenzene derivatives have been measured by different authors [102, 103]. It was observed that the thermal *cis* → *trans* isomerization of 4-amino and 4-hydroxyl azobenzenes usually took place in a such rapid rate and the reproducible absorption spectra could only be obtained by the aid of a rotating shutter [94]. On the basis of the results of five *p*-substituted azobenzene [24], Jaffé claimed that the isomerization rates follow the Hammett '$\sigma\rho$' equation [104]. This claim was later challenged by the study on the thermal isomerization with a much large number of substituents and starting from pure *cis* isomers [105]. The result showed that all *p*-substituted azobenzenes isomerize faster than the parent compound, regardless of the nature of the substituent. Because of the absence of large solvent or substituent effects, the authors proposed that the thermal isomerization with the low activation energy (21–24 kcal/mol) was through the inversion pathway. Recent study has shown that the thermal isomerization can be affected by the surrounding environment. The thermal isomerization of nitro-substituted azobenzenes can be accelerated up to 13 times when they are doped in nematic low molar liquid crystals [106]. The isomerization acceleration can reach 10^3 times when the azo chromophores are covalently linked to a nematic siloxane polymer.

Intensive investigations on the thermal isomerization mechanism have been focused on the *push–pull*-type azo chromophores. Thermal isomerization of the 4-donor–4′-acceptor-substituted azobenzenes has been thoroughly investigated by different groups. However, for a long time, interpretations of the results have been hotly debated following the mode of the rotation versus inversion disputation of azobenzene. This type azobenzenes, such as 4-(diethylamino)-4′-nitroazobenzene (4-NEt_2-4′-NO_2-AB), show much faster rates of thermal isomerization and lower energy of the transition state [107, 108]. The isomerization rate increases and activation energy decreases with the increase in the polarity of solvents [109, 110]. The isomerization reaction is accelerated by hydrogen-bonding solvents relative to aprotic solvents with similar polarity [111]. These observed solvent and substituent effects imply the existence of a polar or dipolar transition state. Considering the reduced NN double bond strength and torsional barrier to rotation due to the contribution from the dipolar resonance, Whitten et al. proposed that the thermal isomerization is through a rotation mechanism [107]. This rotation mechanism of thermal isomerization is contradictory to the inversion mechanism proposed for azobenzene [34, 41, 42] and 4-donor–4′-acceptor-substituted azobenzenes [109]. Therefore, this new theory has aroused a series of subsequent studies and hot discussions.

Through the study of the pressure effects on the thermal *cis* → *trans* isomerization of 4-(dimethylamino)-4′-nitroazobenzene (4-NMe_2-4′-NO_2-AB), Asano proposed that the isomerization mechanism changes from inversion in hexane to rotation in benzene [112]. One of the important evidences to support this inversion–rotation competition mechanism is that the activation volume $\Delta V^{\neq}$ decreases with an increase of solvent polarity and electron-donating power of the *push* substituent. The same change in mechanism takes place for 4-anilino-4′-nitroazobenzene, but not for 4-methoxy-4′-nitroazobenzene, for which the mechanism does not change obviously with the solvent polarity [113]. A strong electron-donating dialkylamino group makes the rotational transition state stable enough to compete with the inversion, while a strong acceptor like nitro increases the stability of the inversion transition state more than that for the rotational one [114].

On the other hand, based on their study on the solvent and substituent effects, Marcandalli et al. indicated that the substitution in the 4′-position invariably leads to an acceleration of the thermal *cis–trans* isomerization regardless of the nature of the substituent [115]. The authors believed that these results support the inversion mechanism rather than rotation mechanism. Through the investigation on the solvent, substituent, and pressure effects of *push–pull* azobenzenes, Nishimura et al. have also concluded that their results are in better agreement with the inversion mechanism rather than the rotation or inversion–rotation competition mechanism [116]. According to the inversion mechanism, the electronic distribution of the π-electrons in the transition state should become similar to that of the *trans* isomer as recovering the coplanarity. An important result to support this viewpoint is that the activation volumes are linearly correlated to the partial molar volumes in solvents with different polarities [117, 118].

Although these observations made by different researchers agree that the transition states of this type of azobenzenes are more polar than that of the *cis* isomer, the interpretations of the results are contradictory. These results have led to the proposition of two different mechanisms concerning the *cis–trans* thermal isomerization. According to the Whitten–Asano inversion–rotation competition mechanism, the increased polarity is caused by the increased proportion of the rotational isomerization [119]. In the Marcandalli–Nishimura inversion mechanism, it is attributed to the contribution of the inversional dipolar resonance structure [118].

In order to resolve the contradiction, more kinetic experiments have been carried out on the *push–pull* azobenzenes. The first-order rate constant for the thermal *cis–trans* isomerization of 4-NMe_2-4′-NO_2-AB was measured in various solvents at different temperatures and pressures. The results show that the temperature dependence of the activation volume is qualitatively different in different solvents, which indicates that there are two competing mechanisms in the process [119]. The solvent effects on the activation free energy ($\Delta G^{\neq}$) were studied for the *cis–trans* isomerization of *push–pull* azobenzenes, and the results were correlated with the Kamlet–Taft π^* solvent parameter [111, 120, 121]. The sharp break in the plots of $\Delta G^{\neq}$ versus π^* suggests that there exist both inversion and rotational isomerization mechanisms. The exact pathway is determined by the extent of solvent-mediated donor–acceptor coupling [121]. Later calculation based on the ab initio density functional theory shows that the preferred mechanism of isomerization in the ground state is the inversion of the NNC angle on the same side as the NO_2 substituent [16]. However, as the solvent effects have not been taken into account in the calculation, the result cannot be directly compared with the experimental results.

2.5.3 Photoisomerization of Substituted Azobenzenes

The photoisomerization of *substituted* azobenzenes has been a subject of considerable interest for many years. The quantum yields and their correlations with solvent and temperature for the photoisomerization were studied by spectrophotometric methods [122, 123]. However, it was observed that the lifetime of the *cis* isomers of azobenzenes bearing donor and/or acceptor substituents was too short to be investigated by these conventional methods [94, 122]. One possible method used at that time was to irradiate the azo compounds at a temperature low enough to prevent thermal isomerization [124]. In recent years, the rapid developments in the time-resolved spectroscopic methods and computational facilities have paved the way to investigate the photoisomerization of these azobenzene derivatives. To facilitate the discussion, the substituted azobenzenes are classified into the donor-, acceptor-, and donor–acceptor-substituted azobenzenes below.

Deep understanding of the photoisomerization dynamics of donor-substituted azobenzenes has been achieved by using the time-resolved spectroscopic methods [125, 126]. The subpicosecond pump-probe spectroscopic investigation was carried out on 4-(dimethylamino)azobenzene [126]. It showed that the rapid isomerization follows the excitation at 400 nm, which induces the lowest lying $\pi \rightarrow \pi^*$ transition. The excited-state internal conversion occurs with a time constant of ~0.8 ps in acetone and ~1.4 ps in fluorotrichloromethane as the solvent. The internal conversion is followed by ground state vibrational relaxation with the ~10 ps timescale. In order to clarify the photoisomerization mechanism, ultrafast spectroscopic investigation using chirp-controlled short pulses has been performed on 4-(dimethylamino)-azobenzene [127]. The study shows that there exists at least one more mode other than the torsional mode, which is coupled to either or both of the N=N and C–N stretching modes. Therefore, the *trans–cis* photoisomerization cannot be simply described by the typical rotation or inversion mechanisms.

This isomerization process of 4-aminoazobenzene has been investigated with static and transient absorption spectroscopy [128]. For this compound, the $\pi \rightarrow \pi^*$ transition band of the *trans* isomer appears at ~390 nm in ethanol and its long wavelength wing overlapped with the $n \rightarrow \pi^*$ band. After photoexcitation, the excited S_2 (π, π^*) state decays to the S_1 (n, π^*) state with the time constant of 0.2 ps. The relaxation of the S_1 state undergoes two pathways with time constants of 0.6 and 1.9 ps. The final vibrational cooling of the ground state takes ~15 ps. As observed by the authors, the S_2 state decay dynamics of 4-aminoazobenzene is similar to that of azobenzene. On the other hand, the dynamics of the following decay from S_1 state of the donor-substituted azobenzene is obviously different with that of azobenzene, but is similar to the process after directly exciting the latter to the S_1 state. It means that the energy gap between the states is an important factor to control the photoisomerization pathways and dynamics. The results have been explained by using the potential energy diagram given in Fig. 2.12, which was obtained based on the previous calculations [43, 48]. Due to the substituted effect, the energy gap between the (π, π^*) state and the (n, π^*) state is rather small. Therefore, the relaxation pathway of the (π, π^*) excited state after the internal conversion to the (n, π^*) state corresponds well to that of the *trans*-azobenzene directly excited to the (n, π^*) state.

Few investigations have been devoted to the photoisomerization dynamics of acceptor-substituted azobenzenes. *p*-Nitroazobenzene as a typical acceptor-substituted azobenzene has been studied by using resonance Raman spectroscopy [129]. This compound shows a strong absorption band around 335 nm corresponding to the S_2 ($\pi \rightarrow \pi^*$) electronic transition and a weak band at 437 for S_1 ($n \rightarrow \pi^*$) transition. This spectral characteristic is similar to that of azobenzene. The results are modeled using Heller's time-dependent formalism for Raman scattering. It shows that upon $n \rightarrow \pi^*$ excitation, N=N and C–N bonds are significantly distorted within 20 fs. Correspondingly, the unsubstituted phenyl ring and nitro stretching vibrations are also affected by the photoexcitation. The authors suggested that the isomerization in the S_1 state occurs through inversion mechanism.

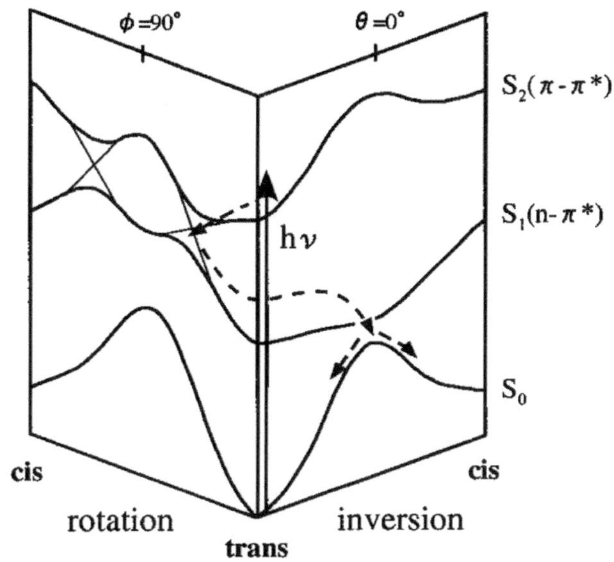

Fig. 2.12 Schematic representation of the potential energy diagram used to explain the results (Reprinted with permission from Ref. [128]. Copyright (2002) American Chemical Society)

The 4-donor–4′-acceptor-substituted azobenzenes (*push–pull* azobenzenes) play important roles as functional structures in various azo polymers. The photoisomerization behavior of this type of molecules has been investigated with femtosecond time-resolved spectroscopic methods. For 4-(4′-aminophenylazo) benzoic acid sodium salt, the excited *trans* isomer reaches the electronic ground state within 1 ps followed by the ground state vibrational cooling in the 10 ps timescale [130]. The resonance Raman spectroscopic study has proved that the solvent polarity has a significant influence on the excited-state energy surface of 4-nitro-4′-dimethylaminoazobenzene [131]. The locally excited state is more favorable in n-hexane, while the charge-transfer state is more stable in benzene. By analysis of the distortions along the N=N and C–N stretching vibrations of the molecule, the mechanism variation in the solvents, i.e., isomerization via rotation in n-hexane and inversion in benzene, has been suggested by the authors. After the excitation to the charge-transfer state, the excited-state evolution occurs along N–O, N=N, C–N, and C–C stretching vibration within 20 fs [132].

From the investigation by femtosecond fluorescence and absorption spectroscopy, the two-stage isomerization process is proposed [133]. It occurs within 100 fs for the large-amplitude motion on the excited-state PES away from the Franck–Condon region. To complete the isomerization, the "search" of the conical intersection (CI) takes ~1 ps. Three stages of the isomerization were proposed to be: (i) extremely rapid (<100 fs) inter-/intramolecular modes to reach the new charge distribution caused by the charge-transfer (CT) excitation, (ii) a large-amplitude motion (within 100 fs) on the excited-state potential away from the Franck–Condon region, and (iii) a "search" of the conical intersection (CI) to the ground state taking ~1 ps. The last two stages are very similar to what was observed in azobenzene by the same group [60]. It indicates that a *push–pull* substitution of azobenzene has little impact on the photoisomerization dynamics and only a moderate acceleration is observed for the photoisomerization 4-nitro-4′-dimethylaminoazobenzene. The authors proposed that the main difference only exists in the initially excited states, which is the n-π* state for azobenzene and CT state for the strong *push–pull* substituted azobenzene. Based on this investigation, the charge-transfer (CT) state is an important characteristic for the photoisomerization of *push–pull* azo chromophores, which is consistent with the results form resonance Raman scattering [101].

In above discussion, the isomerization of low-molecular-weight azo compounds is presented to offer basic understanding of the processes related to the configuration variation. For azo polymers discussed in this book, after light absorption, the polymeric system will undergo a wide variety of transformations related to the photoisomerization and other energy dissipation processes. The photoisomerization rate and degree of the azo chromophores depend on the factors such as free volume, temperature, mobility, and polarity of the polymer systems. Therefore, the azobenzene and its derivatives have been used as sensitive photochemical probes to detect the local environments in the systems. For such applications, we refer to the interested reader to some reviews previously published [134, 135]. Various photoresponsive variations related to the functional applications, which are triggered by the *trans–cis* photoisomerization, will be discussed in the following chapters in detail.

References

1. Barltrop, J.A., Coyle, J.D.: Principles of Photochemistry. John Wiley & Sons Ltd, New York (1978)
2. Turro, N.J.: Modern Molecular Photochemistry. The Benjamin/Cummings Publishing Co. Inc., Menlo Park (1978)
3. Böckmann, M., Marx, D., Peter, C., Site, L.D., Kremer, K., Doltsinis, N.L.: Multiscale modeling of mesoscopic phenomena triggered by quantum events: light-driven azo-materials and beyond. Phys. Chem. Chem. Phys. **13**, 7604–7621 (2011)
4. Hartley, G.S.: The *cis*-form of azobenzene. Nature **140**, 281–282 (1937)
5. Hartley, G.S.: The *cis*-form of azobenzene and the velocity of the thermal *cis* → *trans* conversion of azobenzene and some derivatives. J. Chem. Soc. 633–642 (1938)
6. Wyman, G.M.: The *cis-trans* isomerization of conjugated compounds. Chem. Rev. **55**, 625–657 (1955)
7. Tsuji, T., Takashima, H., Takeuchi, H., Egawa, T., Konaka, S.: Molecular structure and torsional potential of trans-azobenzene. A gas electron diffraction study. J. Phys. Chem. A **105**, 9347–9353 (2001)
8. de Lange, J.J., Robertson, J.M., Woodward, I.: X-ray crystal analysis of *trans*-azobenzene. Proc. R. Soc. London Ser A **171**, 398–410 (1939)
9. Harada, J., Ogawa, K., Tomoda, S.: Molecular motion and conformational interconversion of azobenzenes in crystals as studied by X-ray diffraction. Acta. Crystallogr. B **53**, 662–672 (1997)
10. Traetteberg, M., Hilmo, I., Hagen, K.: Gas electron-diffraction study of molecular-structure of *trans*-azobenzene. J. Mol. Struct. **39**, 231–239 (1977)

11. Armstrong, D.R., Clarkson, J., Smith, W.E.: Vibrational analysis of *trans*-azobenzene. J. Phys. Chem. **99**, 17825–17831 (1995)
12. Biswas, N., Umapathy, S.: Density functional calculations of structures, vibrational frequencies, and normal modes of *trans*- and *cis*-azobenzene. J. Phys. Chem. A **101**, 5555–5566 (1997)
13. Kurita, N., Tanaka, S., Itoh, S.: *Ab Initio* molecular orbital and density functional studies on the stable vibrational properties of *trans*- and *cis*-azobenzene. J. Phys. Chem. A **104**, 8114–8120 (2000)
14. Fliegl, H., Köhn, A., Hättig, C., Ahlrichs, R.: Ab initio calculation of the vibrational and electronic spectra of *trans*- and *cis*-azobenzene. J. Am. Chem. Soc. **125**, 9821–9827 (2003)
15. Cembran, A., Bernardi, F., Garavelli, M., Gagliardi, L., Oriandi, G.: On the mechanism of the *cis-trans* isomerization in the lowest electronic states of azobenzene: S_0, S_1, T_1. J. Am. Chem. Soc. **126**, 3234–3243 (2004)
16. Crecca, C.R., Roitberg, A.E.: Theoretical study of the isomerization mechanism of azobenzene and disubstituted azobenzene derivatives. J. Phys. Chem. A **110**, 8188–8203 (2006)
17. Kurita, N., Ikegami, T., Ishikawa, Y.: *Ab Initio* study of the minimum-energy structure of *trans*-azobenzene. Chem. Phys. Lett. **360**, 349–354 (2002)
18. Tiago, M.L., Ismail-Beigi, S., Louie, S.G.: Photoisomerization of azobenzene from first-principles constrained density-functional calculations. J. Chem. Phys. **122**(1–7), 094311 (2005)
19. Briquet, L., Vercauteren, D.P., Perpète, E.A., Jacquemin, D.: Is solvent *trans*-azobenzene twisted of planer. Chem. Phys. Lett. **417**, 190–195 (2006)
20. Mostad, A., Rømming, C.: A refinement of the crystal structure of *cis*-azobenzene. Acta Chem. Scand. **25**, 3561–3568 (1971)
21. Cattaneo, P., Granucci, G., Persico, M.: Simulations of condensed phase photochemistry: cage effect and internal conversion in azoalkanes and nitrosamines. J. Phys. Chem. **103**, 3364–3371 (1999)
22. Ishikawa, T., Noro, T., Shoda, T.: Theoretical study on the photoisomerization of azobenzene. J. Chem. Phys. **115**, 7503–7512 (2001)
23. Adamson, A.W., Vogler, A., Kunkely, H., Wachter, R.: Photocalorimetry. Enthalpies of photolysis of *trans*-azobenzene, ferrioxalate and cobaltioxalate ions, chromium hexacarbonyl, and dirhenium decarbonyl. J. Am. Chem. Soc. **100**, 1298–1300 (1978)
24. Le Fèvre, R.J.W., Northcott, J.: The effects of substituents and solvents on the *cis* → *trans* change of azobenzene. J. Chem. Soc. 867–870 (1953)
25. Fischer, E., Frankel, M., Wolovsky, R.: Wavelength dependence of photoisomerization equilibria in azocompounds. J. Chem. Phys. **23**, 1367–1368 (1955)
26. Zimmerman, G., Chow, L.Y., Paik, U.J.: The photochemical isomerization of azobenzene. J. Am. Chem. Soc. **80**, 3528–3531 (1958)
27. Yamashida, S., Ono, H., Toyama, O.: The *cis-trans* photoisomerization of azobenzene. Bull. Chem. Soc. Jpn. **35**, 1849–1853 (1962)
28. Bortolus, P., Monti, S.: *Cis-trans* photoisomerization of azobenzene. Solvent and triplet donor effect. J. Phys. Chem. **83**, 648–652 (1979)
29. Rau, H.: Photoisomerization of azobenzenes. In: Rabek, J.F. (ed.) Photochemistry and Photophysics, vol. 2, pp. 119–141. CRC Press, Boca Raton (1990)
30. Rau, H.: Azo compounds. In: Dürr, H., Bouas-Laurent, H. (eds.) Photochromism: Molecules and Systems, pp. 165–192. Elsevier, Amsterdam (1990)
31. Bandara, H.M.D., Burdette, S.C.: Photoisomerization in different classes of azobenzene. Chem. Soc. Rev. **41**, 1809–1825 (2012)
32. Fischer, E.: Temperature dependence of photoisomerization equilibria. Part I. Azobenzene and the azonaphthalenes. J. Am. Chem. Soc. **82**, 3249–3252 (1960)
33. Malkin, S., Fischer, E.: Temperature dependence of photoisomerization. Part II. Quantum yields of *cis*↔ *trans* isomerizations in azo-compounds. J. Phys. Chem. **66**, 2482–2486 (1962)

34. Gegiou, D., Muszkat, K.A., Fischer, E.: Temperature dependence of photoisomerization. VI. The viscosity effect. J. Am. Chem. Soc. **90**, 12–18 (1968)
35. Jones, L.B., Hammond, G.S.: Mechanisms of photochemical reactions in solution. XXX. Photosensitized isomerization of azobenzene. J. Am. Chem. Soc. **87**, 4219–4220 (1965)
36. Fischer, E.: Photosensitized isomerization of azobenzene. J. Am. Chem. Soc. **90**, 796–797 (1968)
37. Ronayette, J., Arnaud, R., Lebourgeois, P., Lemaire, J.: Isomérisation photochimique de l'azobenzèneen solution. I. Can. J. Chem. **52**, 1848–1857 (1974)
38. Keanrs, D.R.: The temperature dependence of the *cis-trans* photoisomerization of azo compounds: theoretical considerations. J. Phys. Chem. **69**, 1062–1065 (1965)
39. Stegemeyer, H.: On the mechanism of photochemical cis↔trans isomerization. J. Phys. Chem. **66**, 2555–2560 (1962)
40. Ronayette, J., Arnaud, R., Lemaire, J.: Isomérisation photosensibilisée par des colorants at photoréduction de l'azobenzène en solution. II. Can. J. Chem. **52**, 1858–1867 (1974)
41. Curtin, D.Y., Grubbs, E.J., McCarty, C.G.: Uncatalyzed *syn-anti* isomerization of imines, oxime ethers and haloimines. J. Am. Chem. Soc. **88**, 2775–2786 (1966)
42. Ljunggren, S., Wettermark, G.: CNDO/2 study of the mechanism of isomerization and conformation of azobenzene. Acta Chem. Scand. **25**, 1599–1606 (1971)
43. Cattaneo, P., Persico, M.: An *ab initio* study of the photochemistry of azobenzene. Phys. Chem. Chem. Phys. **1**, 4739–4743 (1999)
44. Magee, J.L., Shand, W., Erying, H.: Non-adiabatic reactions. Rotation about the double bond. J. Am. Chem. Soc. **63**, 677–688 (1941)
45. Gimarc, B.M.: The shapes of simple polyatomic molecules and ions. I. the series HAAH and BAAB. J. Am. Chem. Soc. **92**, 266–275 (1970)
46. Camp, R.N., Epstein, I.R., Steel, C.: Theoretical studies of the photochemistry of acyclic azoalkanes. J. Am. Chem. Soc. **99**, 2453–2459 (1977)
47. Rau, H.: Further evidence for rotation in the π, π^* and inversion in the n, π^* photoisomerization of azobenzene. J. Photochem. **26**, 221–225 (1984)
48. Monti, S., Orlandi, G., Palmieri, P.: Features of the photochemically active state surfaces of azobenzene. Chem. Phys. **71**, 87–99 (1982)
49. Rau, H., Lüddecke, E.: On the rotation-inversion controversy on photoisomerization of azobenzenes. Experimental proof of inversion. J. Am. Chem. Soc. **104**, 1616–1620 (1982)
50. Rau, H., Shen, Y.Q.: Photoisomerizaton of sterically hindered azobenzenes. J. Photochem. Photobio. A Chem. **42**, 321–327 (1988)
51. Bortolus, P., Monti, S.: Cis↔trans photoisomerization of azobenzene-cyclodextrin inclusion complexes. J. Phys. Chem. **91**, 5046–5050 (1987)
52. Struve, W.S.: Emission from the ^{1}(n, π^*) state of azobenzene: spectrum and ultrashort decay time. Chem. Phys. Lett. **46**, 15–19 (1977)
53. Morgante, C.G., Struve, W.S.: $S_2 \rightarrow S_0$ fluorescence in *trans*-azobenzene. Chem. Phys. Lett. **68**, 267–271 (1979)
54. Nägele, T., Hoche, R., Zinth, W., Wachtveitl, J.: Femtosecond photoisomerization of *cis*-azobenzene. Chem. Phys. Lett. **272**, 489–495 (1997)
55. Lednev, I.K., Ye, T.Q., Hester, R.E., Moore, J.N.: Femtosecond time-resolved UV-visible absorption spectroscopy of *trans*-azobenzene in solution. J. Phys. Chem. **100**, 13338–13341 (1996)
56. Lednev, I.K., Ye, T.Q., Matousek, P., Towrie, M., Foggi, P., Neuwahl, F.V.R., Umapathy, S., Hester, R.E., Moore, J.N.: Femtosecond time-resolved UV-visible absorption spectroscopy of *trans*-azobenzene: dependence on excitation wavelength. Chem. Phys. Lett. **290**, 68–74 (1998)
57. Lednev, I.K., Ye, T.Q., Abbott, L.C., Hester, R.E., Moore, J.N.: Photoisomerization of a capped azobenzene in solution probed by ultrafast time-resolved electronic absorption spectroscopy. J. Phys. Chem. A **102**, 9161–9166 (1998)

58. Fujino, T., Arzhantsev, S.Y., Tahara, T.: Femtosecond time-resolved fluorescence study of photoisomerization of *trans*-azobenzene. J. Phys. Chem. A **105**, 8123–8129 (2001)
59. Lu, Y.C., Diau, E.W., Rau, H.: Femtosecond fluorescence dynamics of rotation-restricted azobenzenophanes: new evidence on the mechanism of *trans* → *cis* photoisomerization of azobenzene. J. Phys. Chem. A **109**, 2090–2099 (2005)
60. Satzger, H., Spörlein, S., Root, C., Wachtveitl, J., Zinth, W., Gilch, P.: Fluorescence spectra of *trans*- and *cis*-azobenzene – emission from the Franck-Condon state. Chem. Phys. Lett. **372**, 216–223 (2003)
61. Satzger, H., Root, C., Braun, M.: Excited-state dynamics of *trans* and *cis*-azobenzene after UV excitation in the $\pi\pi^*$ band. J. Phys. Chem. **108**, 6256–6271 (2004)
62. Hamm, P., Ohline, S.M., Zinth, W.: Vibrational cooling after ultrafast photoisomerization of azobenzene measured by femtosecond infrared spectroscopy. J. Chem. Phys. **106**, 519–529 (1997)
63. Schultz, T., Quenneville, J., Levine, B., Toniolo, A., Martinez, T.J., Lochbrunner, S., Schmitt, M., Shaffer, J.P., Zgieski, M.Z., Stolow, A.: Mechanism and dynamics of azobenzene photoisomerization. J. Am. Chem. Soc. **125**, 8098–8099 (2003)
64. Chowdary, P.D., Umapathy, S.: Ultrafast dynamics and photochemistry of $\pi - \pi^*$ excited *trans*-azobenzene – a comprehensive analysis of resonance Raman intensities. J. Raman Specrosc. **39**, 1538–1555 (2008)
65. Fujino, T., Tahara, T.: Picosecond time-resolved Raman study of *trans*-azobenzene. J. Phys. Chem. A **104**, 4203–4210 (2000)
66. Pancur, T., Renth, F., Temps, F., Harbaum, B., Krüger, A., Herges, R., Näther, C.: Femtosecond fluorescence up-conversion spectroscopy of a rotation-restricted azobenzene after excitation to the S_1 state. Phys. Chem. Chem. Phys. **7**, 1985–1989 (2005)
67. Chang, C.W., Lu, Y.C., Wang, T.T., Diau, E.W.: Photoisomerization dynamics of azobenzene in solution with S_1 excitation: a femtosecond fluorescence anisotropy study. J. Am. Chem. Soc. **126**, 10109–10118 (2004)
68. Bandara, H.M.D., Friss, T.R., Enriquez, M.M., Isley, W., Incarvito, C., Frank, H.A., Gascon, J., Burdette, S.C.: Proof for the concerted inversion mechanism in the *trans* → *cis* isomerization of azobenzene using hydrogen bonding to induce isomer locking. J. Org. Chem. **75**, 4817–4827 (2010)
69. Stuart, C.M., Froutiera, R.R., Mathies, R.A.: Excited-state structure and dynamics of *cis*- and *trans*-azobenzene from resonance Raman intensity analysis. J. Phys. Chem. A **111**, 12072–12080 (2007)
70. Cimiraglia, R., Hofmann, H.J.: Rotation and inversion states in thermal E/Z isomerization of aromatic compounds. Chem. Phys. Lett. **217**, 430–435 (1994)
71. Liu, R.F., Cui, Q., Dunn, K.M., Morokuma, K.: Ab initio molecular orbital study of the mechanism of photodissociation of transazomethane. J. Chem. Phys. **105**, 2333–2345 (1996)
72. Altoè, P., Bernardi, F., Conti, I., Garavelli, M., Negri, F., Orlandi, G.: Light driven molecular switches: exploring and tuning their photophysical and photochemical properties. Theor. Chem. Acc. **117**, 1041–1059 (2007)
73. Diau, E.W.: A new *trans*-to-*cis* photoisomerization mechanism of azo benzene on the S_1 (n, π^*) surface. J. Phys. Chem. A **108**, 950–956 (2004)
74. Ciminelli, C., Granucci, G., Persico, M.: Are azobenzenophanes rotation-restricted? J. Chem. Phys. **123**, 174317 (10 pages) (2005)
75. Nonnenberg, C., Gaub, H., Frank, I.: First-principles simulation of the photoreaction of a capped azobenzene: the rotational pathway is feasible. ChemPhysChem **7**, 1451–1461 (2006)
76. Böckmann, M., Doltsinis, N.L., Marx, D.: Azobenzene photoswitches in bulk materials. Phys. Rev. E. **78**, 036101 (4 pages) (2008)
77. Böckmann, M., Doltsinis, N.L., Marx, D.: Nonadiabatic hybrid quantum and molecular mechanic simulation of azobenzene photoswitching in bulk liquid environment. J. Phys. Chem. A **114**, 745–754 (2010)

78. Tiberio, G., Muccioli, L., Berardi, R., Zannoni, C.: How does the *trans-cis* photoisomerization of azobenzene take place in organic solvents? ChemPhysChem **11**, 1018–1028 (2010)
79. Gagliardi, L., Orlandi, G., Bernardi, F., Cembran, A., Garavelli, M.: A theoretical study of the lowest electronic states of azobenzene: the role of torsion coordinate in the *cis-trans* photoisomerization. Theor. Chem. Acc. **111**, 363–372 (2004)
80. Conti, I., Garavelli, M., Orlandi, G.: The different photoisomerization efficiency of azobenzene in the lowest $n\pi^*$ and $\pi\pi^*$ singlets: The role of a phantom state. J. Am. Chem. Soc. **130**, 5216–5230 (2008)
81. Ciminelli, C., Granucci, G., Persico, M.: The photoisomerization mechanism of azobenzene: a semiclassical simulation of nonadiabatic dynamics. Chem. Eur. J. **10**, 2327–2341 (2004)
82. Toniolo, A., Ciminelli, C., Persico, M., Martínez, T.J.: Simulation of the photodynamics of azobenzene on its first excited state: comparison of full multiple spawning and surface hopping treatment. J. Chem. Phys. **123**, 234308 (10 pages) (2005)
83. Granucci, G., Persico, M.: Excited state dynamics with the direct trajectory surface hopping method: azobenzene and its derivatives as a case study. Theor. Chem. Acc. **117**, 1131–1143 (2007)
84. Cusati, T., Granucci, G., Persico, M.: Photodynamics and time-resolved fluorescence of azobenzene in solution: a mixed quantum-classical simulation. J. Am. Chem. Soc. **133**, 5109–5123 (2011)
85. Sauer, P., Allen, R.E.: Multiple steps and multiple excitations in photoisomerization of azobenzene. Chem. Phys. Lett. **450**, 192–195 (2008)
86. Sauer, P., Allen, R.E.: Influence of laser pulse parameters on dynamical processes during azobenzene photoisomerization. J. Phys. Chem. A **112**, 11142–11152 (2008)
87. Yuan, S., Dou, Y.S., Wu, W.F., Hu, Y., Zhao, J.S.: Why does *trans*-azobenzene have a small isomerization yield for $\pi\pi^*$ excitation than for $n\pi^*$ excitation. J. Phys. Chem. A **112**, 13326–13334 (2008)
88. Shao, J.F., Lei, Y.B., Wen, Z.Y., Dou, Y.S., Wang, Z.S.: Nonadiabatic simulation study of photoisomerization of azobenzene: detailed mechanism and load-resisting capacity. J. Chem. Phys. **129**, 164111 (9 pages) (2008)
89. Ootani, Y., Satoh, K., Nakayama, A., Noro, T., Taketsugu, T.: Ab initio molecular dynamics simulation of photoisomerization in azobenzene in the $n \rightarrow \pi^*$ state. J. Chem. Phys. **131**, 194306 (10 pages) (2009)
90. Burawoy, A.: Studies in the light absorption of organic compounds. Part VIII. azo-compounds. J. Chem. Soc. 1865–1869 (1937)
91. Funkhouser, J.A., Brode, W.R.: The relationship between the absorption spectra and the chemical constitution of dyes. VI. The influence of chromophore position in disazo dyes. J. Am. Chem. Soc. **56**, 2172–2173 (1934)
92. Piper, J.D., Brode, W.R.: The relationship between the absorption spectra and the chemical constitution of dyes. VII. The separation of chromophores in symmetrical disazo dyes. J. Am. Chem. Soc. **57**, 135–138 (1935)
93. Gore, P.H., Wheeler, O.H.: Absorption spectra of aromatic azo and related compounds. III. Substituted azobenzenes. J. Org. Chem. **26**, 3295–3298 (1961)
94. Brode, W.R., Gould, J.H., Wyman, G.M.: The relation between the absorption spectra and the chemical constitution of dyes. XXV. Phototropism and cis-trans isomerization in aromatic azo compounds. J. Am. Chem. Soc. **74**, 4641–4646 (1952)
95. Sheppard, S.E., Newsome, P.T.: The effect of solvents on the absorption spectra of dyes. II. some dyes other than cyanines. J. Am. Chem. Soc. **64**, 2937–2946 (1942)
96. Brode, W.R.: The effect of solvents on the absorption spectrum of a simple azo dye. J. Phys. Chem. **30**, 56–69 (1926)
97. Griffiths, J.: Photochemistry of azobenzene and its derivatives. Chem. Soc. Rev. **1**, 481–493 (1972)

98. Cataliotti, R.S., Murgia, S.M., Paliani, G., Poletti, A., Zgierski, M.Z.: Resonance Raman scattering in dyes derived from azobenzene. J. Raman Spectrosc. **15**, 251–257 (1985)
99. Cataliotti, R.S., Murgia, S.M., Paliani, G., Poletti, A., Zgierski, M.Z.: The analysis of unresolved Raman excitation profiles. J. Raman Spectrosc. **15**, 258–264 (1985)
100. Rava, R.P.: Deconvoluting the visible absorption spectrum of methyl orange using inverse Raman transform techniques. J. Chem. Phys. **87**, 3758–3765 (1987)
101. Cataliotti, R.S., Morresi, A., Paliani, G., Zgierski, M.Z.: Resonance Raman scattering in azo dyes. J. Raman Spectrosc. **20**, 601–604 (1989)
102. Cook, A.H., Jones, D.G., Polya, J.B.: *cis*-Azo-compounds. Part III. Absorption spectra. J. Chem. Soc. 1315–1320 (1939)
103. Birnbaum, P.P., Linford, J.H., Style, D.W.G.: The absorption spectra of azobenzene and some derivatives. Trans. Faraday Soc. **49**, 735–744 (1953)
104. Jaffé, H.H.: A reexamination of the Hammett equation. Chem. Rev. **53**, 191–261 (1953)
105. Talaty, E.R., Fargo, J.C.: Thermal *cis-trans*-isomerization of substituted azobenzenes: a correction of the literature. Chem. Commun. 65–66 (1967)
106. Garcia-Amorós, J., Finkelmann, H., Velasco, D.: Increasing the isomerization kinetics of azo dyes by chemical bonding to liquid-crystalline polymers. Chem. Eur. J. **17**, 6518–6523 (2011)
107. Wildes, P.D., Pacifici, J.G., Irick, G., Whitten, D.G.: Solvent and substituent effects on the thermal isomerization of substituted azobenzenes. A flash spectroscopic study. J. Am. Chem. Soc. **93**, 2004–2008 (1971)
108. Gabor, G., Fischer, E.: Spectra and *cis-trans* isomerization in highly bipolar derivatives of azobenzene. J. Phys. Chem. **75**, 581–583 (1971)
109. Nishimura, N., Sueyoshi, T., Yamanaka, H., Imai, E., Yamamoto, S., Hasegawa, S.: Thermal cis-to-trans isomerization of substituted azobenzenes II. substituent and solvent effects. Bull. Chem. Soc. Jpn. **49**, 1381–1387 (1976)
110. Gille, K., Knoll, H., Quitzsch, K.: Rate constants of the thermal *cis-trans* isomerization of azobenzene dyes in solvents, acetone/water mixtures, and in microheterogeneous surfactant solutions. Int. J. Chem. Kinet. **31**, 337–350 (1999)
111. Schanze, K.S., Mattox, T.F., Whitten, D.G.: Solvent effects upon the thermal *cis-trans* isomerization and charge-transfer absorption of 4-(diethylamino)-4′-nitroazobenzene. J. Org. Chem. **48**, 2808–2813 (1983)
112. Asano, T.: Pressure effects on the thermal *cis-trans* isomerization of 4-dimethylamino-4′-nitroazobenzene. Evidence for a change of mechanism with solvent. J. Am. Chem. Soc. **102**, 1205–1206 (1980)
113. Asano, T., Yano, T., Okada, T.: Mechanistic study of thermal *Z-E* isomerization of azobenzenes by high-pressure kinetics. J. Am. Chem. Soc. **104**, 4900–4904 (1982)
114. Asano, T., Okada, T.: Thermal *Z-E* isomerization of azobenzenes. The pressure, solvent, and substituent effects. J. Org. Chem. **49**, 4387–4391 (1984)
115. Marcandalli, B., Liddo, L.P., Fede, C.D., Bellobono, I.R.: Solvent and substituent effects on thermal *cis-trans*-isomerization of some 4-diethylaminoazobenzenes. J. Chem. Soc. Perkin Trans. **I1**, 589–593 (1984)
116. Nishimura, N., Kosako, S., Sueishi, Y.: The thermal isomerization of azobenzenes. III. substituent, solvent, and pressure effects on the thermal isomerization of *push-pull* azobenzenes. Bull. Chem. Soc. Jpn. **57**, 1617–1625 (1984)
117. Nishimura, N., Tanaka, T., Sueishi, Y.: Evidence of an inversion mechanism for the thermal *cis-trans* isomerization of push-pull azobenzenes. A volumetric study. J. Chem. Soc., Chem. Commun. 903–904 (1985)
118. Nishimura, N., Tanaka, T., Asano, M., Sueishi, Y.: A volumetric study to the thermal *cis*-to-*trans* isomerization of 4-(dimethylamino)-4′-nitroazobenzene and 4,4′-bis-(dialkyl-amino) azobenzenes: evidence of an inversion mechanism. J. Chem. Soc. Perkin Trans II, 1839–1845 (1986)

119. Asano, T., Okada, T.: Further kinetic evidence for the competitive rotational and inversional *Z-E* isomerization of substituted azobenzenes. J. Org. Chem. **51**, 4454–4458 (1986)
120. Sigman, M.E., Leffler, J.E.: Supercritical carbon dioxide. The cis to trans relaxation and π, π* transition of 4-(diethylamino)-4′-nitroazobenzene. J. Org. Chem. **52**, 3123–3126 (1987)
121. Shin, D.M., Whitten, D.G.: Solvent-induced mechanism change in charge-transfer molecules. Inversion versus rotation paths for the $Z \rightarrow E$ isomerization of donor-acceptor substituted azobenzene. J. Am. Chem. Soc. **110**, 5206–5208 (1988)
122. Birnbaum, P.P., Style, D.W.G.: The photo-isomerization of some azobenzene derivatives. Trans. Faraday Soc. **50**, 1192–1196 (1954)
123. Brode, W.R., Gould, J.H., Wyman, G.M.: The relation between the absorption spectra and the chemical constitution of dyes. XXVI. Effect of solvent and of temperature on the cis-trans isomerization of azo dyes. J. Am. Chem. Soc. **75**, 1856–1859 (1953)
124. Fischer, E., Frei, Y.: Photoisomerization equilibrium in azodyes. J. Chem. Phys. **27**, 328–330 (1957)
125. Azuma, J., Tamai, N., Shishido, A., Ikeda, T.: Femtosecond dynamics and stimulated emission from the S_2 state of a liquid crystalline *trans*-azobenzene. Chem. Phys. Lett. **288**, 77–82 (1998)
126. Mayer, S.G., Thomsen, C.L., Philpott, M.P., Reid, P.J.: The solvent-dependent isomerization dynamics of 4-(dimethylamino)azobenzene (DMAAB) studied by subpicosecond pump-probe spectroscopy. Chem. Phys. Lett. **314**, 246–254 (1999)
127. Saito, T., Kobayashi, T.: Conformational change in azobenzene in photoisomerization process studied with chirp-controlled sub-10-fs pulse. J. Phys. Chem. A **106**, 9436–9441 (2002)
128. Hirose, Y., Yui, H., Sawada, T.: Effect of potential energy gap between the n-π* and the π-π* state on ultrafast photoisomerization dynamics of an azobenzene derivative. J. Phys. Chem. A **106**, 3067–3071 (2002)
129. Biswas, N., Abraham, B., Umapathy, S.: Investigation of short-time isomerization dynamics in *p*-nitroazobenzene from Raman intensity analysis. J. Phys. Chem. A **106**, 9397–9406 (2002)
130. Wachtveitl, J., Nägele, T., Puell, B., Zinth, W., Krüger, M., Rudolph-Böhner, S., Oesterhelt, D., Moroder, L.: Ultrafast photoisomerization of azobenzene compounds. J. Photochem. Photobiol. A Chem **105**, 283–288 (1997)
131. Biswas, N., Umapathy, S.: Study of solvent effects on the molecular structure and the reorganization energies of 4-nitro-4′-dimethylamino-azobenzene using resonance Raman intensities. J. Raman Spectrosc. **32**, 471–480 (2001)
132. Biswas, N., Umapathy, S.: Resonance Raman study of the solvent dynamics for ultrafast charge transfer transition in 4-nitro-4′-dimethylamino-azobenzene. J. Chem. Phys. **118**, 5526–5536 (2003)
133. Schmidt, B., Sobotta, C., Malkmus, S., Laimgruber, S., Braun, M., Zinth, W., Gilch, P.: Femtosecond fluorescence and absorption dynamics of an azobenzene with a strong *push-pull* substitution. J. Phys. Chem. A **108**, 4399–4404 (2004)
134. Williams, J.L.R., Daly, R.C.: Photochemical probes in polymers. Prog. Polym. Sci. **5**, 61–93 (1977)
135. Kumar, G.S., Neckers, D.C.: Photochemistry of azobenzene-containing polymers. Chem. Rev. **89**, 1915–1925 (1989)

Chapter 3
Azo Polymer Syntheses

Abstract This chapter discusses different ways to obtain azo polymers. Like other functional polymers, azo polymers can be synthesized by polymerization and copolymerization of monomers that contain azo functional groups or by chemical modification of a suitable precursor polymer to introduce the azo functional groups after polymerization. The chain polymerization and step polymerization are discussed as the two typical ways to obtain azo homopolymers and random copolymers. Controlled radical polymerization is introduced as a new way to synthesize azo polymers with well-controlled molecular weight and distribution. Then, the ways to obtain azo block and graft copolymers, azo dendritic and star polymers, and azo hyperbranched polymers are discussed separately in each section in detail. Finally, syntheses of azo polymers with π-conjugated backbone and azobenzene-containing polypeptide are presented as separate sections due to their unique characteristics.

Keywords Azo homopolymer • Azo random copolymer • Azo block copolymer • Azo graft copolymer • Azo dendritic and star polymer • Azo polymers with π-conjugated backbone • Azo polypeptide

Azo polymers are macromolecules composed of a large number of structural units, which contain different azo functional groups designed to exhibit various functions. To obtain the azo polymers, the functional structural units can be connected together through chemical bonds in any conceivable and feasible manner. The polymeric backbones are built up by a variety of polymerization methods. The azo functional groups can be introduced both by using functional monomers or incorporating them through chemical reactions after polymerizations. As common cases for polymers, the number of structural units in a polymer molecule, i.e., the degree of polymerization (DP), can be very different depending on the synthetic methods. As a polymer is always polydispersed in its molecular weight, the degree of polymerization and molecular weight are usually given in the number-average and weight-average values. The ratio of the weight-average molecular weight (M_w) and the number-average molecular weight (M_n), termed polydispersity index, is often used to characterize the molecular weight distribution. The degree of functionalization (DF) is often used to indicate the proportion of the structural units containing azo chromophores to the total.

X. Wang, *Azo Polymers*, Soft and Biological Matter,
DOI 10.1007/978-3-662-53424-3_3

Azo polymers with diverse compositions and structures have been synthesized in the past decades. According to their compositions and bonded sequences, azo polymers can be classified into homopolymer, random copolymer, and block and graft copolymers. Depending on their topological architecture, azo polymers can be classified into linear, branched, starlike, hyperbranched, dendritic, and cross-linked polymers. A variety of methods has been used to synthesize these azo polymers, which has developed into a very active research frontier for azo polymer investigation and will be discussed in this chapter in detail.

3.1 Synthetic Method Classification

The polymerizations as a typical way to obtain a polymer can be classified as the addition and condensation polymerizations [1, 2]. Addition polymerization is carried out without the loss of small molecules during the formation of a polymer. The major type in this category is the polymerization of monomers containing the carbon–carbon double bond. In the process, the chain polymerization is initiated by an initiator, which is typically a free radical or ion. On the other hand, the condensation polymerizations form polymers through condensation reactions of the monomers with the elimination of some small molecules as by-products. More accurate classification should be made according to the polymerization mechanism. The polymerizations are then classified into chain and step polymerizations [1, 2]. Chain polymerization occurs through the propagation of a reactive center (free radical, cation, or anion) by the successive addition of a large number of monomer molecules. On the other hand, step polymerization proceeds by the stepwise reaction between the reactive groups of monomers.

Azo polymers can be considered as a specific type of functional polymers, to which the azo functional groups are attached. Integration of the azo functional groups into a polymeric architecture is necessary to obtain an azo polymer. Typically, the azo functional groups can be incorporated into the polymer main chain or, much more frequently, be linked to the backbone as side-chain pendant groups either directly or via a spacer group. Although azo polymers can include more complicated molecular architectures, the azo chromophores are either directly incorporated in a backbone or tethered to the backbone as pendant groups. There are two typical ways to synthesize azo polymers: (a) polymerization and copolymerization of monomers that contain azo functional groups and (b) chemical modification of a suitable precursor polymer to introduce the azo functional groups after polymerization. These two approaches are typically used for functional polymer syntheses, and their advantages and disadvantages have been discussed in previous monographs [3, 4]. In real applications, the structures of the target polymers and preparation conditions should be considered to answer the question which approach is a better choice than the other.

Generally speaking, azo polymers can be prepared by many different methods. In practice, some of the methods might be preferred for the preparation of azo polymers with specified structures, and others could be totally unfeasible. In many cases, a suitable method to obtain azo polymers with special functions is a topic intensively investigated by chemists. Instead of discussing all these possibilities, this chapter will present some commonly used methods for synthesizing azo polymers with different structures. The synthetic methods are discussed below according to the structural characteristics of azo polymers. Azo homopolymer and random copolymers are usually prepared by conventional addition and condensation polymerizations, which are discussed in Sect. 3.2. Azo block and graft copolymers are mainly prepared by controlled radical polymerization, which is the theme of Sect. 3.3. Dendritic polymers include dendrons, dendrimers, and hyperbranched polymers [5–7]. Dendrimers as well as dendrons are prepared by stepwise chemical reactions through convergent and divergent schemes. Hyperbranched azo polymers are typically prepared by using AB_x monomer through one-pot synthesis. The syntheses of the dendritic polymers are discussed in Sect. 3.4. Azo polymers with conjugated backbone and azo polypeptides are usually prepared by some specific methods, which are presented in Sects. 3.5 and 3.6, respectively.

3.2 Azo Homopolymer and Random Copolymer

Azo homopolymer and *random copolymers* are azo polymers with relatively simple structures and can be prepared by conventional addition and condensation polymerizations. The azo polymers are often prepared by polymerization and copolymerization of monomers containing the azo functional groups. In some cases, the azo polymers are also synthesized through chemical modification of a suitable precursor polymer. The early explorations in this area have been reviewed by Nuyken [8], Kumar and Neckers [9], and Kumar [4]. Progresses before 2003 have been summarized in some comprehensive reviews [10–12].

3.2.1 Chain Polymerization

The most widely used *chain polymerization* is addition polymerization of vinyl monomers initiated by free radicals. In the radical polymerization, polymers are formed by conversion of the double bonds of vinyl monomers into saturated linkages as represented by Fig. 3.1, where X is a side-chain substituent containing azo functional groups. A random copolymer can be prepared by this reaction using two or more types of monomers. Radical polymerization has been widely used to prepare various types of azo polymers. This versatile method can be applied to different types of vinyl monomers containing azo functional groups under mild reaction conditions. The radical polymerization can be performed in a broad temperature range (−20–200 °C) under different process conditions such as bulk

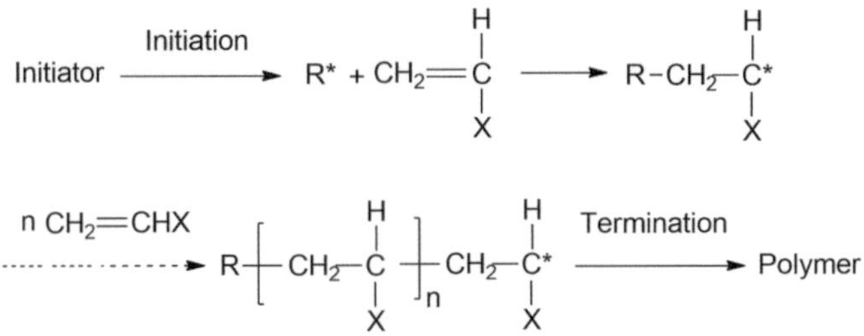

Fig. 3.1 Scheme of the radical polymerization to obtain side-chain azo polymers

polymerization, solution polymerization, emulsion polymerization, and heterogeneous polymerization. The only requirement for a radical polymerization is the absence of oxygen and other inhibitors. As many types of monomers can be copolymerized by radical polymerization, this method is also widely used to prepare azo random copolymers.

The syntheses of azo polymers through radical polymerization can be traced back to the 1960s and 1970s [13–15]. However, due to the purpose of these syntheses, such as studying the photochemical and thermal isomerization of aromatic azo residues in polymer systems, the content of the azo functional groups was very low for many of the polymers. The main component is from common monomers, such as acrylic and methacrylic acids [13] and methacrylate and styrene [15]. The monomers were polymerized in sealed ampules under vacuum or in an inert atmosphere. Typically, polymerizations were carried out through solution polymerization in a suitable solvent and initiated by azobisisobutyronitrile (AIBN). After the early stage of the development, radical polymerization has been widely used to prepare various azo polymers. Two typical types of polymers are discussed below to illustrate the synthetic methods.

Natansohn et al. have synthesized a series of acrylate-/methacrylate-based amorphous homopolymers and copolymers bearing electronic donor–acceptor azo chromophores [16–21]. Figure 3.2 shows some representatives with the chemical structure and the generic names, which are abbreviations of the corresponding names of the disperse azo dyes. The polymerizations are usually performed in anhydrous toluene with 10 % AIBN as initiator. This initiator concentration is obviously higher than the value used for conventional radical polymerization, but it has been found to produce the highest yield (50 %) of the polymers without unfavorably affecting the molecular weight. The polymerizations require a relatively long time, such as 4 days at 60 °C. The polymers are finally obtained by precipitation in methanol and tetrahydrofuran (THF)/methanol mixture. The number-average molecular weights of polymers are not high, which are about 4000 as measured by gel permeation chromatography (GPC) [16]. The yields and molecular weights depend on the monomers used in the polymerizations. For instances, the yields were 50 % and 56 % for pNDR1M and pDR1M, and the GPC-measured molecular weights are 3400 and 9700 for pNDR1M and pDR1M, respectively. As a general rule, the relatively low yield and low molecular weight

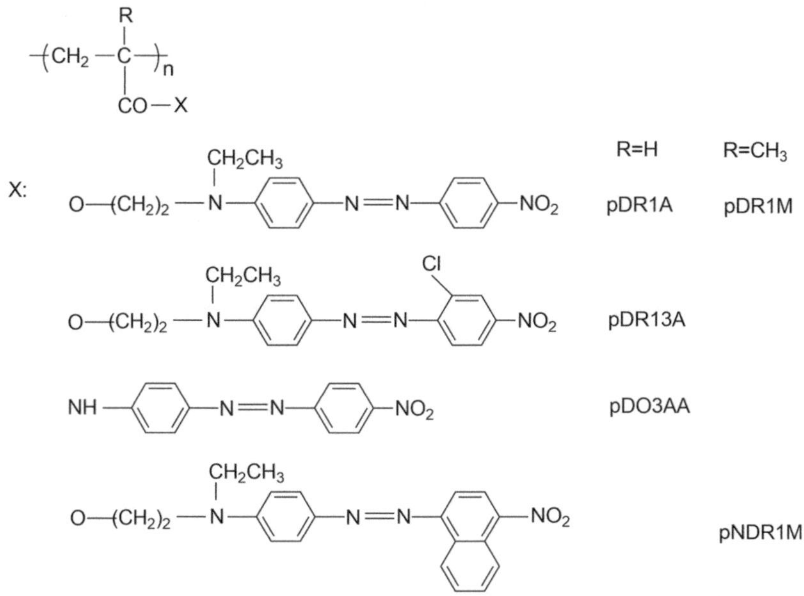

Fig. 3.2 Chemical structure of acrylate-/methacrylate-based amorphous homopolymers and their generic names

are typically observed for the polymerization of monomers bearing strong electronic donor–acceptor azo chromophores, which has been attributed to the inhibition and retardation effect of azo groups to radical.

Another well-known type of azo polymers obtained from the radical polymerization is liquid crystalline (LC) homopolymer and copolymers containing azobenzene side-chain groups [22–24]. Figure 3.3 shows the chemical structures and the generic names of a series of LC copolymers synthesized by Ikeda et al. [24]. For a typical preparation condition, the radical polymerization is carried out in tetrahydrofuran (THF) by using azobisisobutyronitrile (AIBN) as the initiator. Benzene can also be used as the solvent for some cases when high-molecular-weight polymers are required. Compared with the polymerization of monomers with strong *push–pull* chromophores, the polymerization conditions are less demanding. To control the sequential structure of copolymers, the conversion is usually kept low (<15 %). The polymers can be well purified by precipitation from chloroform solution into cold ether.

3.2.2 Step Polymerization

Step Polymerization has been widely used to prepare azo polymers (Fig. 3.4). Many among them are synthesized through condensation reactions of azo

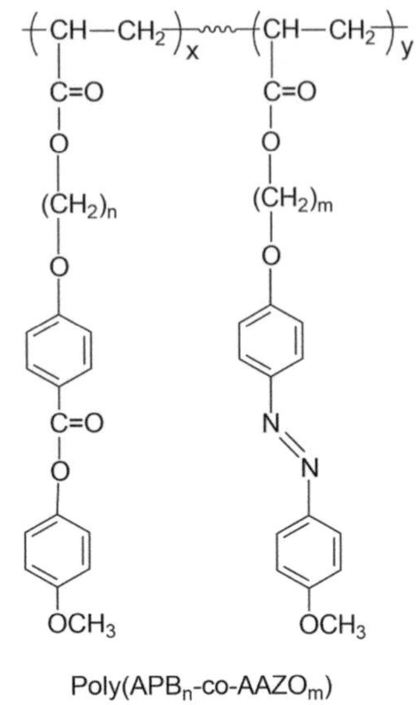

Fig. 3.3 Chemical structure of a series of the acrylate-based LC copolymers

monomers with elimination of some small molecules. A polymer obtained from this way is a typical condensation polymer as its synthesis involves the elimination of small molecules. On the other hand, step polymerizations also include those without the elimination of small molecules. In this case, although the composition of the polymer is the same as the monomer, it can also be classified into a condensation polymer due to the nature of the linking groups [2]. According to the linking groups that join the repeating units together, the azo polymers can be classified into polyamide, polyester, polyurethane, and others. Depending on the position of azo functional groups in polymeric structures, the azo polymers are further divided into main-chain and side-chain azo polymers. In the past decades, many types of main-chain condensation azo polymers have been synthesized by different types of reactions of the reactive groups in the monomers.

In one of the earliest reports, azo polymers were synthesized by azo coupling of bisdiazonium salts to polyphenols [25]. Different polyphenols and aromatic diamines were used as monomers for the polymerizations, which produced polymers with number-average molecule weights ranging from below 2000 to over 20,000. In 1970s and 1980s, copolymers containing a small number of azobenzene residues were prepared to study the rate of conformation transitions and local free volume. The azo polymers synthesized for this purpose include polyamides [15, 26, 27], poly(ester urethane)s [28], and polyurethanes [29, 30].

Since then, many types of *main-chain* azo polymers containing azo monomer units as the major components have been synthesized for different purposes. The azo polymers can be categorized as polyesters [31–33], polyimides [34], polyamides [35–41], polybenzimidazoles [42], polyhydrazides [43], polyamide-hydrazides [44], polyureas [45], polyurethanes [46–48], and poly(aryl ether ketone

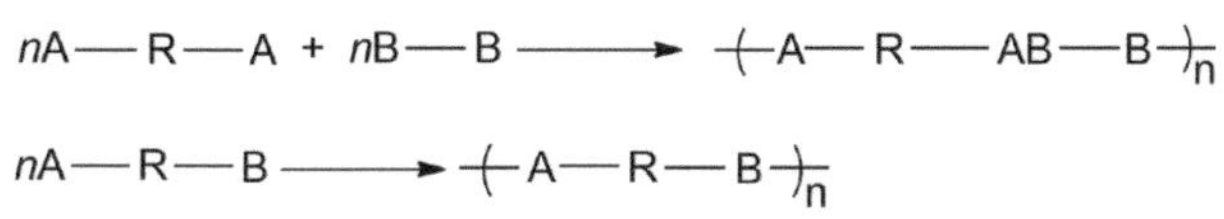

Fig. 3.4 Scheme of step polymerization to obtain polymers, where R is the structures to contain main-chain or side-chain azo chromophores

amide)s [49]. Figure 3.5 summarizes some types of the main-chain azo polymers prepared by the step polymerization.

Typically, main-chain azo polymers are prepared by condensation reactions between bifunctional and/or polyfunctional monomers through conventional condensation polymerization scheme. The polymerizations can be carried out under different process conditions such as solution polymerization [34], interfacial polycondensation [35], low-temperature solution, and interfacial polycondensation [39]. The incorporation of azoaromatic groups into the polymer main chain imparts the inherent color and backbone rigidity to the polymers. Reversible photocontractile and conformation changes have been observed for some of the azo polymers [34, 37, 38]. To obtain a polymer with sufficient high molecular weight, a very high conversion of the reactive groups is required, and a polar organic solvent is usually needed to dissolve the produced polymers. A unique series of main-chain azo polymers containing chiral binaphthyl bends or spirobiindane turns have been synthesized by the solution polycondensation [50]. The typical structure of the polymers is shown in Fig. 3.6. The stimuli-responsive polymers include polyaramides [51, 52], polyesters [53, 54], and poly(ester-amide)s [53, 55]. Light-, heat-, and solvent-regulated optical rotatory function has been demonstrated for the polymers.

Step polymerization has also been widely used to prepare *side-chain* azo polymers and end-bonded azo polymers. For an end-bonded azo polymer, one end of azo chromophores is directly bonded to polymeric backbone without flexible spacers. As the azoaromatic groups have one free end, the properties of the end-bonded azo polymers are analogous to those of side-chain polymers. The condensation azo polymers with side-chain and end-bonded structures include polyesters [56–59], polyurethanes [60–62], polyimides [63–69], and polyimides with siloxane linkage [70]. Figure 3.7 summarizes some types of the side-chain azo polymers prepared by the step polymerization. Most of the polymers have been developed as the second-order nonlinear optical (NLO) polymers. The polymerizations typically proceed by the reactions between the corresponding functional groups, such as the hydroxyl and acyl chloride, hydroxyl and isocyanate, and amine and anhydride in suitable solvents. The polyester with high molecular weight can also be prepared by polytransesterification of monomers [57].

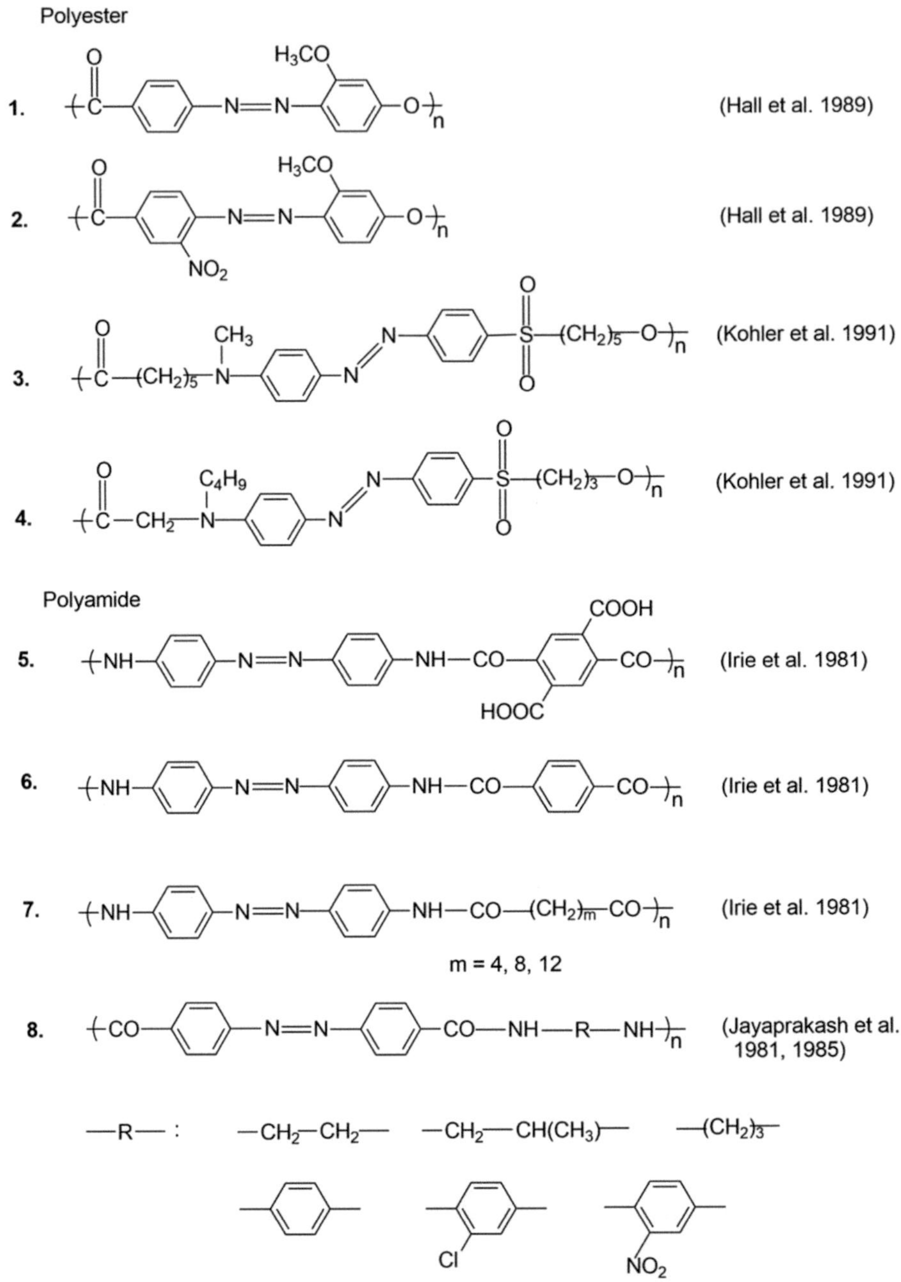

Fig. 3.5 Main-chain azo polymers prepared by the step polymerization

Polybenzimidazole

9. 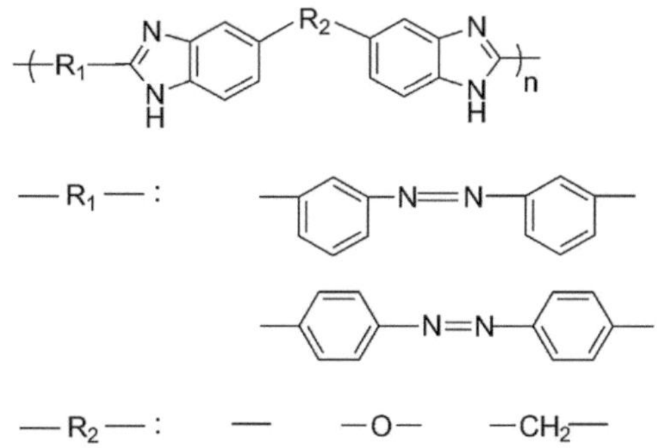

(Srinivasan et al. 1982)

Polyureas

10. 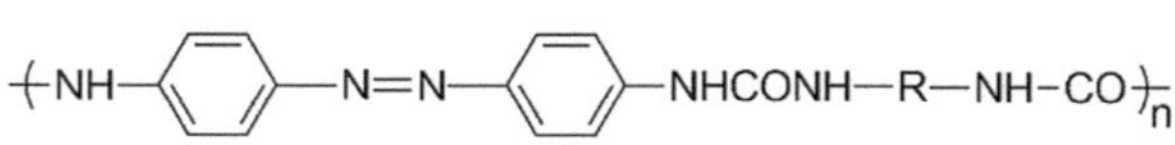

(Kumar et al. 1984)

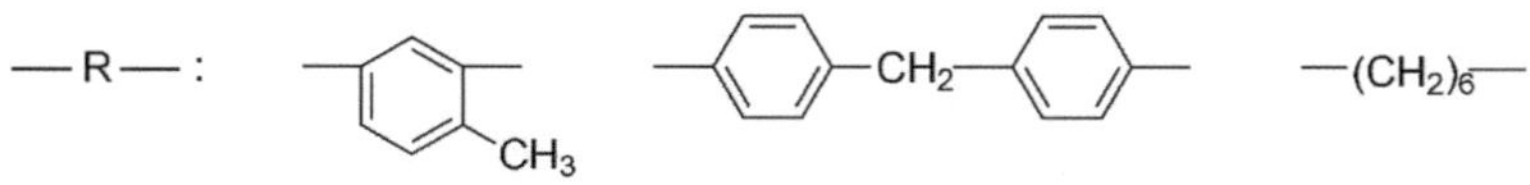

Polyurethane

11. 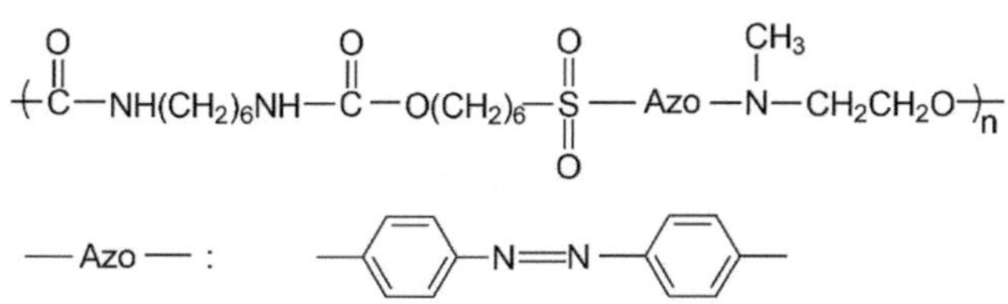

(Xu et al. 1992)

Poly(aryl ether ketone amide)

12. 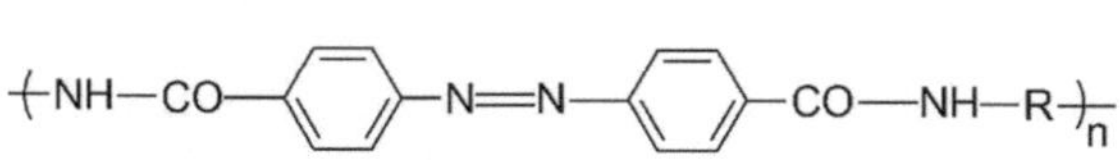

(Jaycox 1998)

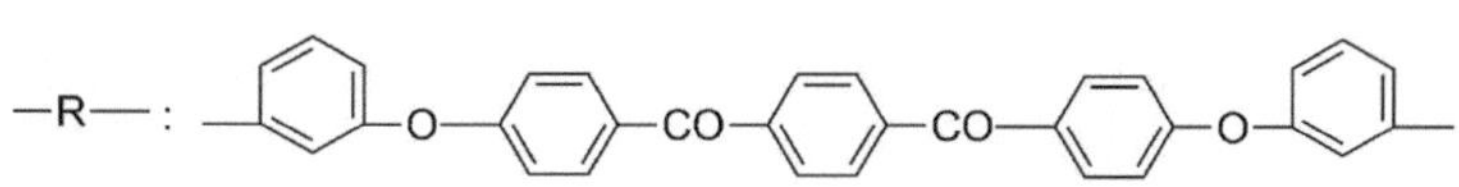

Fig. 3.5 (continued)

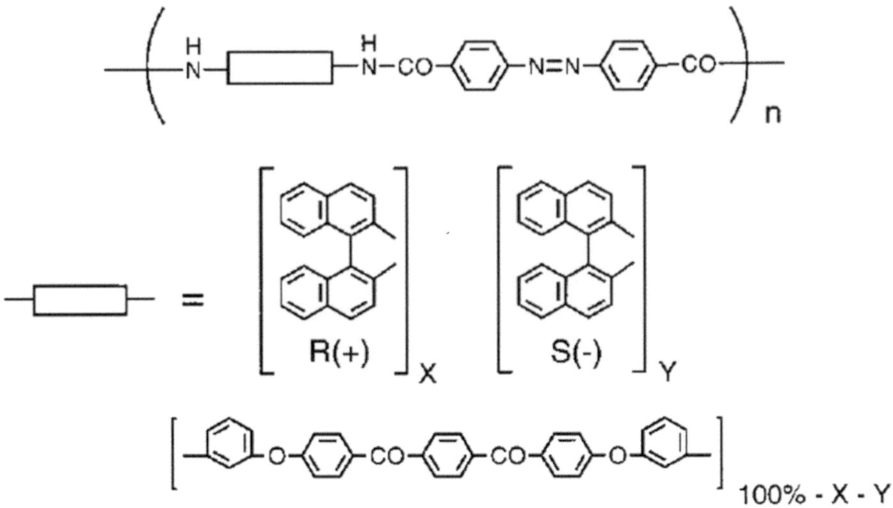

polymer	X (%)	Y (%)
1(R)	100	0
1(S)	0	100
2(S)	0	80
3(S)	0	60
4(S)	0	40
5(S)	0	20
6	0	0

Fig. 3.6 Azo polymers containing *trans*-1(R) and *trans*-1(S) (Reprinted with permission from Ref. [51]. Copyright (2001) American Chemical Society)

3.2.3 Post-polymerization Modification

Chemical modification of a suitable precursor polymer is another way to introduce the azo functional groups. Synthesis through polymerization of functional monomers is a straightforward approach to obtain the azo polymers. However, there are some drawbacks related to the direct polymerization methods. Due to the strong inhibition of the azo chromophores to the radical polymerization, the polymerization of the monomers may suffer from the low efficiency of the initiators and extremely long polymerization time to obtain high-molecular-weight polymers. Moreover, some side reactions could occur for the azo chromophores in the polymerization processes, especially for polycondensations with the high reaction temperature. To avoid those problems, *post-polymerization functionalization* has been developed as an alternative method. In this case, precursor polymers for

Polyester

1. 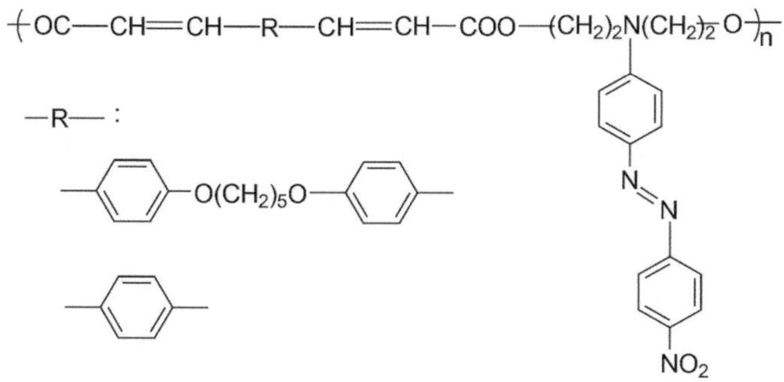

(Chen et al. 1991)

2. 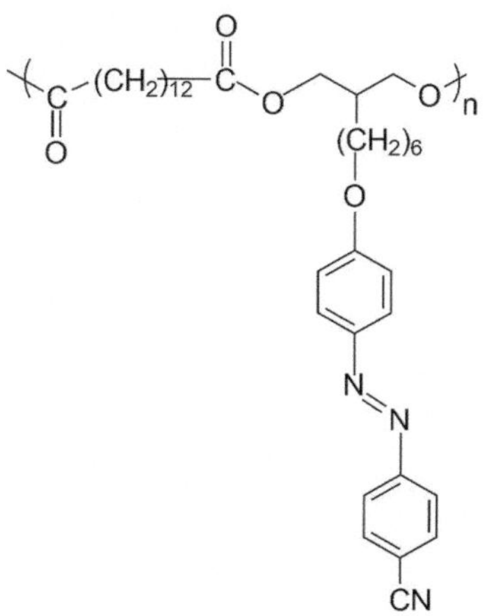

(Hvilsted et al. 1992)

Polyurethane

3. 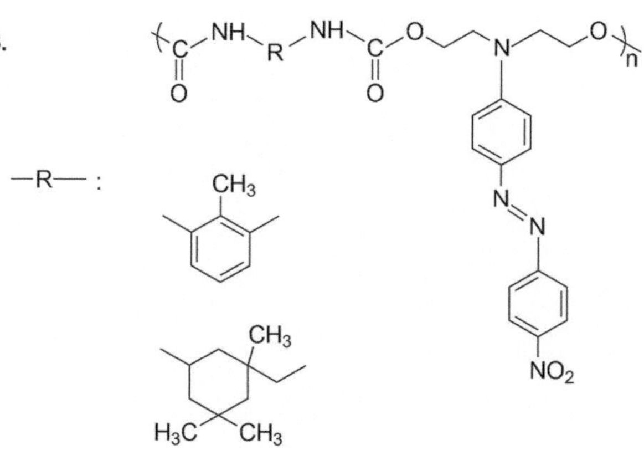

(Wang et al. 1995)

Fig. 3.7 Side-chain azo polymers prepared by the step polymerization

Polyimide

4.

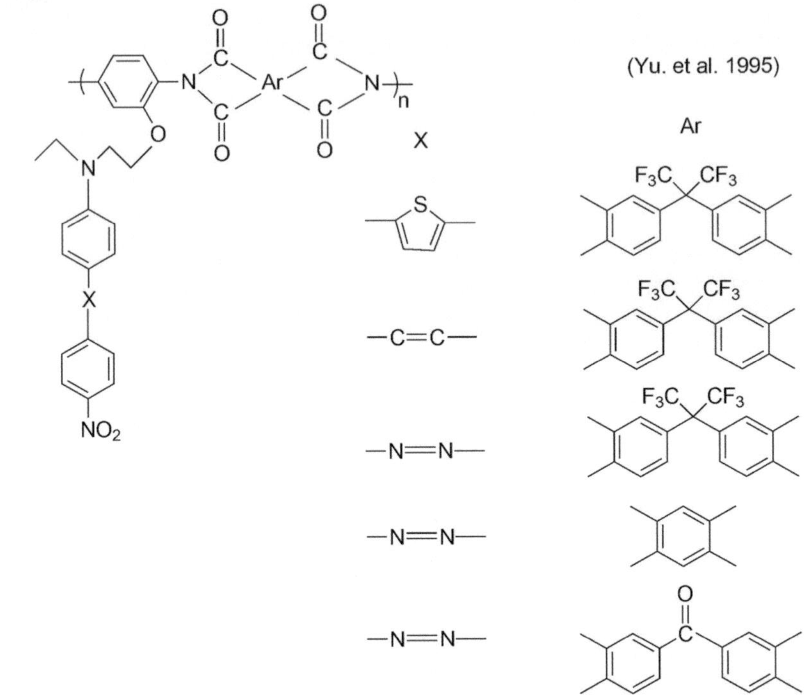

(Yu. et al. 1995)

5.

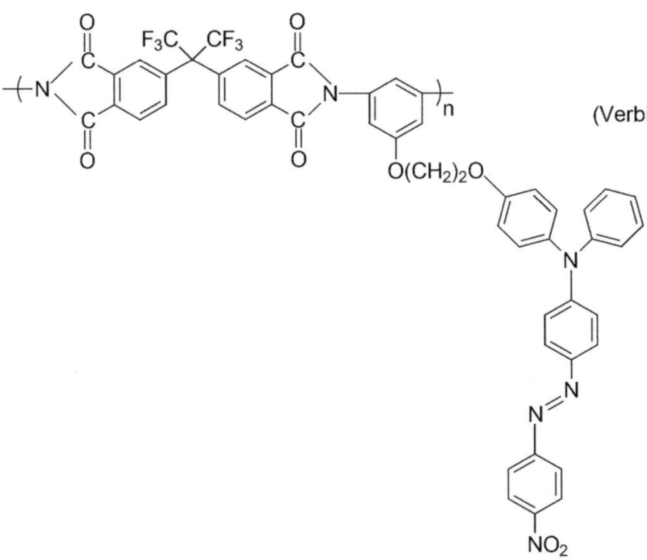

(Verbiest et al. 1995)

Fig. 3.7 (continued)

further reactions are prepared first, and then functionalization is carried out by converting these precursors into the azo polymers.

The precursor polymers can be prepared by both chain polymerization and step polymerization. As a critical requirement, the high efficiency is necessary for the reactions between precursor polymers and reactants to introduce the azo functional groups. Polypeptide is one unique type of the precursor polymers, which will be discussed in Sect. 3.6. Aromatic azo functional groups can be introduced through nucleophilic reactions of the sodium salts of alcohol (or phenol) groups of different azo compounds with poly(dichlorophosphazene) [71, 72]. Esterification between fully saponified polyvinyl alcohol (PVA) and azo compounds bearing carboxylic acid groups is another way used to synthesize a series of amphiphilic random copolymers [73]. Mitsunobu condensation is proved to be a highly efficient reaction to modify the precursor polymers bearing phenol group through the reaction with hydroxyl group on azo chromophores [74]. In a typical case, the hydroxyl-containing polyimides are obtained by the condensation polymerization of hydroxyl diamine and dianhydride monomers. The post-Mitsunobu condensation can reach a loading level from 0 to 50 wt% for the side-chain chromophores of the NLO polyimides.

A very reactive precursor polymer, poly(acryloyl chloride) (PAC), has been synthesized and used to prepare amphiphilic azo copolymers via the post-polymerization modification method [75]. Typically, PAC is prepared by the radical polymerization of acryloyl chloride at 60 °C under vacuum for 48 h. It is necessary to carry out the precipitation, washing, and drying of the product under a strictly anhydrous condition. Azo polymers are then obtained by reaction of PAC with azobenzene derivatives bearing hydroxyl end groups through the Schotten–Baumann reaction (Fig. 3.8). A certain amount of the unreacted acyl chloride groups will remain on the polymeric chain, which are then hydrolyzed to yield the hydrophilic COOH groups. This synthetic scheme is feasible for the preparation of amphiphilic azo random copolymers with different azo chromophores [75, 76].

Azo-coupling reaction is a widely used method to prepare various azo dyes. This reaction has been used as a way to prepare various azo polymers through post-polymerization modifications. In one early report, polymeric diazonium salt was prepared by directly diazotizing polysulfanilamide, and water-soluble polymeric azo dyes were prepared by azo-coupling reaction with 2-naphthol-6-sulfonate in cold aqueous alkali solution [77]. In a separate report, cyanovinyl-substituted benzenediazonium salt has been prepared and used to react with a methacrylate copolymer containing aniline moieties in an AcOH/NaOAc solution [78].

Later study shows that the post-polymerization azo-coupling reactions can be carried out in polar organic solvents with high efficiency [79, 80]. Figure 3.9 shows the synthetic routes and chemical structure of a series of epoxy-based azo polymers obtained from the post-polymerization azo-coupling reaction. The precursor polymer was prepared by the polymerization between the diglycidyl ether of bisphenol A and aniline. Polymerization was carried out at a relatively low temperature (110 °C) to avoid the reaction between the secondary OH groups formed during the polymerization and the unreacted epoxide rings. The precursor polymer was

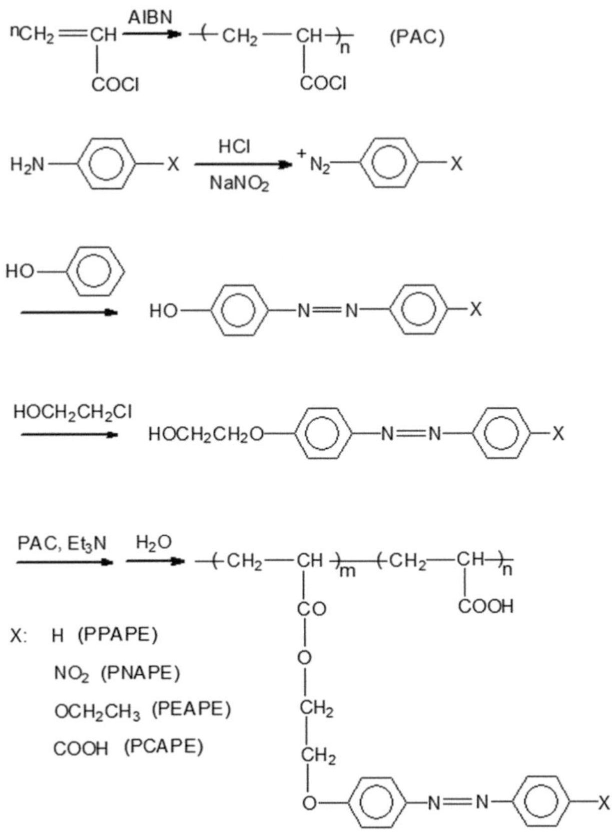

Fig. 3.8 Azo polymers obtained by reaction of poly(acryloyl chloride) (PAC) with azobenzene derivatives bearing hydroxyl end groups (Reprinted with permission from Ref. [75]. Copyright (2001) American Chemical Society)

postfunctionalized by azo-coupling reaction and tricyanovinylation to form a number of NLO polymers with different conjugation lengths and electron acceptors. The results showed that by using a slightly excessive amount of the diazonium salts, the conversion of the side-chain anilino moieties to azo chromophores can reach near 100 %. As many types of precursor polymers can be well dissolved in polar organic solvent, this post-polymerization azo-coupling scheme can be used to obtain different types of azo polymers. Various types of azo polymers containing a variety of azo chromophores have been synthesized by this method [81–87].

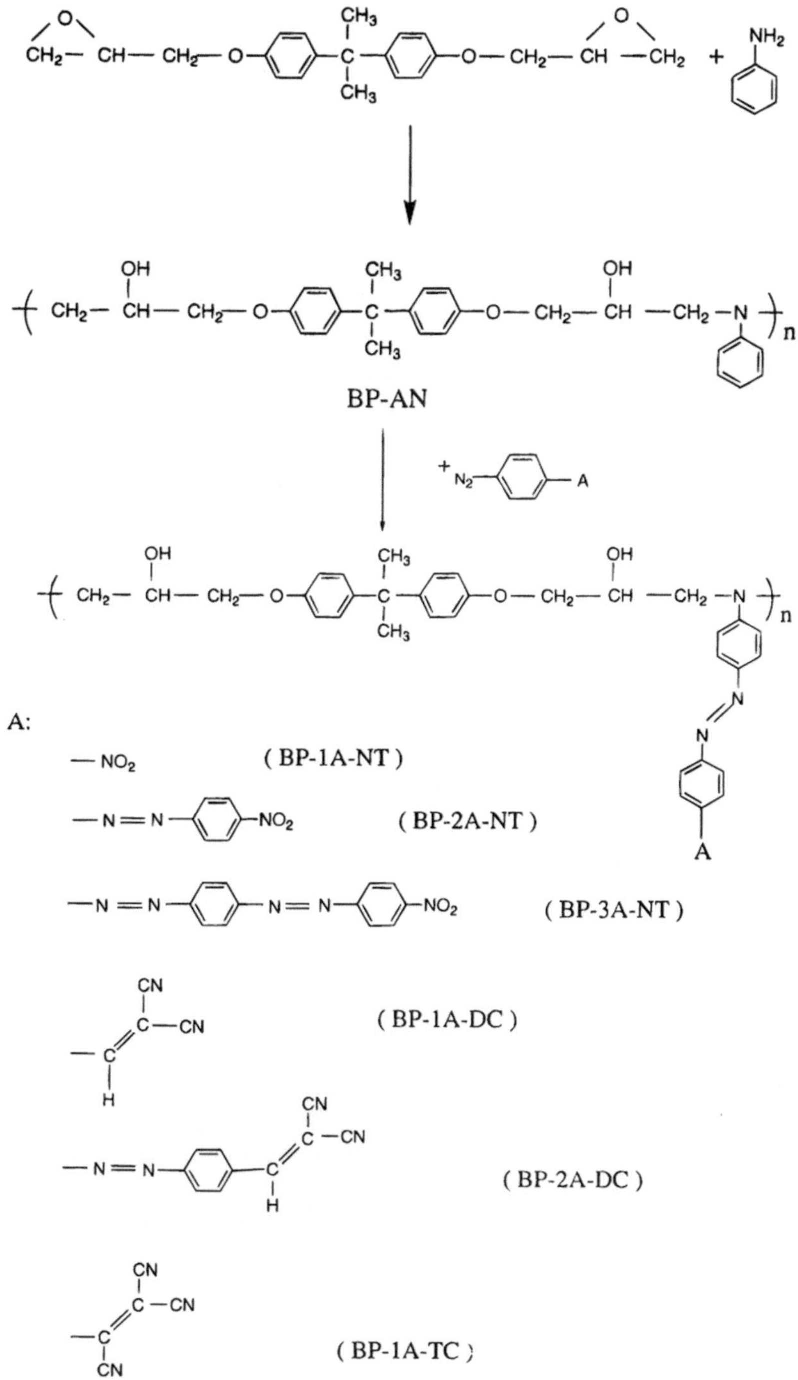

Fig. 3.9 Synthetic route and chemical structure of a series of epoxy-based azo polymers obtained from the post-polymerization azo-coupling reaction (Reprinted with permission from Ref. [80]. Copyright (1997) American Chemical Society)

3.2.4 Controlled Radical Polymerization

The abovementioned methods can be used to prepare a variety of polydispersed azo homopolymers and random copolymers. Some interesting photoresponsive properties of the polymers will be discussed in the following chapters. However, the conventional radical polymerization and condensation polymerization are usually unsatisfied for precise control of polymer structures and compositions, such as the degree of polymerization, polydispersity, structure of the end group, chain sequence structure, and compositions of the copolymers. Over the past decades, polymer chemistry has made substantial progresses in the methods of *"living"/controlled radical polymerization* [88–93]. Azo polymers with well-defined and novel architectures can be prepared by these new methods. In recent years, these methods have been used to prepare azo polymers with the controlled average molecular weight, narrow molecular weight distribution, and defined end groups [94–96]. "Living"/ controlled radical polymerizations are widely applied to synthesize various block copolymers, which is discussed in Sect. 3.3.

3.2.5 Cross-Linked Azo Polymer

For this category of azo polymers, the polymer molecules are linked to each other to form the network structure. The cross-linking can be formed during the polymerization process by using appropriate monomers, where at least one must have functionality greater than two to work as the branching unit. It can also be introduced after polymerization by various chemical modifications. According to the number of cross-links and rigidity of the segments, the azo polymers can show different properties, which have been prepared for different purposes. When the number of cross-links is sufficiently large, the formed rigid three-dimensional azo polymer networks are thermosetting polymers. Most of the polymers in this category have been prepared for NLO applications. In this case, the thermal and temporal stability are imparted to the polymers through the cross-linking. The cross-linked azo NLO polymers, with different main-chain compositions and linking groups, include polyimide/inorganic composite [97], sol–gel system [98], polymethacrylates with cross-linkable chromophore pendants [99, 100], epoxy-based interpenetrating polymer network [101], and polyurethanes [102–105].

When an appropriate number of cross-links are introduced, resulted polymer systems can show completely different properties compared with those of the thermosetting polymers. The polymer network with the light cross-linking density is referred to as elastomer. *Liquid crystalline elastomers* (*LCEs*) containing azobenzene moieties belong to a unique type of the photoresponsive polymers, which endow photoresponsive properties to a system with anisotropic orientation of LCs and rubbery elasticity of polymer networks. This type of cross-linked azo polymer network will be discussed in Chap. 6 in detail.

Hydrophilic polymers with space network structure have been prepared as photoresponsive *hydrogels* and hydrogels with other functions. A suitable number of cross-links are required to ensure the proper swelling and recovery properties of the systems. Azo hydrogels have been prepared by copolymerization of different monomers, such as copolymerization of *N*-isopropylacrylamide with 4-methacryloylaminoazobenzene and 4,4′-di(methacryloylamino)azobenzene [106], and copolymerization of methacryloyloxy azobenzene with hydroxyethyl methacrylate [107]. Hydrogels with more than two components have been synthesized by copolymerization of *N*,*N*-dimethylacrylamide, *t*-butylacrylamide, acrylic acid, 4,4′-di(methacryloylamino)azobenzene, and *N*-alkanoyl and *O*-methacryloylhydroxyl amines [108]. The hydrogel composed of konjac glucomannan has been prepared by graft copolymerization with acrylic acid and cross-linking with *bis*(methacryloylamino)azobenzene [109]. As an alternative approach, a pH-sensitive hydrogel has also been synthesized by a "one-step" polymer–polymer reaction of two precursors [110]. The polymeric hydrogels were prepared for photo-controlled release of solutes [106] and colon-targeted drug delivery [107, 109].

3.3 Azo Block and Graft Copolymer

The block copolymers are copolymers with long sequences of each monomer in the copolymer architecture. Distinct from a *block copolymer* as a linear copolymer, a *graft copolymer* is a branched copolymer with different compositions for its backbone and side chain. Block copolymers can be synthesized by sequential monomer addition, transformation reactions, telechelic polymer routes, and coupling reactions [2]. The most effective way to prepare block copolymers is sequential addition of monomers to a living polymerization system (a polymerization without termination) such as anionic polymerization of nonpolar monomers. However, it is very difficult to find azo monomers that are stable against the attack of the anions during the chain growth process. Therefore, for a long period, block copolymers with azobenzene side groups could only be prepared by post-polymerization modification of precursor block copolymers obtained from anionic polymerization [111]. Few azo block copolymers have been prepared by sequential addition of monomers to living cationic polymerization [112].

Although conventional radical polymerization can be used to polymerize a variety of azo monomers, it is almost impossible for the polymerization to undergo persistent propagation without termination. Therefore, to obtain block polymers, the basic requirement of "living"/controlled criterion needs to be satisfied, i.e., the chain transfer and termination are effectively suppressed. Additional requirements to be fulfilled are that the initiation should be completed at early stage of polymerization and the exchange between species of various reactivities is fast in comparison with propagation [89]. For controlled radical polymerization, these requirements are met by exchange between active and dormant species during chain propagation through some well-designed processes, such as reversible homolytic cleavage of covalent species catalyzed by transition-metal complexes and the degenerative transfer through addition–fragmentation process. The most widely applied methods based on these mechanisms are *atom transfer radical polymerization* (*ATRP*) and *reversible addition–fragmentation chain transfer* (*RAFT*) polymerization. Many types of azo block copolymers with well-defined structures have been prepared by ATRP and RAFT polymerization as discussed below. The controlled radical polymerization methods also provide a convenient way to prepare azo graft copolymers.

3.3.1 Atom Transfer Radical Polymerization (ATRP)

Atomic Transfer Radical Polymerization (ATRP) employs atom transfer from an organic halide to a transition-metal complex to generate active radicals for initiation and chain propagation [88, 89]. In the process, the concentration of active radicals is always kept low by effective back-transfer from the transition metal to the active site, which is a necessary step to minimize the termination. The polymerization is usually initiated by a suitable alkyl halide, and the atomic transfer process is catalyzed by a complex of a copper (I) halide with 2,2′-bipyridyl derivatives, 1,1,4,7,10,10-hexamethyl-triethylenetetramine (HMTETA), *N,N,N′*, *N″,N″*-pentamethyl-diethylenetriamine (PMDETA), and others. The initiator and catalyst systems generate the organic radical for polymerization and copper (II) complex, where the reversible process is the formation of dormant species. Due to the "living" nature, the activated alkyl halide at a polymer chain end can initiate the polymerization of the suitable monomers to synthesize di-, tri-, or even multiblock azo copolymers [91]. Typically, azo block copolymers can be synthesized through two schemes, i.e., using an available *macroinitiator* and *sequential ATRP steps* of different types of monomers.

Well-defined block copolymers containing azo functional groups can be feasibly synthesized by ATRP through macroinitiator approach, where polymerization is initiated by a macromolecule with the activated alkyl halide end group referred to as macroinitiator. A macroinitiator is mainly obtained by modification of a commercially available macromolecule. In the syntheses through sequential addition of monomers to a typical ATRP system with small molecule initiators, the intermediate isolated in the preparation process can also act as a macromolecular initiator for next step polymerization. In this case, there is no substantial difference between the macroinitiator and sequential addition approaches. Besides ATRP, a macroinitiator can be prepared by other types of polymerizations. A macromolecule synthesized by one type of the polymerizations can be used as an ATRP macroinitiator either directly or after a transformation reaction.

The approach to obtain macroinitiator from commercially available polymer is a convenient way to prepare various azo block copolymers. Poly(ethylene glycol)

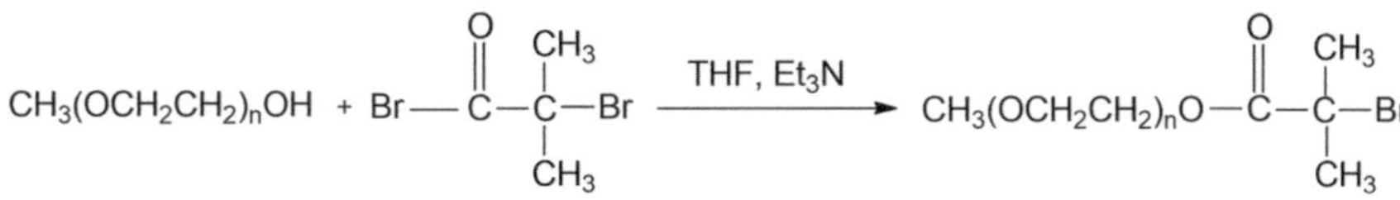

Fig. 3.10 Synthetic route of the macroinitiator (PEGBr)

(PEG) has been used to prepare the macroinitiator and synthesize azo block copolymers [113]. This method is later proved to be one of the most effective ways to prepare amphiphilic block polymers through relatively simple routes. The macroinitiator (PEGBr) is prepared via Schotten–Baumann reaction between PEG monomethyl ether and 2-bromo-2-methylpropionyl chloride in the presence of trimethylamine (Fig. 3.10). The ATRP of hydrophobic polymethacrylates containing azobenzene moieties on side chains is initiated by the PEGBr/CuCl/HMTETA initiating system in anisole at 80 °C. This system shows a nearly linear relationship of $\ln([M]_0/[M])$ versus reaction time, which evidences the "living"/controlled character of the polymerization. Narrow polydispersities in the range from 1.08 to 1.11 can be achieved for the diblock copolymers. It is able to control the composition of the PEGBr and azo blocks by adjusting the polymerization degree of the azo block relative to the PEGBr. This approach has been used to prepare block copolymers composed of PEG and side-on azo liquid crystalline blocks [114]. Such type of amphiphilic azo block polymers can show a variety of self-assembling properties in a selective solvent.

By using macroinitiators obtained from PEG, triblock and multiblock copolymers containing azobenzene moieties can be prepared. The block copolymers are prepared either through chain propagation from both ends of a telechelic PEG macroinitiator with two bromide functionalized terminals [115, 116] or successive ATRP processes initiated by PEGBr [117]. The approach using a telechelic PEG macroinitiator can be used to prepare ABA-type triblock copolymers, where A and B correspond to azobenzene-containing polymethacrylate and PEG blocks. The telechelic PEG macroinitiator with two bromide functionalized terminals is prepared by the esterification of the terminal hydroxyl group of commercial PEG with 2-bromoisobutyryl bromide. The macroinitiator can initiate the ATRP of methacrylate monomers with azobenzene moieties pending in the side chains. Under proper conditions, the polymerization of the azo methacrylates propagates from both ends of the PEG chains to yield the ABA triblock copolymers. The approach using PEGBr as a macroinitiator can be applied to prepare ABC triblock copolymers and multiblock polymers as well. As a typical example, the successive ATRP of 11-(4-cyanobiphenyloxy)undecyl methacrylate and 11-(4-(4-butylphenylazo)phenoxyl)-undecyl methacrylate has been initiated by PEGBr [117]. Azobenzene-containing ABC triblock copolymers have also been obtained by sequential ATRP of styrene and 6-(4-methoxy-4′-oxyazobenzene)hexyl methacrylate by using PEGBr as macroinitiator and CuBr/PMDETA as the catalyst [118].

This PEGBr-based approach can be used to prepare block copolymers functionalized with strong *push–pull* azo chromophores through post-polymerization azo-coupling reactions. In this case, the ATRP of anilino-containing monomers is initiated by PEGBr to obtain the precursor polymer, and then post-polymerization azo-coupling reaction is used to introduce the functional groups [119, 120]. The synthetic route and chemical structure of the block copolymer are shown in Fig. 3.11. For the synthesis, the precursor polymer PEG-b-PEMA is synthesized by the ATRP reaction of 2-(ethyl(phenyl)amino)ethyl methacrylate (EMA) by using PEGBr as the macroinitiator and CuBr/HMTETA catalyst system. The amphiphilic azo block copolymers (PEG-b-PCN) are obtained by the azo-coupling reaction between PEG-b-PEMA and the diazonium salt of 4-aminobenzonitrile. The azo-coupling reaction is very efficient and nearly all the anilino moieties of the precursor polymer can be converted to the azo

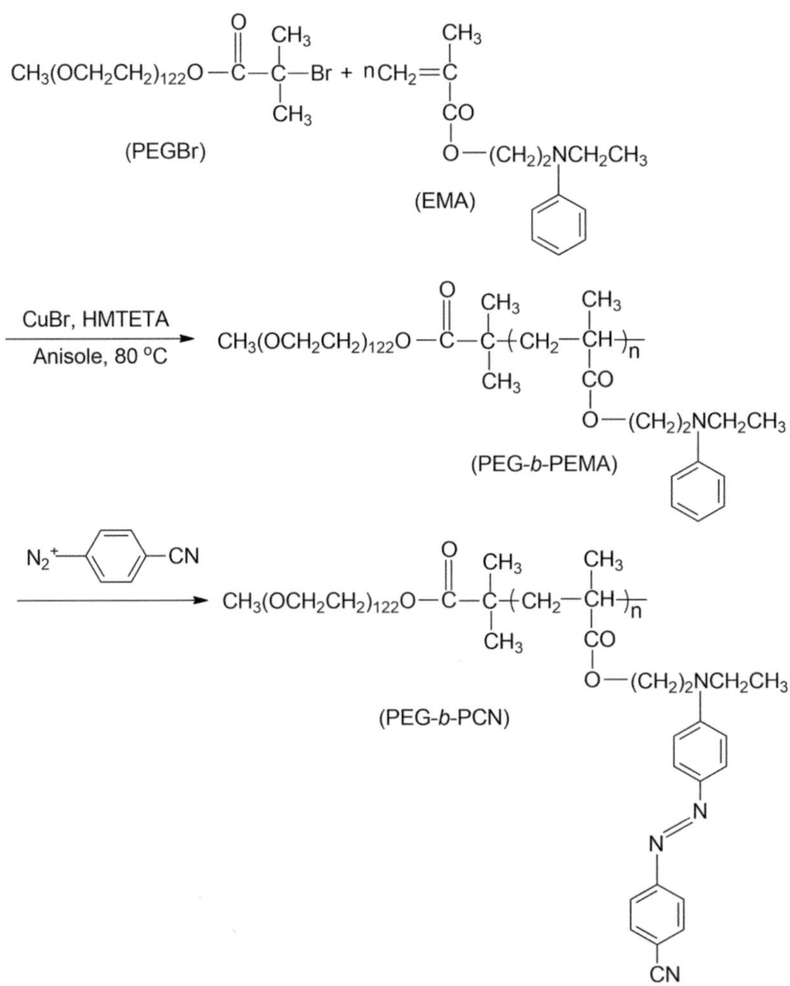

Fig. 3.11 Synthetic route of the azo block copolymer PEG-b-PEMA

chromophores. The ratios of the two blocks can be controlled by adjusting the molar feed ratio of EMA to the macroinitiator in the ATRP step, and the polydispersity indexes of PEG-b-PCN with different compositions are usually low (<1.3).

Azo block copolymers can also be prepared with a small molecule initiator through sequential ATRP steps of different types of monomers. By using ATRP in this way, to transfer from one block to another may sometimes be difficult. Although there are certain rules that can be followed [91], it is no guarantee for the propagation centers to keep "living" in the processes. As the strong inhibition of some azo chromophores to the radical polymerization, the polymerization of some monomers is hard to produce a polymer with high molecular weight. For this approach, the sequential order to add the monomers is important, and the polymerization of azo monomers is usually the last step for the "living" propagation process. The monomers that can be feasibly used for ATRP include styrenes, (meth)acrylates, and acrylonitrile. ATRPs of some functional monomers, such as (meth)acrylic acid and 4-vinylpyridine, may sometimes face challenging problems such as the poison of the catalysts by coordinating to the transition metal [91].

Styrene is one of the monomers suitable for preparing azo block copolymers. Diblock copolymers composed of a polystyrene block and azobenzene-containing polymethacrylate block have been synthesized by the two-step ATRP [121, 122]. In this approach, a polystyrene (PS) macroinitiator is first synthesized by ATRP using ethyl bromoisobutylate (EBB)/CuBr/2,2-dipyridine (dPy) initiating system. Then, monofunctional macroinitiator (PS-Br) is used to further initiate the polymerization of the azo monomer, 6-[4-(4-methoxyphenylazo)phenoxy]hexyl methacrylate. Diblock copolymers with various contents (or volume fractions) of the LC azobenzene methacrylate block can readily be obtained by controlling the conversion and/or the amount of azobenzene monomer added for the extension of polymerization. Most block copolymers prepared have a polydispersity similar to that of the PS macroinitiator (~1.25).

(Meth)acrylate is another type of the monomers that can be feasibly used to copolymerize with azo monomers. Block copolymers composed of azo block and poly(*n*-butyl acrylate) (PBA) can be prepared by a two-step ATRP process [94]. The mono- and difunctional PBA macroinitiators are synthesized by the first-step ATRPs (Fig. 3.12). Monofunctional macroinitiator (PBA-Br), which has one bromine group at the end of the PBA chain, is synthesized by ATRP of BA using 2-bromopropionate, CuBr, and PMDETA at 70 °C. Difunctional macroinitiator (Br-PBA-Br), which has bromine groups at both ends of PBA chain, is prepared by ATRP of BA using dimethyl 2,6-dibromoheptanedioate as the initiator under similar conditions. The number-average molecular weights of PBA-Br and Br-PBA-Br are 42,100 and 24,700 with low polydispersities of 1.13 and 1.14, respectively. The diblock and triblock copolymers containing *p*-cyanoazobenzene moieties are prepared by using the macroinitiators through the second-step ATRP. The compositions of the copolymers can be adjusted by controlling the molar feed ratio of the azo monomers to the mono- or difunctional macroinitiators. The similar approach has been used to prepare diblock copolymer with methyl methacrylate (MMA) and azo blocks [123], azobenzene-containing

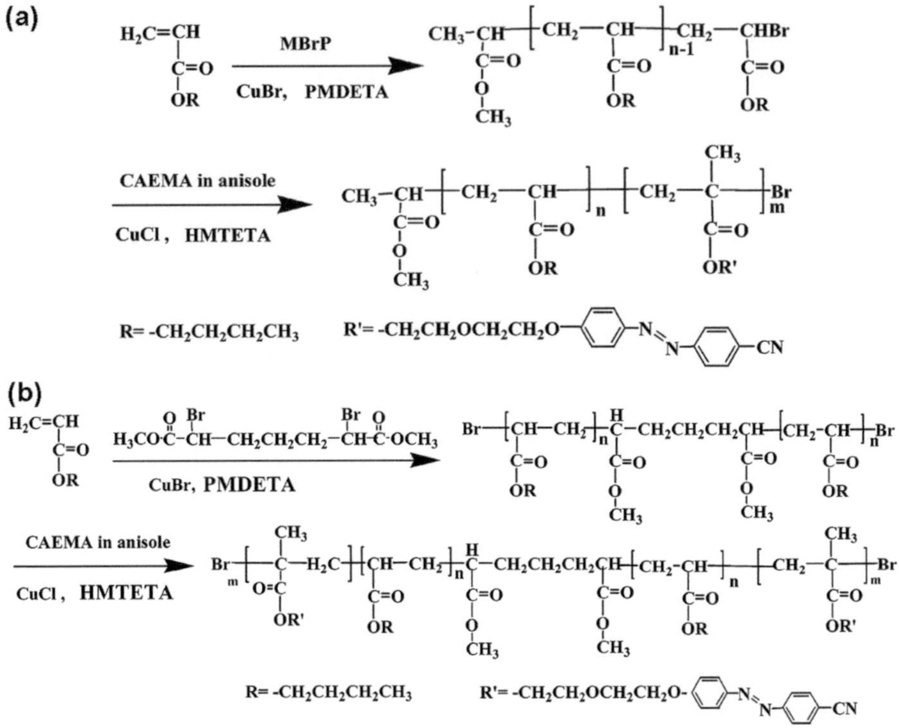

Fig. 3.12 Synthetic routes for diblock and triblock azo copolymers (Reprinted with permission from Ref. [94]. Copyright (2004) American Chemical Society)

ABA triblock copolymers with poly(n-butyl acrylate) (PBA) as the midblock [124], liquid crystalline triblock copolymers with azobenzene-containing block as the midblock [125].

This method has been applied to prepare an amphiphilic diblock azo copolymer (PAA-b-PAzoMA) composed of poly(acrylic acid) (PAA) and PAzoMA blocks [126]. Figure 3.13 shows the chemical structure of PAzoMA-*b*-PAA. As acrylic acid is difficult to be used for ATRP, tert-butyl acrylate (tBA) is used as the monomer in the first step of the polymerization. It is polymerized by using ethyl bromoisobutylate/CuBr/PMDETA initiating system to obtain the macroinitiator PtBA-Br. After that, the PAzoMA block is introduced by ATRP initiated with PtBA-Br to obtain precursor (PAzoMA-*b*-PtBA). Finally, the precursor is converted to PAzoMA-*b*-PAA via selective hydrolysis of the tBA groups by using trimethylsilyl iodide (TMSI) as the catalyst. The hydrolysis degree of tert-butyl ester groups is in the range from 32 to 90 %. After the hydrolysis, the average molecular weights of the diblock copolymers decrease slightly, and their polydispersity indexes increase from 1.28–1.31 to 1.39–1.92. By this approach, a series of amphiphilic diblock copolymers (P(CNAZOm-b-MAAn)), composed of 2-(*N*-ethyl-*N*-(4-(4′-cyanophenylazo)phenyl)amino)ethyl methacrylate (CNAZO) and

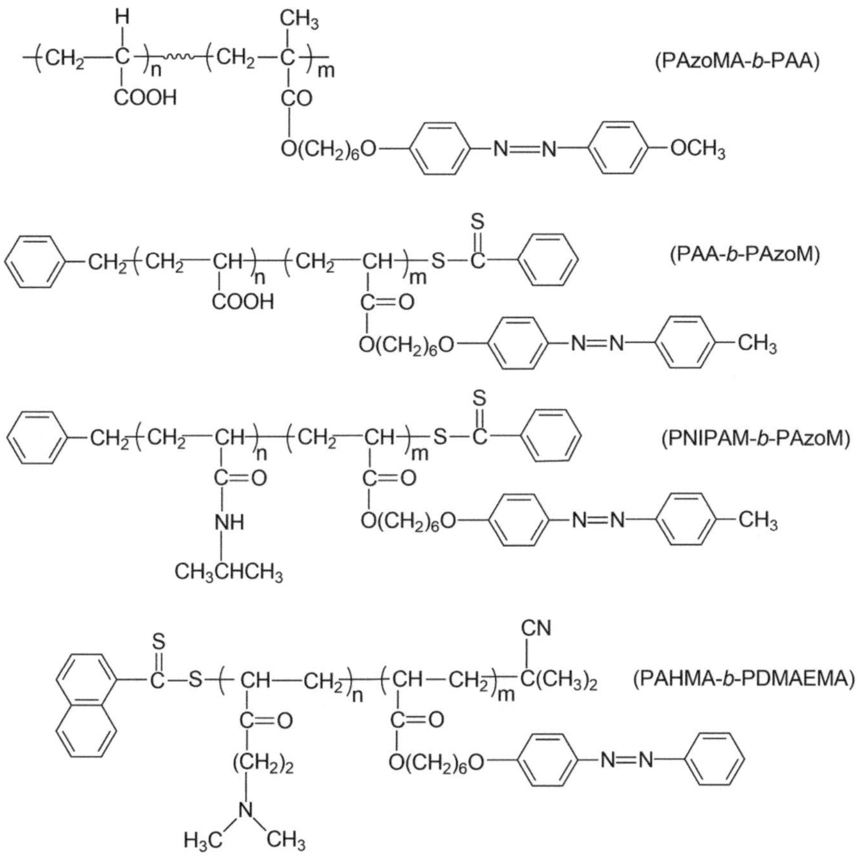

Fig. 3.13 Chemical structure of different types of amphiphilic azo block copolymers

methacrylic acid (MAA) blocks, has been synthesized by the sequential ATRP route of methacrylates and followed by the selective hydrolysis [127]. The strong *push–pull* azo chromophores are introduced through the post-polymerization azo-coupling reaction. One of the major purposes of synthesizing the amphiphilic azo block copolymers is used to prepare photoresponsive micelles or colloidal particles through the self-assembling processes.

For those monomers suitable for ATRP, amphiphilic block copolymers can also be directly synthesized by the sequential ATRP route. The amphiphilic azo diblock copolymer containing poly(4-vinylpyridine) as the hydrophilic block has been prepared by the two-step sequential ATRP reactions of 4-vinylpyridine and the azo monomer [128]. A series of amphiphilic azo diblock copolymers composed of poly(2-(dimethylamino)ethylmethacrylate) (PDMAEMA) as hydrophilic segments has been synthesized through this two-step sequential ATRP route [129, 130]. In this approach, the macroinitiator PDMAEMA is obtained in the first step of the polymerization and used to initiate the polymerization of methacrylate monomers containing azo chromophores. In this reported case, the polydispersity indexes of the block copolymers were in the range of 1.12–1.20, but the DPs of the azo blocks were relatively low.

3.3.2 Reversible Addition–Fragmentation Chain Transfer Polymerization

The *reversible addition–fragmentation chain transfer (RAFT)* polymerization is another type of controlled radical copolymerization that has been widely applied to prepare block copolymers. In this process, the requirements for controlled radical process are met by the degenerative transfer through addition–fragmentation process [90, 93]. A major advantage of the RAFT polymerization is that it is compatible with a wide range of monomers, especially those containing functional groups. The "living"/controlled process is achieved by a well-designed reversible addition–fragmentation sequence between active and dormant chains. RAFT polymerization achieves the degenerative chain transfer relying on a RAFT chain transfer agent (CTA). The typical types of the RAFT-CTAs include dithioesters, xanthates, dithiocarbamates, and trithiocarbonates. Among them, 2-(2-cyanopropyl) dithiobenzoate (CPDB), benzyl dithiobenzoate (BDTB), and 2-phenylprop-2-yl dithiobenzoate (PPDTB) are some frequently used CTAs. The key structural components include the so-called Z and R groups [93]. The Z group can be adjusted to determine the reactivity. The R group is a good free radical (homolytic) leaving group, for which the radical generated must be able to initiate polymerization (small molecule CTA) and keep propagation (macro-CTA). Azo block copolymers with different structures have been prepared by a well-designed RAFT polymerization as discussed below.

Two types of amphiphilic azo block copolymers have been prepared by RAFT polymerization [131, 132]. The chemical structures of the diblock copolymers, PAA-*b*-PAzoM and PNIPAM-*b*-PAzoM, are shown in Fig. 3.13. The macro-CTA of hydrophilic poly(acrylic acid) (PAA) or poly(*N*-isopropylacrylamide) (PNIPAM) is prepared through RAFT polymerization by using dithiobenzoate derivatives (BDTB or PPDTB) as the small molecule CTA. The macro-CTA is obtained by the RAFT polymerization of the monomers initiated by AIBN in the presence of the dithiobenzoate derivatives. The macro-CTA is then used in the RAFT polymerization of hydrophobic azobenzene-containing methacrylate monomer to obtain the amphiphilic diblock copolymers. In this two-step process, the macro-CTA is separated and characterized before the second-step polymerization. The RAFT polymerization can be used to synthesize amphiphilic block copolymers with more complicated structures [133].

In many cases, azo block copolymers can be prepared from a macro-CTA of either monomer. Azo block copolymers are also synthesized by first obtaining macro-CTA of the azo block and then carrying out the second-step RAFT in the presence of the macro-CTA. An azobenzene-containing macro-CTA has been obtained by the RAFT polymerization, and its copolymerization ability with methyl methacrylate (MMA) and styrene (St) has been tested [134]. A linear increase of the molecular weights versus the monomer consumption is observed to be maintained until 86.6 % conversion for MMA and 39.5 % conversion for St, respectively. Amphiphilic azobenzene-containing diblock copolymers (PAHMA-*b*-PDMAEMA, Fig. 3.13) are also obtained by using macro-CTA of the azo block through the two-step RAFT polymerization [135]. The azo block copolymers prepared by the RAFT method possess the molecular weight distributions in a range from 1.2 to 1.5, and the enhanced fluorescence from the aggregation of the block copolymer in dispersions has been observed.

Double side-chain liquid crystalline (SCLC) block copolymers are synthesized by the two-step RAFT polymerization [136]. In this process, the macro-CTA (PAzoMA-CTA) is obtained through RAFT polymerization of the azobenzene-containing monomer using CPDB as the starting chain transfer agent and AIBN as the initiator (Fig. 3.14). The same experimental condition is also used to prepare the other macromolecular chain transfer agent, PBiPMA-CTA, which contains side-chain biphenyl mesogens. The double SCLC block copolymers are synthesized through the RAFT chain extension reaction using the macromolecular CTAs, which can be either PAzoMA-CTA for the growth of the second block of PBiPMA or PBiPMA-CTA for the growth of the PAzoMA block. RAFT polymerization is effective to produce the double SCLC block copolymers no matter the order of polymerization of the two monomers.

Due to the inhibition and retardation effect, it is favorable to polymerize the monomers containing strong electronic donor–acceptor azo chromophores in the last step of the RAFT polymerization to synthesize the block copolymers. Two series of diblock copolymers bearing strong *push–pull* azo chromophores and mesogenic cholesteryl or biphenyl groups have been synthesized by the two-step

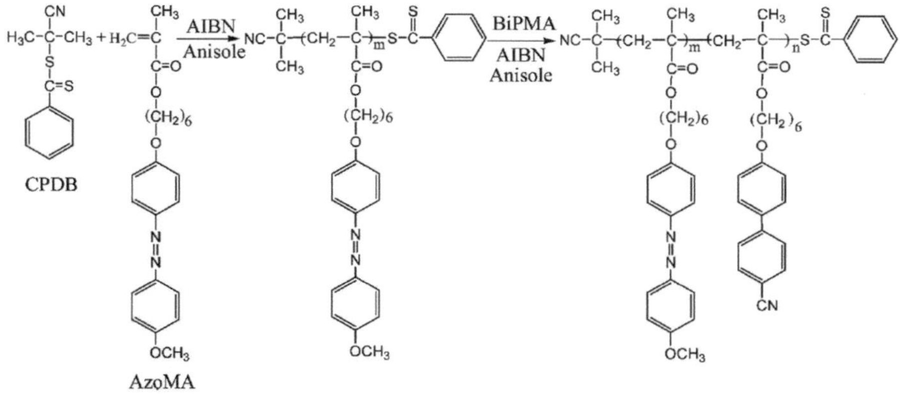

Fig. 3.14 Double side-chain liquid crystalline (SCLC) block copolymers obtained by the two-step RAFT polymerization (Reprinted with permission from Ref. [136]. Copyright (2008) American Chemical Society)

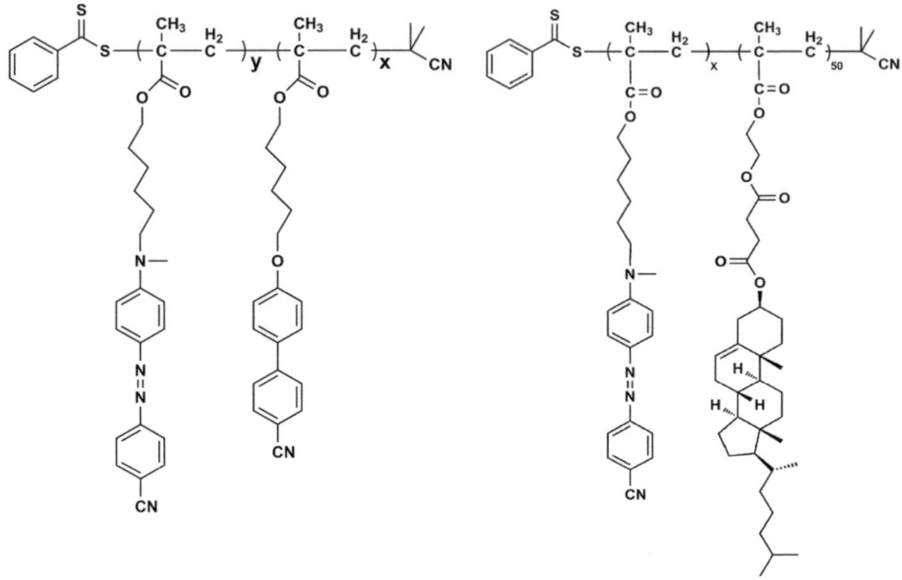

Fig. 3.15 Two series of liquid crystalline azo block copolymers obtained by two-step RAFT polymerization

RAFT polymerization [137, 138]. The chemical structure of the block copolymers obtained by this approach is given in Fig. 3.15. In the process, two macro-CTAs are synthesize through the RAFT polymerization of 6-(4-(4′-cyanophenyl)phenoxy) hexyl methacrylate (BiPMA) and 2-(3-((cholesteryl)oxycarbonyl)propionyloxy) ethyl methacrylate (ChEMA). The polymerizations are initiated by AIBN in the presence of CPDB. PChEMA-CTA and PBiPMA-CTA are then used in the second-step RAFT copolymerization of 6-(*N*-methyl-*N*-(4-(4′-cyanophenylazo)phenyl) amino)hexyl methacrylate.

3.3.3 Azo Graft Copolymer Synthesis

Differing from a block copolymer, a *graft copolymer* possesses a branched structure with a backbone of one monomer, to which one or more side chains of a different monomer (monomers) are attached [2]. Typically, there are three methods to synthesize graft copolymers, which are grafting through, grafting onto, and grafting from. Grafting through refers to the copolymerization of a monomer (monomers) with a macromonomer bearing the grafting segment. Grafting onto and grafting from mean the methods to attach growing polymeric chains onto a polymer backbone or initiate chain growth from a polymer backbone, respectively. Generally speaking, a variety of azo graft copolymers can be synthesized by these

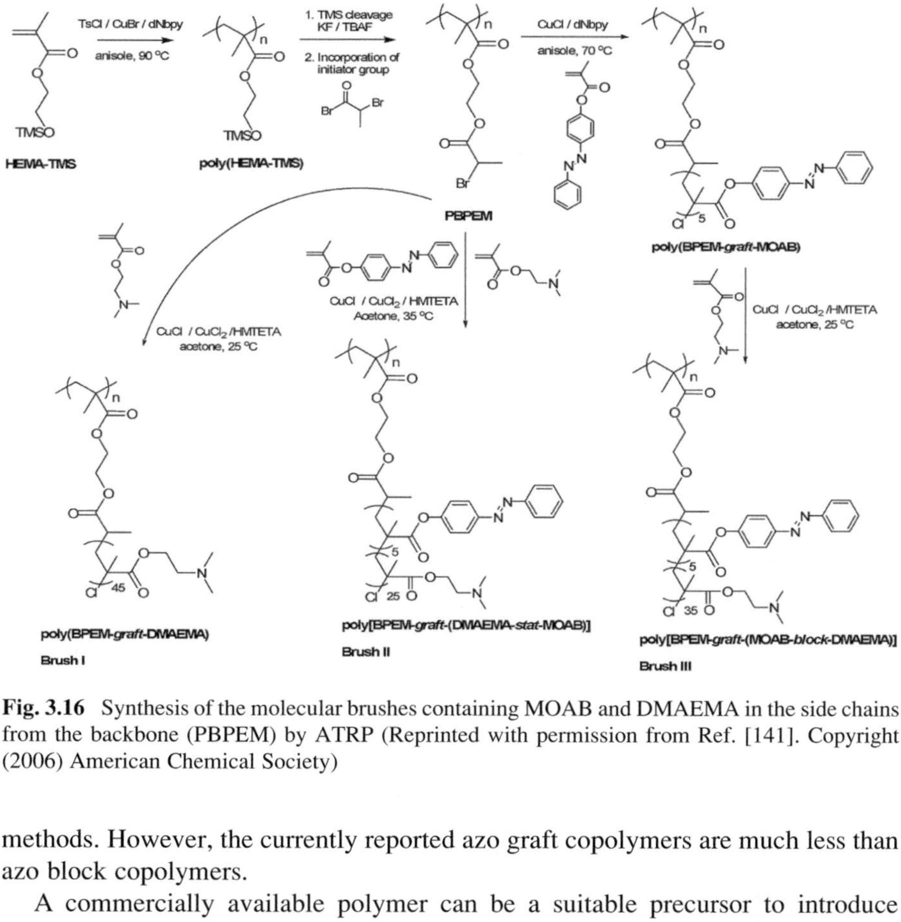

Fig. 3.16 Synthesis of the molecular brushes containing MOAB and DMAEMA in the side chains from the backbone (PBPEM) by ATRP (Reprinted with permission from Ref. [141]. Copyright (2006) American Chemical Society)

methods. However, the currently reported azo graft copolymers are much less than azo block copolymers.

A commercially available polymer can be a suitable precursor to introduce the grafts. Styrene–butadiene–styrene (SBS) triblock copolymer is a well-known thermoplastic elastomer. SBS can be used to prepare graft copolymers with azobenzene-containing grafts onto the central PB block [139, 140]. In this process, the graft copolymers are obtained by polymerization of azobenzene-containing acrylate monomers in solutions of SBS with benzoyl peroxide as the initiator. Azobenzene-containing thermoplastic elastomers with tunable optical properties can be obtained by this method.

Controlled radical polymerization can provide a more versatile way to prepare graft copolymers. Copolymers with a high density of grafts, so-called molecular brushes, have been synthesized through atom transfer radical polymerization (ATRP) [141]. The molecular brushes are synthesized by using poly (2-(2-bromopropionyloxy)ethyl methacrylate) (PBPEM) as a macroinitiator. Under a typical ATRP condition, the bromine in the PBPEM macroinitiator is replaced by chlorine from CuCl to initiate the grafting-from reaction, which can initiate a propagation of side chains from the backbone (Fig. 3.16). A progressive increase in the molecular weight is observed during the polymerization of grafting chains from the PBPEM backbone. The number-average molecular weight (M_n) of the brushes ranges from 4.7×10^5 to 1.1×10^6 depending on molecular architecture. The molecular weight distribution of the brushes is narrow ($M_w/M_n = 1.23$–1.36). In either chloroform or aqueous solution, azobenzene units in the side chains of the brush copolymers can show reversible isomerization upon irradiation with UV (365 nm) or visible light (442 nm).

3.4 Azo Dendritic and Star Polymers

Highly branched macromolecules and star polymers have attracted considerable attention in recent years. *Dendritic polymers* are highly branched macromolecules, which include dendrons, dendrimers, and hyperbranched polymers [6, 142, 143]. Dendrimers are macromolecules with regular extension of the branched structure along each dendron. A dendrimer structure can be divided into a core, framework (backbone), and peripheral or terminal groups. Hyperbranched polymers are highly branched macromolecules with less regular structures [7]. The structural units in a hyperbranched polymer can be classified into dendritic, linear, and terminal ones. The degree of branching (DB) is a factor describing the relative amount of the dendritic and terminal units in total units. Typically, dendrimers are synthesized by stepwise chemical reactions through divergent route [144–146] and convergent route [147]. The process of the growth is extended to different scales, which form dendrimers with different generations. Dendrons with a focal point can be considered as an intermediate during the convergent approach to prepare a dendrimer. A dendrimer can then be prepared by the reaction between the reactive group at the focal point and core molecules [147]. Hyperbranched polymers are typically prepared by using AB_x monomer through one-pot synthesis [7].

Star Polymer is another type of the macromolecules with a nonlinear and precisely controlled architecture, which contains several linear polymer chains connected at one central core [148]. Macromolecules with more complicated nonlinear structures include the dendrimer–polymer hybrids among others [5]. In recent years, azo dendrimers, hyperbranched polymers, star polymers, and other nonlinear macromolecules have been synthesized by different methods, which are discussed below according to their structural characteristics.

3.4.1 *Azo Dendron and Dendrimer*

Azo Dendrimers have been prepared by the divergent and convergent approaches to contain azo chromophores at the core, in the internal structure, and on the periphery [142, 149]. Accordingly, dendrons with azo chromophores at the different positions have also been prepared. The reactive groups at the focal point of a

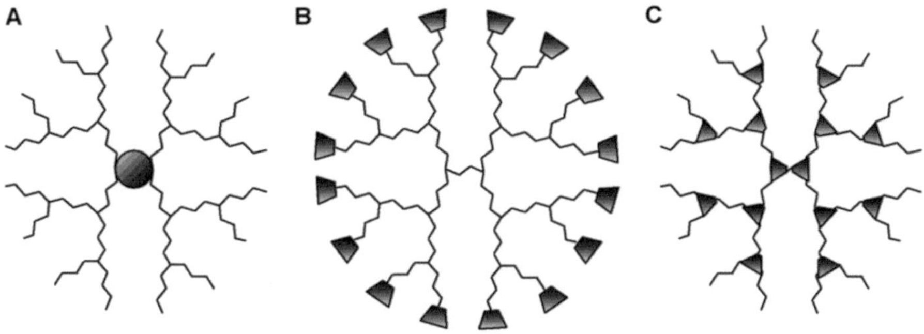

Fig. 3.17 Location of the functional groups in monofunctionalized dendrimers (Reprinted with permission from Ref. [142]. Copyright (2000) Elsevier Science Ltd)

dendron or on the periphery of a dendrimer and dendron can be directly employed to introduce azo chromophores at the core or on the periphery. For other types of azo dendrimers, some specifically designed monomers are usually required. The photo-switchable dendrimer with six azobenzene units on the periphery was first synthesized by Vögtle et al. more than 20 years ago [150]. In the past 20 years, many types of azo dendrons and dendrimers have been synthesized. As summarized by Vögtle et al. for general cases of functional dendrimers [142], there are three typical locations for the functional groups in monofunctionalized dendrimers (Fig. 3.17). Following this classification, the azo dendrons and dendrimers are discussed below according to the positions of azo chromophores in the dendritic structures, i.e., at the core, on the periphery, and in the internal locations.

The dendrimers with azobenzenes at the cores can be synthesized by the coupling reactions between an azobenzene derivative and a dendron bearing reactive group at the focal point. In a typical way, the dendrimers are obtained through coupling reactions between benzyl aryl ether dendrons and azobenzenes with functional groups. The dendritic polyether macromolecules are synthesized through the convergent approach by using 3,5-dihydroxy-benzyl alcohol as the monomer [147]. The formation of benzyl ethers is realized through the Williamson reaction between phenols and benzylic halides, which are then reacted with halogenating agents to restore the bromomethyl functionality at the focal point of the growing dendrons. By using the dendritic benzyl bromides, azobenzene-containing aryl ether dendrimers have been synthesized by alkaline-catalyzed Williamson coupling with 3,3′,5,5′-tetra-hydroxyazobenzene [151] or 4, 4′-bis(hydroxyl)azobenzene derivatives [152, 153]. The coupling reactions of azobenzene derivatives with the appropriate dendritic bromides can be efficiently carried out in the presence of K_2CO_3 in acetone. The dendritic structures show the ability to harvest low-energy photons to trigger isomerization [151].

Other specially designed dendrons bearing reactive groups at focal points have also been synthesized and used for this core and wedge coupling approach. Polyamidoamine (PAMAM) is a class of the starburst-dendritic macromolecules developed by Tomalia et al. [145]. The dendrimers are prepared by (a) Michael addition of an acrylate ester to an amino core and (b) amidation of the resulting ester moieties with alkylene diamine to yield the first-generation dendrimer. Dendrimers with different generations can be synthesized by repeating this synthetic procedure. PAMAM-based dendrimers with an azobenzene core have been synthesized by using ethanolamine in the first-step Michael addition to obtain the PAMAM dendritic wedge with one hydroxyl group at the focal point [154, 155]. The targeted dendrimer is then synthesized from the dendritic wedge by coupling reactions with azobenzene dicarboxylic acid or 4,4′-bis(chlorocarbonyl)azobenzene.

The second type according to above classification is dendrimers with azo chromophores on the periphery. This type of dendrimer can be synthesized by using an available dendrimer through the reactions of the terminal reactive groups with various azobenzene derivatives. Typically, for a divergent approach, the functional groups in each generation grow exponentially because of the multifunctionality of monomers. Dendrimers obtained by this method have a large amount of peripheral reactive groups in outmost generation, which can be used to introduce different functional groups [142].

Poly(propylene imine)s (PPIs) are a series of dendrimers prepared by the divergent method to possess numerous terminal amino groups. Typically, the dendrimers are prepared by two alternating reactions, i.e., the Michael addition of a primary amino group to acrylonitrile and hydrogenation of the nitrile group to regenerate the amino group [144]. A variety of PPI-based azo dendrimers have been prepared by reactions of the peripheral amino groups of PPIs. The azobenzene derivatives used for the reactions include *N*-hydroxysuccinimide esters of 3- and 4-(phenylazo)benzoic acid [156], sulfonic acid chloride of methyl orange [157], 4-bromo-methylazobenzene [158], pentafluorophenyl 11-[4-(4-hexyloxyphenylazo)phenyloxy)]-undecanoate [159], pentafluorophenyl 11-[4′-cyanophenylazophenyloxy]undecanoate [160], and pentafluorophenyl 6-{4-[4′-(S)-methylpropyloxyphenylazo]phenyloxy}hexanoate [161]. Photoactivity and pH sensitivity have been demonstrated for the dendrimers. Besides PPIs, dendrimers with azo chromophores on the periphery have also been prepared on the basis of other types of dendrimers, such as PAMAM [162] and carbosilane dendrimers [163].

The third type according to above classification is dendrons and dendrimers with azobenzene moieties in the backbone structure. For synthesizing this type of dendritic molecules, the monomers with two different types of reactive groups are generally required. The azobenzenes moieties can be located at a single layer or at several selected layers. This type of azo dendrimers can have many different subtypes with diversified architecture. Some typical types of dendrons and dendrimers containing azobenzene groups in different locations have been summarized by Deloncle and Caminade as shown in Fig. 3.18 [149].

A typical way to synthesize this type of azo dendritic macromolecules is through AB_2-type coupling reactions with a proper protection–deprotection scheme. The generation growth can be an esterification reaction with coupling reagent of dicyclohexylcarbodiimide (DCC) catalyzed by dimethylaminopyridine [164, 165]. A convergent synthetic approach has been developed to synthesize

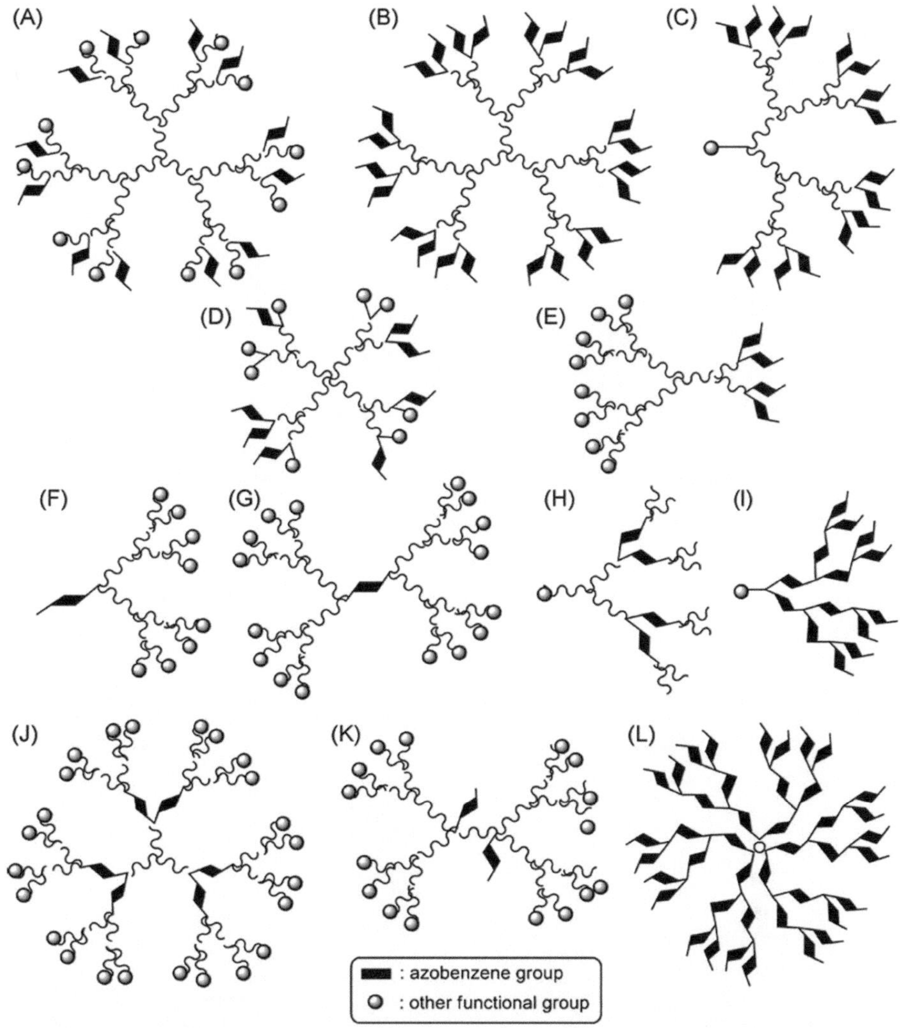

Fig. 3.18 Typical types of dendrimers (**a**, **b**, **d**, **e**, **g**, **j**, **k**, **l**) and dendrons (**c**, **f**, **h**, **i**) containing azobenzene group(s) in their structure (Reprinted with permission from Ref. [149]. Copyright (2010) Elsevier Science Ltd)

poly(benzyl ester) monodendrons with azobenzene groups throughout the molecular architecture [166]. The azo dendrimers can then be prepared by coupling different generation monodendrons with the four-branched azobenzene core AB_4 through esterifications mediated by DCC/DPTS (4-(dimethylamino)pyridinium 4-toluenesulfonate) at room temperature. Figure 3.19 shows the structures of the three generations of azo dendrimers. An improved approach without the protection–deprotection steps has been developed to synthesize azo dendrons through the 1,3-dipolar cycloaddition between azides and alkynes (click reaction) [167]. The

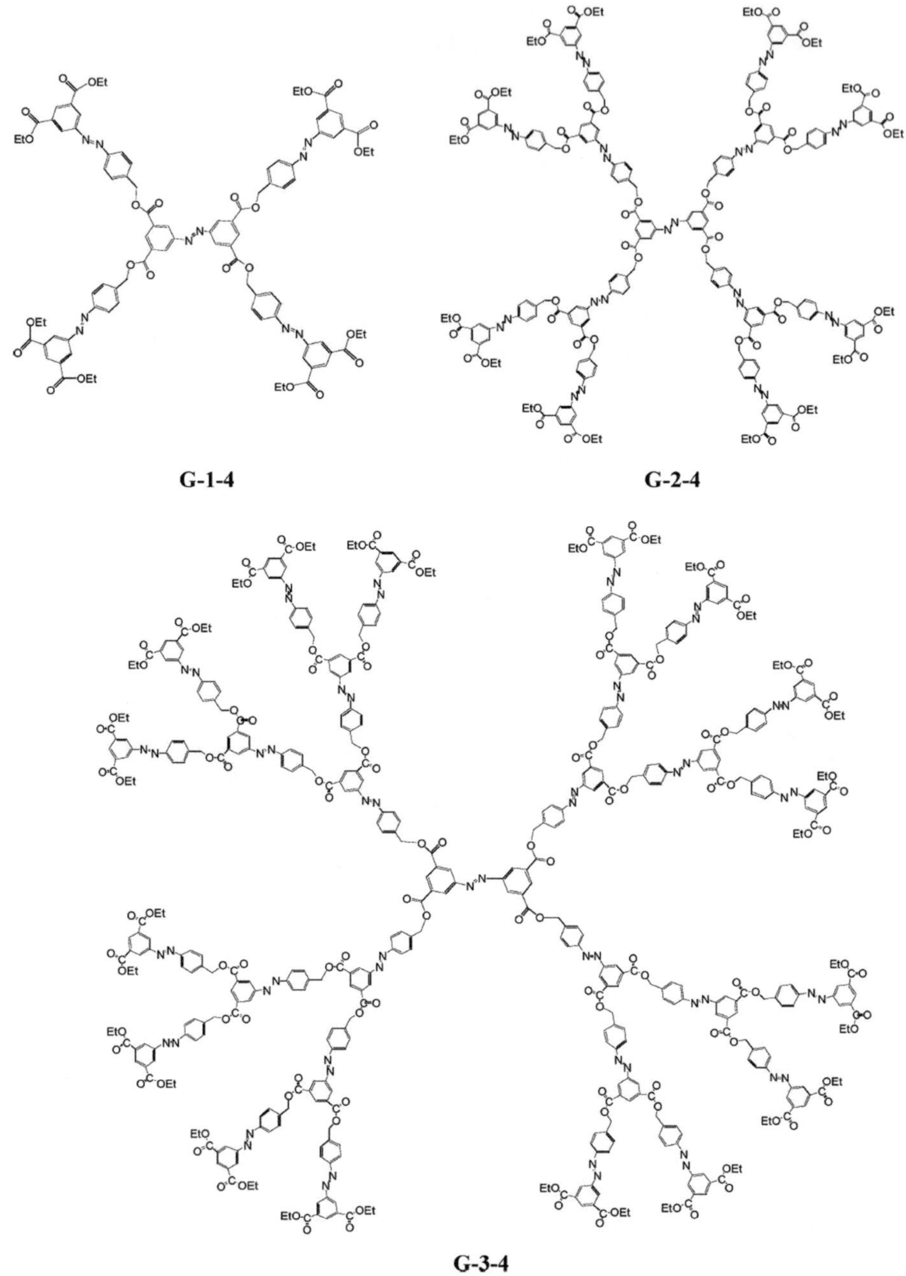

G-1-4

G-2-4

G-3-4

Fig. 3.19 Three generations of azo dendrimers G-1-4, G-2-4, and G-3-4 (Reprinted with permission from Ref. [166]. Copyright (2004) American Chemical Society) growth of dendrons is achieved via repetitive click reactions between the alkyne and azide reactive groups of monomers.

In addition to those with azobenzene moieties throughout the dendritic structures, the azobenzene moieties can be located in different interior layers or other positions. Those dendrimers contain three interior azobenzene moieties radially configured about the core unit and dendritic benzyl ether wedges linked to the branches [153]. The dendrimers were synthesized by reactions between 1,3,5-tricarboxybenzene and corresponding dendrons through esterification. Phosphorus-containing dendrimers having azobenzene derivatives specifically located at some generations in the interior and/or on the surface have been synthesized [168]. An unsymmetrical azobenzene derivative, 4-(4′-hydroxyphenylazo) benzaldehyde, was used for the synthesis. The reaction of $P(S)Cl_2$ reactive groups in the core with the azo compound in the presence of cesium carbonate can afford the dendrimer with six aldehyde groups on the surface. By using the aldehyde groups, higher generations of dendrimers are obtained by the continuation of the synthetic procedure.

3.4.2 Azo Hyperbranched Polymer

Hyperbranched Polymers are dendritic macromolecules with less regular structures and a broader molecular weight distribution. Hyperbranched polymers can be prepared through some relatively simple methods, most of them are one-pot reactions, which shows the attractive advantage for large-scale preparation. The typical methods to synthesize hyperbranched polymers include (a) step polymerization of AB_x-type monomers (where x is usually two or three) or $A_x + B_y$ monomer pairs (where x and y are usually two and three), (b) self-condensing vinyl polymerization of AB* monomers (where B* is an initiator), and (c) multibranching ring-opening polymerization [7]. Some modified methods, such as coupling monomer approach, have also been used to prepare hyperbranched polymers with specially designed structures [169]. These methods can all be used to synthesize hyperbranched polymers containing azo chromophores. Although the polymers have less regular structures, hyperbranched polymers can also be classified into those with azo chromophores at the core, in the internal structure, and on the periphery. Typical types of hyperbranched azo polymers and their syntheses are discussed below according to their structures and synthetic methods.

A variety of hyperbranched polymers with azobenzene moieties in the internal structures has been synthesized, where the azo functional groups can be either part of the backbone or pendant groups linked to the backbone. The hyperbranched azo polymers can be synthesized by the self-condensing atom transfer radical polymerization (ATRP) [170, 171]. In this case, a monomer with ATRP initiator at the side chain is needed, for which the radical polymerization can be consistently initiated in the side chain during the radical polymerization. Other types of the controlled

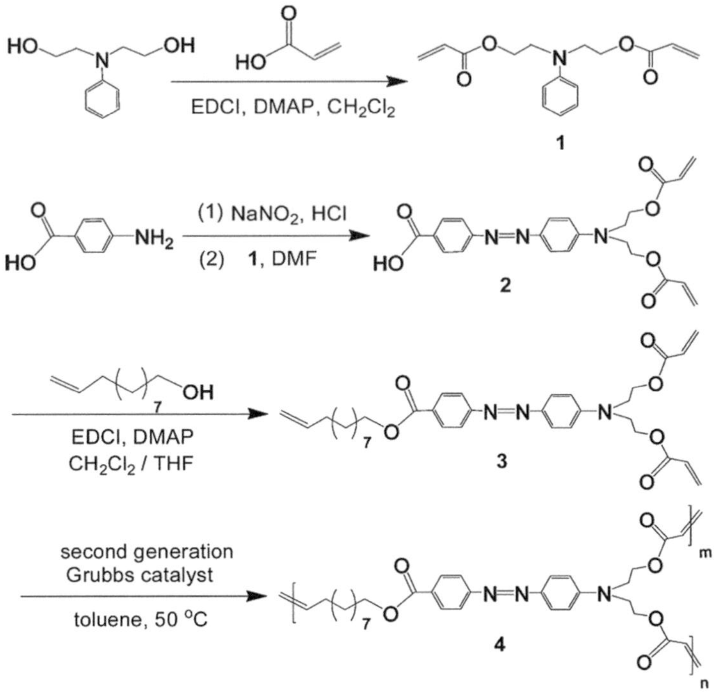

Fig. 3.20 Synthesis of hyperbranched azo polymer through acyclic diene metathesis (ADMET) (Reprinted with permission from Ref. [173]. Copyright (2009) American Chemical Society)

radical polymerizations can also be used for the self-condensing approach. More frequently, the hyperbranched polymers with azobenzene moieties in the internal structures are synthesized by different kinds of step polymerizations using AB_x- or $A_x + B_y$-type monomers. The monomers are prepared to contain both azobenzene moieties and reactive groups. Theoretically, any reactions with high yield and mild reaction conditions can be adopted for this purpose.

As a straightforward way, hyperbranched azo polymers can be prepared by step polymerizations using AB_2-type monomer. A hyperbranched polyester containing pseudo-stilbene-type azo chromophores throughout the hyperbranched architecture has been synthesized by ester-interchange reaction of a AB_2 monomer, ethyl 4-(4′-(bis(2-hydroxyethyl)amino)phenylazo)benzoate [172]. The condensation polymerization was catalyzed by $Zn(OAc)_2$. More recently, a hyperbranched azo polymer has been prepared using acyclic diene metathesis (ADMET) polymerization of an AB_2 monomer [173]. Figure 3.20 shows the synthetic route of the monomer and polymer. The polymerization of the monomer, 4-undecylenyloxycarbonyl-4′-[bis(2-ethyl acrylate)amino]azobenzene, is carried out by using a Grubbs catalyst, i.e., a ruthenium-based catalyst $RuCl_2(PCy_3)$(SIMes)(CHPh).

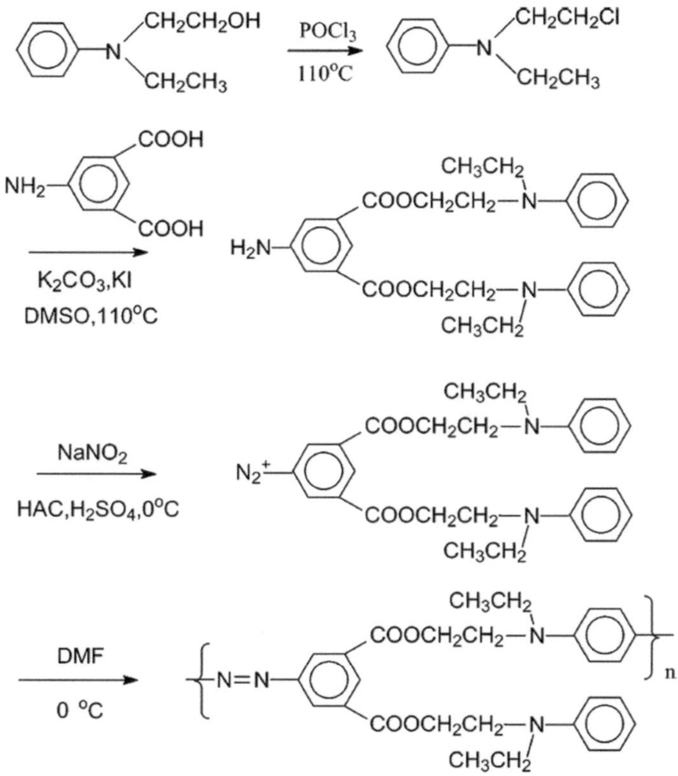

Fig. 3.21 Synthetic route of the precursor hyperbranched azo polymer through azo-coupling reaction (Reprinted with permission from Ref. [175]. Copyright (2005) American Chemical Society)

Azo-coupling reaction has been used to build up hyperbranched azo polymers by using AB_2-type monomers [174, 175]. A unique feature of this method is that the formations of azo chromophores and growth of polymeric backbone occur at the same time. The hyperbranched azo polymers are prepared by the azo-coupling reaction of specially designed monomers with both amino and anilino groups. The polymerization is carried out by using diazonium salt of the monomer in DMF under mild condition. Figure 3.21 shows the synthetic route of the monomer and precursor polymer. The hyperbranched azo polymer obtained by this way can be further modified to introduce other functional groups. For instance, carboxylic groups at the periphery can be obtained by two-stage azo-coupling reactions. In the first step, a precursor hyperbranched polymer is prepared by the step-growth polymerization of the AB_2 monomer through azo-coupling reaction. After that, the dendritic precursor is modified through the azo-coupling reaction with diazonium salt of 4-aminobenzoic acid.

The $A_x + B_y$ approach has been more widely used to prepare hyperbranched polymers with internal azo chromophores in the backbone or as pendant groups.

Hyperbranched polymers containing pendant azo chromophores have been obtained by esterification reaction between dispersed red 19 (DR19) as an A_2 monomer and 1,3,5-benzenetricarbonyl trichloride (BTC) as a B_3 monomer [176]. Hyperbranched poly(aryl ether)s with azobenzene moieties in both the backbone and pendant positions of the branched arms have been prepared by using $A_2 + B_3$ approach through a nucleophilic aromatic substitution polycondensation [177]. Hyperbranched polymers prepared by the $A_2 + B_3$ approach also include polyaspartimides synthesized via Michael addition reaction between azobenzene derivatives (A_2) and trimaleimide (B_3) [178], polyamides synthesized via Schotten–Baumann reaction [179]. This $A_x + B_y$ approach can be used to prepare hyperbranched polymers through other high efficient reactions, such as Sonogashira coupling and click chemistry. For the Sonogashira coupling approach, the hyperbranched azo polymers are synthesized by palladium-catalyzed coupling of diethynylazobenzene (A_2) with triiodoarenes (B_3) [180, 181] and triethynylazobenzene (A_3) with diiodoarenes (B_2) [182]. Figure 3.22 shows the synthetic scheme of the azo polymer. The $A_x + B_y$ approach can also be carried out by using the click chemistry syntheses. The hyperbranched polymer is obtained by reactions between azo-diazide (A_2) and 1,3,5-tris(propynyloxy)benzene (B_3) [183] or 1,3,5-trisazo-triazide (A_3) and siloxane diyne (B_2) [184].

Similar to dendrimers, hyperbranched polymers can be synthesized by incorporation of azobenzene moieties in the periphery of the dendritic structures. A

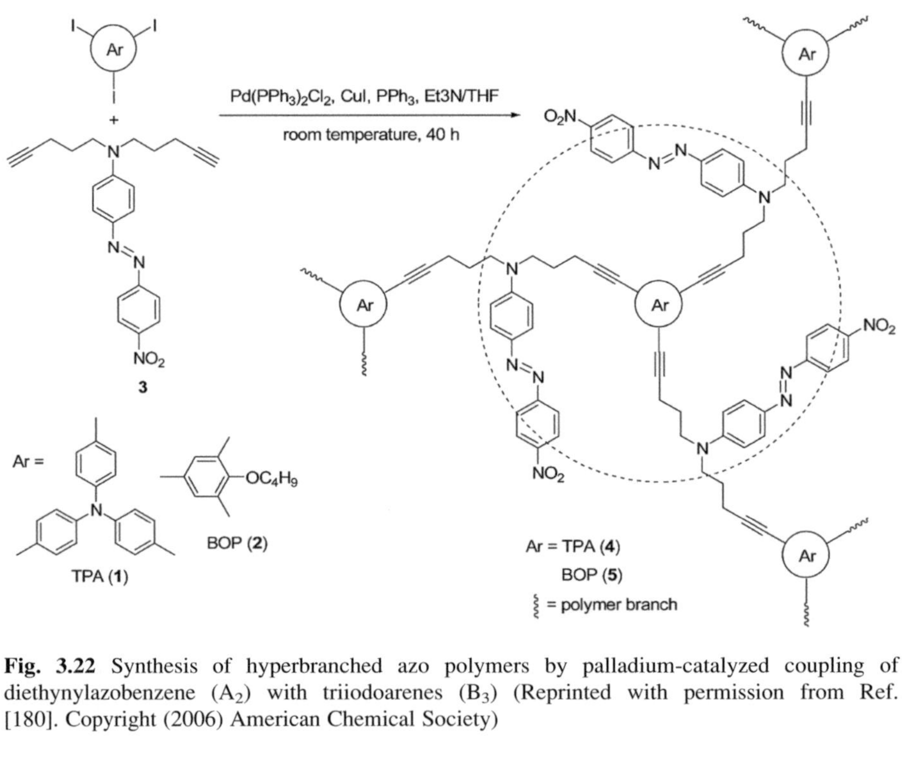

Fig. 3.22 Synthesis of hyperbranched azo polymers by palladium-catalyzed coupling of diethynylazobenzene (A_2) with triiodoarenes (B_3) (Reprinted with permission from Ref. [180]. Copyright (2006) American Chemical Society)

hyperbranched polymer containing azo chromophores on the periphery has been synthesized from transesterification reaction between a hyperbranched polyester bearing peripheral hydroxyl groups and ethyl 4-{4′-[*N*, *N*-di(hydroxyethyl)-aminobutoxy]phenylazo}benzoate [185]. The hyperbranched polymers have also been prepared by ionic interaction between amino-terminated hyperbranched poly (ethyleneimine) (PEI) and an azobenzene derivative with carboxyl group, which is realized in THF through the proton transfer from carboxyl group to amino groups [186]. The peripheral functionalization can be achieved by other highly efficient chemical reactions too. More recently, hyperbranched poly(ether amine)s have been synthesized through nucleophilic addition/ring-opening reaction of diglycidyl ether with amine and functionalized to prepare hyperbranched azo polymers [187]. In this process, the hyperbranched poly(ether amine)s are functionalized through reaction with an azobenzene derivative with epoxy group.

3.4.3 Other Azo Polymers with Nonlinear Architecture

Dendritic polymers discussed above are nonlinear macromolecules with highly branched structures. *Star polymers* are another type of the nonlinear macromolecules, which contain several linear polymer chains connected at one central core. Typically, star polymers can be synthesized by growing arms from a multifunctional initiator (the "core-first" method), cross-linking linear arm precursors using a cross-linker (the "arm-first" method), and attaching linear arm precursors onto a well-defined multifunctional core (the "coupling-onto" method) [148]. Dendrimer–polymer hybrids are another type of the nonlinear macromolecules, which can be prepared by various methods [5]. Some typical methods include growing chains from the peripheral of a dendrimer, polymerization of dendritic monomer, and different coupling reactions between a dendrimer and linear polymer. Similarly, a hyperbranched polymer can be used instead of the dendrimers for the same purpose to obtain the similar hybrid structures.

Methacrylate-based azo polymers with a three-arm star structure have been synthesized by atom transfer radical polymerization (ATRP) [188]. In this approach, azo monomer is polymerized by using 1,3,5-(2′-bromo-2′-methylpropionato)benzene (BMPB) as the initiator under typical ATRP conditions. Asymmetric or heteroarm star polymers (miktoarm star copolymers) are unique type of the star polymers. The central core of a miktoarm star copolymer bears different types of polymer arms, which differ in the molecular weight, the chemical composition, and the functional end group. Star copolymer containing an azobenzene moiety at the core has been synthesized to have poly(methyl methacrylate)$_2$–(polystyrene)$_2$ miktoarms. To obtain the polymer, miktofunctional initiator with an azo core is synthesize to bear tertiary bromide (for ATRP) and 2,2,6,6-tetramethylpiperidin-1-yloxy (TEMPO) (for nitroxide-mediated free radical polymerization, NMP) functionalities. The initiator was used in the controlled radical polymerizations, i.e., ATRP of methyl methacrylate and NMP of styrene, respectively, to give A_2B_2-type miktoarm star copolymer [189]. Miktoarm star polymers composed of two azobenzene-containing arms have been prepared by ATRP [190]. If a hyperbranched polymer is used as a multifunctional initiator to grow arms from the core, a polymeric structure with multi-arms can be synthesized. By this approach, multi-arm star side-chain azo copolymers with hyperbranched cores have been synthesized through ATRP by using a multifunctional hyperbranched polyether as the initiator [191].

Hybrids of dendritic and linear polymers can be synthesized to contain azobenzene moieties at peripheral positions. Linear-dendritic azo diblock copolymers have been synthesized by click chemistry between the dendritic block and an alkyne-functionalized PEG through the Huisgen′s 1, 3-dipolar cycloaddition [192]. The copolymers consist of poly(ethylene glycol) (PEG) and the first four generations of dendritic aliphatic polyesters based on 2,2-bis(hydroxymethyl) propionic acid (bis-MPA) functionalized with cyanoazobenzene units at the periphery. The dendritic block is synthesized through a two-step convergent approach starting from the 2-azidoethyl ester of 2,2-bis(hydroxymethyl)-propionic acid (bis-MPA) as shown in Fig. 3.23. The azo chromophores at the periphery are introduced through an esterification reaction between 11-(4-(4′-cyanophenylazo) phenyloxy)undecanoic acid and dendrons of the different generations. As an alternative approach, linear-dendritic azo diblock copolymers have been prepared from

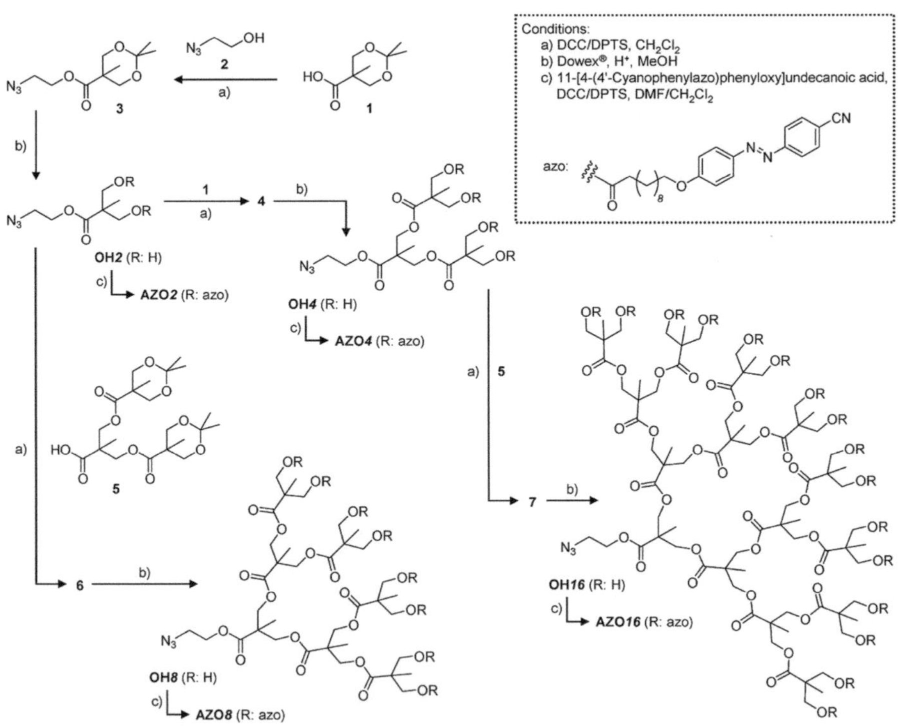

Fig. 3.23 Synthesis of the azo dendrons, AZOn (Reprinted with permission from Ref. [192]. Copyright (2009) American Chemical Society)

linear-dendritic block copolymers with a linear poly(ethylene glycol) (PEG) block and dendritic PAMAM of different generations [193]. The azo copolymers are obtained by Michael addition between the peripheral amine groups of dendritic segments and the azo acrylate bearing octyloxy tail and flexible decylmethylene spacer.

3.5 Azo Polymers with π-Conjugated Backbone

Since the discovery of Shirikawa, Heeger, and MacDiarmid in the late 1970s, *π-conjugated polymers* (*CPs*) have aroused tremendous research interest as conducting polymers and organic semiconductor [194]. Various CPs have been synthesized, which are characterized by rigid main chains, extended π-systems, and many interesting properties. Among them, azo polymers with π-conjugated backbones are unique type of CPs with photoresponsive properties. Typical ways to prepare CPs include transition-metal-catalyzed polymerizations, Lewis acid-induced polymerization, and electrochemical synthesis, among other addition and condensation polymerizations. CPs can also be prepared by an indirect route, where precursor polymers are first obtained by polymerization and then converted to final conjugated polymers by various reactions. This indirect approach can avoid the poor solubility of CPs and other problems of the direct synthesis. Azo polymers with π-conjugated backbone can be synthesized by these methods. According to the positions of the azo functional groups, linear azo CPs can be classified as side-chain and main-chain types. Few dendritic CPs containing azobenzene moieties have also been synthesized. The main types of the azo CPs and their synthetic methods are discussed below.

3.5.1 Conjugated Polymer with Side-Chain Azo Groups

CPs bearing side-chain azo groups have been synthesized by polymerizations of corresponding monomers or post-polymerization modifications. According to the main-chain structures, the polymers can be classified into polythiophene (PTh), poly(1,4-phenylenevinylene) (PPV), polyaniline (PANI), polyacetylene (PA), and poly(diacetylene) (PDA). Among them, polymers with main-chain aromatic rings are obtained by condensation polymerizations, where azo polymers based on PTh and PPV have been most widely investigated. On the other hand, acetylene-based polymers are typically obtained by addition polymerization.

Polythiophene is a type of conjugated polymers with excellent electrical conductivity, which also shows reasonably good mechanical properties and stability toward oxygen and moisture at ambient temperatures. A series of azobenzene-substituted polythiophenes has been synthesized from monomers 3-(2-(4-((4′-ethoxyphenyl)azo)phenoxy)ethoxy)-4-methyl-thiophene, 3-((6-(4-((4′-ethoxyphenyl)azo)phenoxy)hexyl)oxy)-4-methyl-thiophene, and 3-(((6-(4-((4′-ethoxyphenyl)azo)phenoxy)ethyl)tri(ethoxy)oxy)-4-methyl-thiophene [195]. The polymers are obtained by the chemical oxidative polymerization induced by Lewis acid ($FeCl_3$) in chloroform for 24 h. Polythiophene derivatives containing azo chromophore have been prepared via copolymerization of 3-alkylthiophene and 2-(*N*-ethyl-*N*-(4-((4-nitrophenyl)azo) phenyl)amino)ethyl 3-thienylacetate by the similar method [196]. By the chemical oxidative polymerization, poly(4-((4-(phenyl)azo)phenoxy)butyl 3-thienylacetate) and the copolymer of 3-hexylthiophene with 4-((4-(phenyl)azo)phenoxy)butyl 3-thienylacetate have also been synthesized [197].

Poly(1,4-phenylenevinylene) (PPV) is another important type of conjugated polymers with many interesting properties. The PPV-based azo polymers have been synthesized by different methods. A soluble conjugated PPV-based polymer, poly[2-methoxy-5-((4-(4-methoxyphenylazo)phenyloxy)hexyloxy)-1,4-phenylene vinylene], and PPV-based copolymers, composed of 2-methoxy-5((4-(4-methoxyphenylazo)phenyloxy)hexyloxy)-1,4-phenylenevinylene and 2-methoxy-5-*n*-butoxy-1,4-phenylenevinylene units, have been synthesized through the dehydrochlorination route [198, 199]. Conducting PPV derivative containing side-chain azobenzene moieties has also been prepared by Matsui et al. [200]. Fluorescence modulation via isomer-dependent energy transfer has been demonstrated by using the azobenzene-functionalized PPV derivatives [201–203].

Polyacetylenes possessing directly connected azobenzene moieties have been synthesized by polymerization of 3-ethynylazobenzene (3EAB) and 4-ethynylazobenzene (4EAB) [204]. In this approach, polymers are prepared by polymerizations of the monomers catalyzed by suitable transition-metal catalysts. Polydiacetylenes (PDAs) are another class of conjugated polymers possessing unique structures and properties, which are typically obtained from topochemical polymerization of appropriate diacetylene (DA) monomer. PDAs with side-chain azobenzene moieties have been prepared by post-polymerization azo-coupling reactions [205, 206]. The chemical structure and the synthetic route are shown in Fig. 3.24. The monomers are obtained by reaction between 9-(*N*-methyl-*N*-phenylamino)-5,7-nonadiyn-1-ol and appropriate isocyanates. The PDA precursors are obtained by polymerizing the monomers with exposure to ^{60}Co γ radiation. The PDA precursors obtained are soluble in organic solvent, and azo-coupling reactions are then used to introduce the side-chain azo functional groups.

Besides the PTh, PPV, PA, and PDA, polyaniline with side-chain azo chromophores has also been synthesized [207]. In this synthetic route, the emeraldine base (EB) form of polyaniline is synthesized by a chemical oxidation method. From the EB aniline, the fully reduced leucoemeraldine (LEB) form of PANI is obtained by reduction with hydrazine. The side-chain azo chromophores are finally introduced by reaction of LEB PANI with an azo compound with a bromo-terminal group.

3.5.2 Main-Chain Conjugated Azo Polymers

Poly(*p*-phenylene)-based (PPP-based) and poly(phenylenevinylene)-based (PPV-based) azo polymers are two types of the main-chain conjugated azo

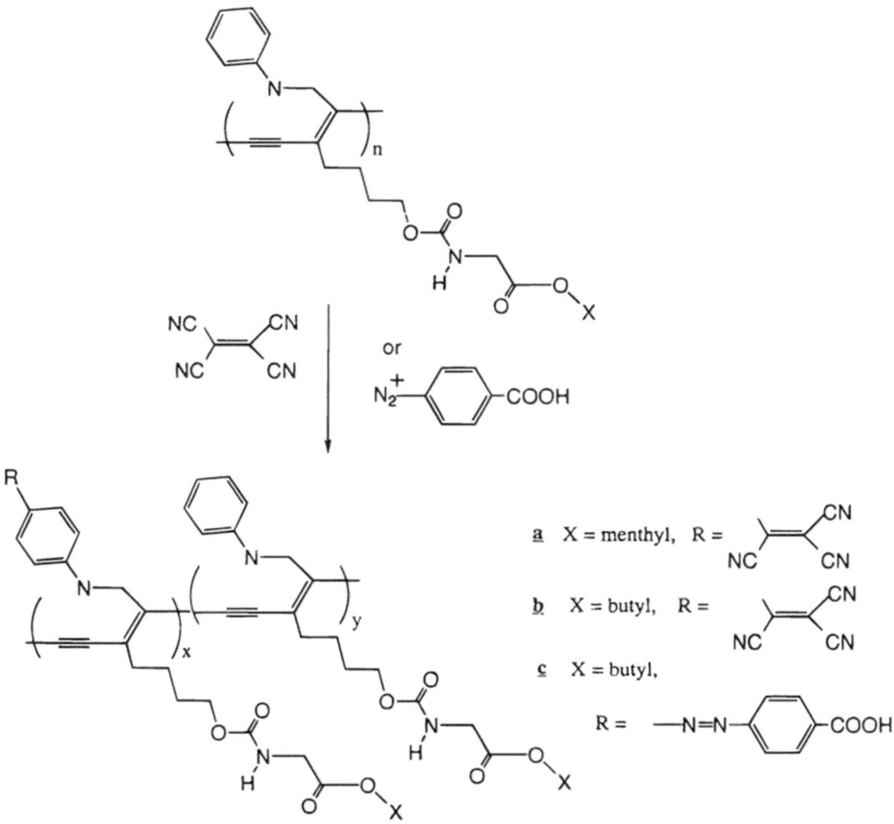

Fig. 3.24 Functionalization of polydiacetylenes with tricyanovinylation and azo-coupling reaction (Reprinted with permission from Ref. [205]. Copyright (1998) American Chemical Society)

polymers among the earliest reported [208, 209]. As shown in Fig. 3.25, PPP-based conjugated azo polymers are synthesized by the palladium-catalyzed Suzuki coupling of 1,3-propanediol diesters of 1,4-phenylenediboronic acids with 4,4′-diiodoazobenzenes [208]. Even the polymers with the rigid conjugated backbone, the photochemical *trans–cis* reversible isomerizations of the azo polymers in both tetrahydrofuran and toluene can be observed upon irradiation with UV light of appropriate wavelengths. These isomerization processes are accompanied by the change in the three-dimensional hydrodynamic volume of the polymer as evidenced by size exclusion chromatography. PPV-based azo polymers are obtained by the P_d-catalyzed coupling polymerization of 4,4′-divinylazobenzene with dihaloarene or divinylbenzenes with 4,4′-dihaloazobenzenes under the Heck reaction conditions [209]. Compared to the monomer, a remarkably red-shifted absorption in the visible region (456 nm) is observed for one of the polymers.

Poly(*p*-phenyleneethynylene)-based(PPE-based) main-chain azo polymer is another type of the conjugated azo polymers. The PPE-based azo polymers have been

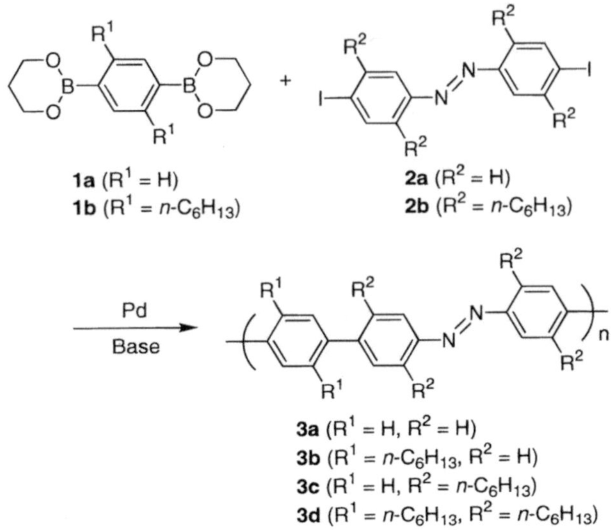

Fig. 3.25 PPP-based conjugated azo polymers synthesized by the palladium-catalyzed Suzuki coupling (Reprinted with permission from Ref. [208]. Copyright (2000) American Chemical Society)

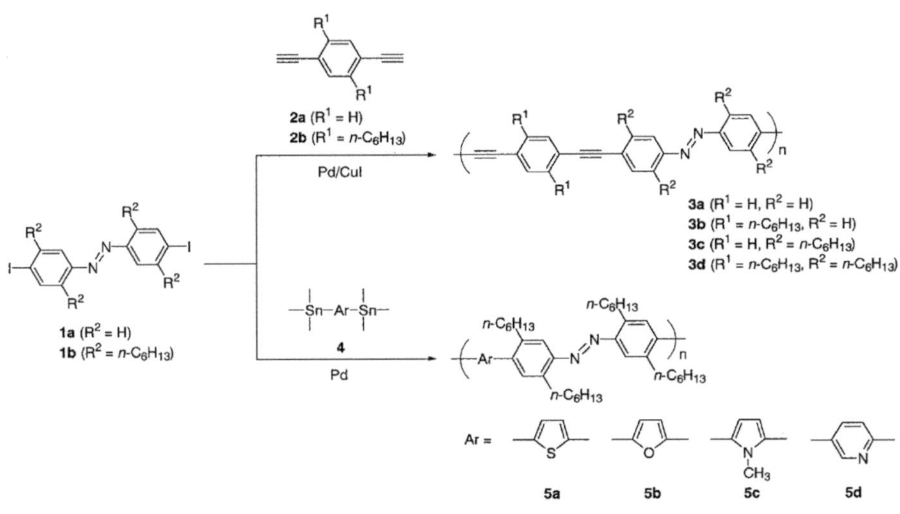

Fig. 3.26 PPE-based azo polymers synthesized by palladium-catalyzed cross-coupling reaction (Reprinted with permission from Ref. [210]. Copyright (2001) American Chemical Society)

synthesized by palladium-catalyzed cross-coupling reaction of 4,4′-diiodoazobenzenes with 1,4-diethynylbenzenes [210]. In this approach, the polymers are obtained under the Sonogashira coupling reaction conditions, i.e., catalyzed by $PdCl_2(PPh_3)_2$ and CuI in *N*,*N*-dimethylformamide (DMF) and triethylamine at 80 °C for 24 h (Fig. 3.26). A series of optically active poly(*m*-phenyleneethynylene)s containing azobenzene moieties has been synthesized by the Sonogashira–Hagihara coupling polymerization of diiodo monomers with 3,5-diethynylazobenzenes, 3,3′-diethynylazobenzene, and 4,4′-diethynylazobenzene [211, 212]. Chiroptical properties and photoresponse of helically folding are most important functions of the polymers.

Other types of main-chain azo polymers include the azo polymers containing heteroaromatics such as thiophene and pyridine. The azo polymers have been obtained by the Stille reaction, i.e., the palladium-catalyzed cross-coupling reaction of dihaloarenes with bis(trialkylstannyl) heteroaromatics [210], and electropolymerizations [213]. Conjugated main-chain azo polymer containing fluorine has recently been synthesized [214]. The polymer is obtained by the Suzuki coupling reaction of 4,4′-dibromoazobenzene and 9,9-dioctylfluorene-2,7-diboronic acid bis(1,3-propanediol)ester. Light-driven chiroptical modulation between chiral aggregation and achiral disaggregation has been demonstrated by using the polymers.

3.5.3 Conjugated Dendrimer with Azobenzene Core

Conjugated dendrimers with an azobenzene core have been synthesized by divergent approach [215]. In this approach, the rigid polyphenylene dendrimers with an azobenzene core is synthesized from 4,4′-bis(4-ethynyl)-azobenzene. The divergent synthesis of the dendrimers can be accomplished on the basis of Diels–Alder cycloaddition of different cyclopentadienones to alkynes in refluxing *o*-xylene at 142 °C or diphenyl ether at 210 °C. The dendrimers with different generations are obtained by repetitive Diels–Alder cycloaddition from the azo core with an A_2B monomer, 3,4-bis (4-triisopropyl-silylethynylphenyl)-2,5-diphenyl-cyclopentadienone, and quantitative cleavage of the protecting groups with ammonium fluoride. For the growth of the generations 2, 3, or 4, the addition/deprotection cycles need to be carried out for one, two, or three times followed by a Diels–Alder addition. It is observed that for the polyphenylene dendrimers, the *cis* isomer content under 364 nm irradiation decreases with increasing dendrimer generation and degree of branching.

3.6 Azobenzene-Containing Polypeptide

Azobenzenes as photoresponsive functional groups have been introduced into the different biomacromolecular systems. *Azobenzene-containing polypeptides* are a major type among them, which is discussed in this section. For other types of azo biomacromolecules, we refer the interested reader to some comprehensive reviews [216, 217]. Polypeptides are composed of α-amino acids as the basic units, which are joined by amide linkages (peptide bonds) to form the polymers. The photochemical effect caused by azo chromophores can result in the helix reversals, coil–helix transitions, and aggregation–disaggregation variations [218–220]. Such properties are distinguished from other types of azo polymers discussed above and have aroused considerable interest. Like side-chain azo polymers, azobenzene-containing polypeptides can be synthesized by both polymerizations of corresponding monomers and post-polymerization modifications as discussed below.

3.6.1 Polymerization of Azo Monomers

Polymerization of *N*-carboxy anhydride (NCA) of an amino acid is a typical way to synthesize azo polypeptides. In a pioneering study, a series of azo polypeptides was prepared from L-*p*-(pheny1azo)phenylalanine and γ-benzyl-L-glutamate [221]. The polypeptides were obtained by homopolymerization of L-*p*-(phenylazo)phenylalanine NCA and copolymerization with varying quantities of γ-benzyl-L-glutamate NCA. The optical rotatory dispersions (ORD) of the polypeptides in trifluoroacetic acid showed large positive Cotton effects and even larger negative Cotton effects before and after irradiation at 425 nm [222]. The NCA polymerization approach has also been used to prepare random copolymers of γ-benzyl-L-glutamate with *m*- and *p*-phenylazobenzyl-L-aspartates [223]. Conformational changes of the azoaromatic polyaspartate can be induced by the light irradiation as revealed by circular dichroism (CD) spectroscopy [224]. Photoinduced *trans–cis* isomerization of the side-chain azobenzene moieties causes helix reversal of these copolymers in appropriate solvent compositions [225].

Series of sequential polypeptides with repeating units Lys(Z)$_m$-azoAla (I, $m = 1–3$), azoAla-Lys(Z)-azoAla-Aib (II, $m' = 1$), and Lys(Z)$_{m'-1}$-azoAla-Aib (II, $m' = 2$ and 3) have been synthesized by the Goodman's method, where Lys(Z), azoAla, and Aib refer to N′-[(benzyloxy)carbonyl]-L-lysin, L-*p*-phenyl azo-phenylalanine, and α-aminoisobutyric acid, respectively [226]. Figure 3.27 shows the chemical structure of the polymers. Caused by the *trans–cis* photoisomerization of azobenzene moieties, the polypeptides show a reversible change of optical rotation at 589 nm. The magnitude and profile of experimental CD spectra of $m(m') = 1$ and 2 polypeptides are consistent with the theoretical calculations [227].

3.6.2 Azo Functionalization of Polypeptide

Various reactive groups in the polypeptides can be used to introduce azo side-chain groups under proper conditions. Poly(L-glutamic acid) with side-chain azobenzene moieties have been prepared by reaction of poly(L-glutamic acid) with *p*-aminoazobenzene [228, 229]. In this approach, the polypeptides containing 13–56 mol% of azo groups are obtained by the reaction in the presence of dicyclohexylcarbodiimide and *N*-hydroxybenzotriazole in dimethylformamide.

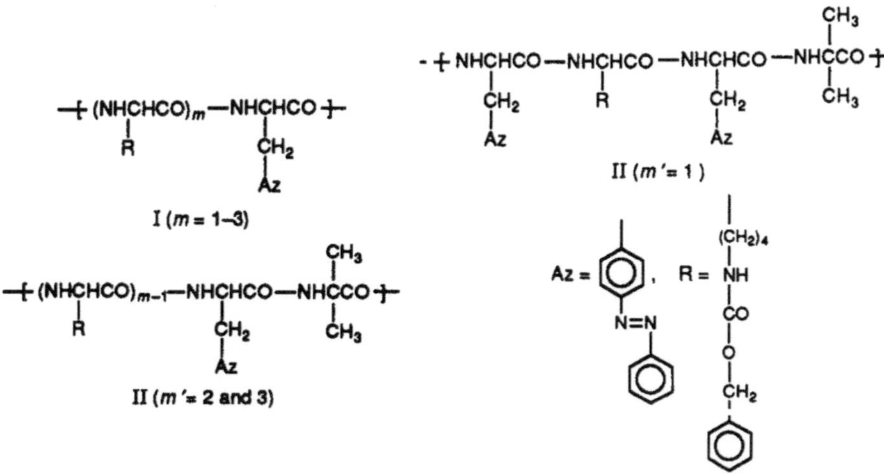

Fig. 3.27 Chemical structure of a series of sequential polypeptides (Reprinted with permission from Ref. [226]. Copyright (1991) American Chemical Society)

For the azo polypeptides, light irradiation can induce reversible conformation changes [229], aggregation change [230], and solubility modification [231]. Poly (L-glutamic acids) containing 1.9, 9.3, and 46.3 mol % azobenzenesulfonate moieties in the side chains have also been prepared by condensation reaction of poly (L-glutamic acid) with 4-amino-l,l′-azobenzene-4′-sulfonic acid sodium salt [232]. The photoinduced conformation transition, from α-helix to coil, can be observed for the polypeptide containing 9.3 mol% azobenzenesulfonate moieties at adequate pH. For the vesicles of azobenzene-modified poly(γ-methyl L-glutamate-*co*-L-glutamic acid), UV light (<330 nm) irradiation can induce intervesicular adhesion [233]. It has been attributed to a transfer of the polypeptide from the hydrophobic bilayer membrane interior to the hydrophilic surface caused by the *trans–cis* photoisomerization of the azobenzene moieties.

Poly(L-lysine) with azo side chains has been synthesized by the reaction of poly (L-lysine) hydrobromide with *p*-(phenylazo)benzoic acid *p*-nitrophenyl ester in the presence of triethylamine in aqueous dimethylformamide [234]. As an alternative approach, the polypeptide has been prepared by the reaction of poly(L-lysine) hydrobromide with *p*-(phenylazo)benzoic acid in the presence of 1-ethyl-3-[3-(dimethylamino)propyl]carbodiimide [235]. Photoresponsive poly(L-lysine) can also be prepared by the reaction of poly(L-lysine) with *p*-phenylazobenzenesulfonyl chloride [236]. For this polypeptide, the azobenzene units are linked to the Lys chain by sulfonamide groups (Fig. 3.28). For the azo poly(L-lysine), irradiation with light at 340 and 417 nm can alternately induce reversible transition between a helical and random coil structure of the macromolecules.

As a more sophisticated synthetic method, diamino acid-N^{α}-substituted oligopeptides (DNOs) bearing azo side chains have been synthesized by Merrifield

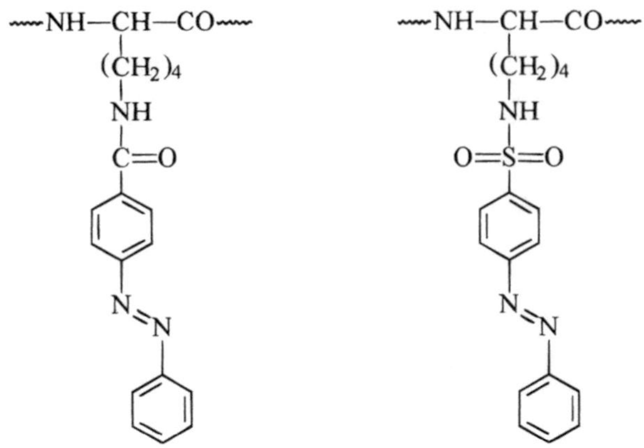

Fig. 3.28 Poly(L-lysine) prepared by the reaction of poly(L-lysine) with p-phenylazobenzenesulfonyl chloride (Reprinted with permission from Ref. [236]. Copyright (1996) American Chemical Society)

solid-phase synthesis [237]. Each cycle of the solid-phase assembly includes two coupling step, i.e., introduction of the backbone unit and the incorporation of the side-chain azobenzene moieties. In the process, the different levels of protection are required to obtain the final product. One advantage of this approach is that the sequential structure of the oligopeptides can be predesigned and well controlled.

3.7 Synthesis, Structures, and Functions

In the past decades, the azo polymers with great diversity have been synthesized by different methods. As a general tendency, the structures of the azo polymer undergo an evolution from simple to complex in the past decades. As a result, almost every type of polymers can find its azo polymer counterparts, which contain azobenzene moieties in the main chain, side chain, or other positions. A variety of polymerization methods, no matter conventional or newly developed, has been used to build up the azo polymers with designed structures. The azo polymers have been synthesized to satisfy the curiosity of the researchers in many cases, but more frequently, the polymers have been designed and synthesized to demonstrate some novel functions.

As a general rule, the functions of azo polymers are determined by hierarchical structures from molecular architectures to condensation states, which can be briefly divided into four levels. The azo chromophores can be considered as the basic and core structure of azo polymers. As discussed in Chap. 2, the azo chromophores can be classified as azobenzene type, aminoazobenzene type, and pseudo-stilbene type, which show different photochemical properties. The primary structure of azo polymers includes the covalently bonded linkages to form the macromolecules as thoroughly discussed in this chapter. Many properties of azo polymers are determined by the azo chromophores and primary structure as discussed above. The secondary structure of azo polymer is the stereochemical structure of azo polymer, such as configuration and conformation, which is often related to the hydrogen bonding in the polymers. Light-induced transition from helix to random coils and the helix reversals, as discussed above for polypeptides, are responses of the secondary structure to light irradiation. Many functions related to azo polymer solutions, especially for dilute solutions, are related to the core, primary, and secondary structures. The tertiary structure of azo polymers can be considered as three-dimensional structure formed from macromolecules. The structure is controlled by hydrophilic/hydrophobic interaction, hydrogen bonding, and electrostatic interaction. In recent years, various types of self-assembling methods have been explored to manipulate the structures. Quaternary structure is the condensation structure of azo polymers, which can be amorphous, liquid crystal, and crystal. Most optical functions and potential applications of azo polymers are related to the bulk structure such as solid films.

In the following chapters (Chaps. 4, 5, and 6), only some most intensively investigated functions of azo polymers will be discussed. Most of them are related to the structures at different levels as discussed above. It should be mentioned that the azo polymers synthesized so far can be much more diversified in structures and functions than those discussed in the following chapters. As it is almost impossible to discuss the diversified structures and functions of azo polymers in detail, we can only refer the interested reader to the literature given in this chapter and other chapters.

References

1. Flory, P.J.: Principles of Polymer Chemistry. Cornell University Press Ltd., Ithaca (1953)
2. Odian, G.: Principles of Polymerization, 2nd edn. Wiley, New York (1981)
3. Akelah, A., Moet, A.: Functionalized Polymers and their Applications. Chapman and Hall, London (1990)
4. Kumar, G.S.: Azo Functional Polymers: Functional Group Approach in Macromolecular Design. Technomic Publishing Co. Inc., Lancaster (1992)
5. Roovers, J., Comanita, B.: Dendrimers and dendrimer-polymer hybrids. In: Roovers, J. (ed.) Advances in Polymer Science, vol. 142, pp. 179–228. Springer, Berlin (1999)
6. Hult, A., Johansson, M., Malmström, E.: Hyperbranched Polymers. In: Roovers, J. (ed.) Advances in Polymer Science, vol. 143, pp. 1–33. Springer, Berlin (1999)
7. Jikei, M., Kakimoto, M.A.: Hyperbranched polymers: a promising new class of materials. Prog. Polym. Sci. **26**, 1233–1285 (2001)
8. Nuyken, O.: Azo polymers. In: Kroschwitz, J.I. (ed.) Encyclopedia of Polymer Science and Engineering, vol. 2, pp. 158–175. Wiley, New York (1985)
9. Kumar, G.S., Neckers, D.C.: Photochemistry of azobenzene-containing polymers. Chem. Rev. **89**, 1915–1925 (1989)
10. Xie, S., Natansohn, A., Rochon, P.: Recent developments in aromatic azo polymers research. Chem. Mater. **5**, 403–411 (1993)
11. Natansohn, A., Rochon, P.: Photoinduced motions in azo-containing polymers. Chem. Rev. **102**, 4139–4175 (2002)

12. Shibaev, V., Bobrovsky, A., Boiko, N.: Photoactive liquid crystalline polymer systems with light-controllable structure and optical properties. Prog. Polym. Sci. **28**, 729–836 (2003)
13. Lovrien, R., Waddington, J.C.B.: Photoresponsive systems. I. Photochromic macromolecules. J. Am. Chem. Soc. **86**, 2315–2322 (1964)
14. Kamogawa, H., Kato, M., Sugiyama, H.: Syntheses and properties of photochromic polymers of the azobenzene and thiazine series. J. Polym. Sci. A **6**, 2967–2991 (1968)
15. Paik, C.S., Morawetz, H.: Photochemical and thermal isomerization of azoaromatic residues in the side chains and the backbone of polymers in bulk. Macromolecules **5**, 171–177 (1972)
16. Natansohn, A., Rochon, P., Gosselin, J., Xie, S.: Azo polymers for reversible optical storage. 1. Poly[4′-[[2-(acryloyloxy)ethyl]ethylamino]-4-nitroazobenzene]. Macromolecules **25**, 2268–2273 (1992)
17. Xie, S., Natansohn, A., Rochon, P.: Microstructure of copolymers containing disperse red 1 and methyl methacrylate. Macromolecules **27**, 1885–1890 (1994)
18. Brown, D., Natansohn, A., Rochon, P.: Azo polymers for reversible optical storage. 5. Orientation and dipolar interactions of azobenzene side groups in copolymers and blends containing methyl methacrylate structural units. Macromolecules **28**, 6116–6123 (1995)
19. Ho, M.S., Natansohn, A., Rochon, P.: Azo polymers for reversible optical storage. 7. The effect of the size of the photochromic groups. Macromolecules **28**, 6124–6127 (1995)
20. Meng, X., Natansohn, A., Barrett, C., Rochon, P.: Azo polymers for reversible optical storage. 10. Cooperative motion of polar side groups in amorphous polymers. Macromolecules **29**, 946–952 (1996)
21. Ho, M.S., Natansohn, A., Rochon, P.: Azo polymers for reversible optical storage. 9. Copolymers containing two types of azobenzene side groups. Macromolecules **29**, 44–49 (1996)
22. Ringsdorf, H., Schmidt, H.W.: Electro-optical effects of azo dye containing liquid crystalline copolymers. Makromol. Chem. **185**, 1327–1334 (1984)
23. Eich, M., Wendorff, J.H.: Erasable holograms in polymeric liquid crystals. Makromol. Chem. Rapid. Commun. **8**, 467–471 (1987)
24. Ikeda, T., Horiuchi, S., Karanjit, D.B., Kurihara, S., Tazuke, S.: Photochemically induced isothermal phase transition in polymer liquid crystals with mesogenic phenyl benzoate side chains. 1. Calorimetric studies and order parameter. Macromolecules **23**, 36–42 (1990)
25. Ravve, A., Fitko, C.: Polymer formation through diazonium coupling. J. Polym. Sci. A **2**, 1925–1940 (1964)
26. Tabak, D., Morawetz, H.: Rates of conformational transitions in solutions of randomly coiled polymers. III. The cis-trans isomerization of azobenzene residues in the backbone of polyamides. Macromolecules **3**, 403–410 (1970)
27. Chen, D.T.L., Morawetz, H.: Photoisomerization and fluorescence of chromophores built into backbones of flexible polymer chains. Macromolecules **9**, 463–468 (1976)
28. Eisenbach, C.D.: *Cis-trans* isomerization of aromatic azo chromophores, incorporated in the hard segments of poly(ester urethane)s. Makromol. Chem. Rapid. Commun. **1**, 287–292 (1980)
29. Sung, C.S.P., Lamarre, L., Tse, M.K.: Molecular motion of polymer main chains in the solid state as studied by photoisomerization. Macromolecules **12**, 666–669 (1979)
30. Sung, C.S.P., Lamarre, L., Chung, K.H.: Use of azochromophoric labels as a molecular probe of physical aging in amorphous polymers. Macromolecules **14**, 1839–1841 (1981)
31. Korshak, V.V., Vinogradova, S.V., Antonova-Antipova, I.P.: Self-colored polyesters based on di(hydroxyalkoxy)azobenzenes. Bull. Acad. Sci. USSR, Div. Chem. Sci. (1968). doi:10.1007/BF00922854.
32. Hall, H.K., Kuo, T., Leslie, T.M.: New AB polyesters and a polymethacrylate containing dipolar *p*-phenyleneazo groups. Macromolecules **22**, 3525–3529 (1989)
33. Köhler, W., Robello, D.R., Willand, C.S., Williams, D.J.: Dielectric relaxation study of some novel polymers for nonlinear optics. Macromolecules **24**, 4589–4599 (1991)
34. Agolini, F., Gay, F.P.: Synthesis and properties of azoaromatic polymers. Macromolecules **3**, 349–351 (1970)

35. Riordan, J.E., Blair, H.S.: Synthesis and characterization of inherently coloured azo polyamides. Polymer **20**, 196–202 (1979)
36. Balasubramanian, M., Nanjan, M.J., Santappa, M.: Synthesis of polyamides containing 4,4′-azodibenamido units. Makromol. Chem. **182**, 853–859 (1981)
37. Irie, M., Schnabel, W.: Photoresponsive polymers. On the dynamics of conformational changes of polyamides with backbone azobenzene groups. Macromolecules **14**, 1246–1249 (1981)
38. Irie, M., Hirano, Y., Hashimoto, S., Hayashi, K.: Photoresponsive polymers. 2. Reversible solution viscosity change of polyamides having azobenzene residues in the main chain. Macromolecules **14**, 262–267 (1981)
39. Jayaprakash, D., Ravikumar, L., Nanjan, M.J.: Synthesis of polyamides containing azo groups. Makromol. Chem. Rapid. Commun. **2**, 611–615 (1981)
40. Sachindrapal, P., Ramasamy, S., Nanjan, M.J.: Synthesis of aromatic-aliphatic polyamides containing azo linkage. Polym. Bull. **5**, 417–424 (1981)
41. Jayaprakash, D., Balasubramanian, M., Nanjan, M.J.: Synthesis and characterization of certain polyamides containing the azo group in the main chain. J. Polym. Sci. A **23**, 2319–2326 (1985)
42. Srinivasan, P.R., Srinivasan, M., Mahadevan, V.: Preparation and properties of polybenzimidazoles containing azo groups. J. Polym. Sci. A **20**, 1145–1150 (1982)
43. Jayaprakash, D., Nanjan, M.J.: Synthesis and characterization of certain polyhydrazides containing azo group in the main chain. J. Polym. Sci. A **20**, 1959–1963 (1982)
44. Nanjan, M.J., Jayaprakash, D.: Synthesis of a new azopolyamide-hydrazide. J. Polym. Sci. A **21**, 1227–1231 (1983)
45. Kumar, G.S., DePra, P., Zhang, K., Neckers, D.C.: Chelating copolymers containing photosensitive functionalities 2. Macromolecules **17**, 2463–2467 (1984)
46. Xu, C.Z., Wu, B., Dalton, L.R., Ranon, P.M., Shi, Y.Q., Steler, W.H.: New random main-chain, second-order nonlinear optical polymers. Macromolecules **25**, 6716–6718 (1992)
47. Xu, C.Z., Wu, B., Becker, M.W., Dalton, L.R., Ranon, P.M., Shi, Y.Q., Steler, W.H.: Main-chain second-order nonlinear optical polymers: random incorporation of amino-sulfone chromophores. Chem. Mater. **5**, 1439–1444 (1993)
48. Li, Q.Q., Li, Z., Zeng, F.X., Gong, W., Li, Z., Zhu, Z.C., Zeng, Q., Yu, S.S., Ye, C., Qin, J.G.: From controllable attached isolation moieties to possibly highly efficient nonlinear optical main-chain polyurethanes containing indole-based chromophores. J. Phys. Chem. B **111**, 508–514 (2007)
49. Jaycox, G.D.: Azobenzene modified poly(aryl ether ketone amide)s. I. Synthesis and physical properties. Polymer **39**, 2589–2596 (1998)
50. Everlof, G.J., Jaycox, G.D.: Stimuli-responsive polymers. 4. Photo- and thermo-regulated chiroptical behavior in azobenzene-modified polymers fitted with main chain spirobiindane turns and chiral binaphthyl bends. Polymer **41**, 6527–6536 (2000)
51. Lustig, S.R., Everlof, G.J., Jaycox, G.D.: Stimuli-responsive polymers. 5. Azobenzene modified polyaramides containing atropisomeric binaphthyl linkages: tuning chiroptical behavior with light and heat. Macromolecules **34**, 2364–2372 (2001)
52. Jaycox, G.D.: Stimuli-responsive polymers. VII. Photomodulated chiroptical switches: periodic copolyaramides containing azobenzene, phenylene, and chiral binaphthylene main-chain linkages. J. Polym. Sci. A **42**, 566–577 (2004)
53. Jaycox, G.D.: Stimuli-responsive polymers. VIII. Polyesters and poly(ester amides) containing azobenzene and chiral binaphthylene segments: highly adaptive materials endowed with light-, heat-, and solvent-regulated optical rotatory power. J. Polym. Sci A **44**, 207–218 (2006)
54. Jaycox, G.D.: Stimuli-responsive polymers. 9. Photo-regulation of optical rotations in chiral polyesters: altering responsive outputs with conformationally flexible backbone elements. Polymer **48**, 82–90 (2007)

55. Lynch, J.G., Jaycox, G.D.: Stimuli-responsive polymers. 10. Photo-regulation of optical rotations in azobenzene modified poly(ester-amide)s containing highly structured, atropisomeric backbone geometries. Polymer **55**, 3564–3572 (2014)
56. Chen, M., Yu, L.P., Dalton, L.R., Shi, Y.Q., Steier, W.H.: New polymers with large and stable second-order nonlinear optical effects. Macromolecules **24**, 5421–5428 (1991)
57. Hvilsted, S., Andruzzi, F., Ramanujam, P.S.: Side-chain liquid-crystalline polyesters for optical information storage. Opt. Lett. **17**, 1234–1236 (1992)
58. Kang, C.S., Winkelhahn, H.J., Schulze, M., Neher, D., Wegner, G.: Synthesis and properties of aromatic main-chain polyesters having disperse red 1 nonlinear optical chromophores in the side chain. Chem. Mater. **6**, 2159–2166 (1994)
59. Nemoto, N., Miyata, F., Nagase, Y., Abe, J., Hasegawa, M., Shirai, Y.: Novel types of polyesters containing second-order nonlinear optically active chromophores with high degree. Macromolecules **29**, 2365–2371 (1996)
60. Chen, M., Dalton, L.R., Yu, L.P., Shi, Y.Q., Steier, W.H.: Thermosetting polyurethanes with stable and large second-order optical nonlinearity. Macromolecules **25**, 4032–4035 (1992)
61. Shi, Y.Q., Steier, W.H., Chen, M., Yu, L.P., Dalton, L.R.: Thermosetting nonlinear optical polymer: polyurethane with disperse red 19 side groups. Appl. Phys. Lett. **60**, 2577–2579 (1992)
62. Wang, N.P., Leslie, T.M., Wang, S.P., Kowel, S.T.: Syntheses of second-order nonlinear optical polyurethanes for electrooptic etalons. Chem. Mater. **7**, 185–191 (1995)
63. Yu, D., Gharavi, A., Yu, L.P.: Novel aromatic polyimides for nonlinear optics. J. Am. Chem. Soc. **117**, 11680–11686 (1995)
64. Chen, T.A., Jen, A.K.Y., Cai, Y.M.: Facile approach to nonlinear optical side-chain aromatic polyimides with large second-order nonlinearity and thermal stability. J. Am. Chem. Soc. **117**, 7295–7296 (1995)
65. Verbiest, T., Burland, D.M., Jurich, M.C., Lee, V.Y., Miller, R.D., Volksen, W.: Exceptionally thermally stable polyimides for second-order nonlinear optical applications. Science **268**, 1604–1606 (1995)
66. Miller, R.D., Burland, D.M., Jurich, M., Lee, V.Y., Moylan, C.R., Thackara, J.I., Twieg, R.J., Verbiest, T., Volksen, W.: Donor-embedded nonlinear optical side chain polyimides containing no flexible tether: materials of exceptional thermal stability for electrooptic applications. Macromolecules **28**, 4970–4974 (1995)
67. Yu, D., Gharavi, A., Yu, L.P.: Highly stable copolyimides for second-order nonlinear optics. Macromolecules **29**, 6139–6142 (1996)
68. Saadeh, H., Gharavi, A., Yu, D., Yu, L.P.: Polyimides with a diazo chromophore exhibiting high thermal stability and large electrooptic coefficients. Macromolecules **30**, 5403–5407 (1997)
69. Tsutsumi, N., Morishima, M., Sakai, W.: Nonlinear optical (NLO) polymers. 3 NLO polyimide with dipole moments aligned transverse to the imide linkage. Macromolecules **31**, 7764–7769 (1998)
70. Jiang, H.W., Kakkar, A.K.: Soluble high-T_g polymers for second-order nonlinear optics from an unusual mix of imide and siloxane linkages in the backbone. Macromolecules **31**, 4170–4176 (1998)
71. Allcock, H.R., Kim, C.: Liquid crystalline phosphazenes. High polymeric and cyclic trimeric systems with aromatic azo side groups. Macromolecules **22**, 2596–2602 (1989)
72. Allcock, H.R., Dembek, A.A., Kim, C., Devine, R.L.S., Shi, Y.Q., Steier, W.H., Spangler, C. W.: Second-order nonlinear optical poly(organophosphazenes: synthesis and nonlinear optical characterization. Macromolecules **24**, 1000–1010 (1991)
73. Seki, T., Sakuragi, M., Kawanishi, Y., Suzuki, Y., Tamaki, T., Fukuda, R.I., Ichimura, K.: "Command surfaces" of Langmuir-Blodgett films. Photoregulations of liquid crystal alignment by molecularly tailored surface azobenzene layers. Langmuir **9**, 211–218 (1993)
74. Chen, T.A., Jen, A.K.Y., Cai, Y.M.: Two-step synthesis of side-chain aromatic polyimides for second-order nonlinear optics. Macromolecules **29**, 535–539 (1996)

75. Wu, L.F., Tuo, X.L., Cheng, H., Chen, Z., Wang, X.G.: Synthesis, photoresponsive behavior, and self-assembly of poly(acrylic acid)-based azo polyelectrolytes. Macromolecules **34**, 8005–8013 (2001)
76. Wang, H.P., He, Y.N., Tuo, X.L., Wang, X.G.: Sequentially adsorbed electrostatic multilayers of branched side-chain polyelectrolytes bearing donor-acceptor type azo chromophores. Macromolecules **37**, 135–146 (2004)
77. Dawson, D.J., Gless, R.D., Wingard Jr., R.E.: Poly(vinylamine hydrochloride). Synthesis and utilization for the preparation of water-soluble polymeric dyes. J. Am. Chem. Soc. **98**, 5996–6000 (1976)
78. Schilling, M.L., Kate, H.E., Cox, D.I.: Synthesis and reactions of cyanovinyl-substituted benzenediazonium salts for nonlinear optics. J. Org. Chem. **53**, 5538–5540 (1988)
79. Wang, X.G., Chen, J.I., Marturunkakul, S., Li, L., Kumar, J., Tripathy, S.K.: Epoxy-based nonlinear optical polymers functionalized with tricyanovinyl chromophores. Chem. Mater. **8**, 45–50 (1997)
80. Wang, X.G., Kumar, J., Tripathy, S.K., Li, L., Chen, J.I., Marturunkakul, S.: Epoxy-based nonlinear optical polymers from post azo coupling reaction. Macromolecules **30**, 219–225 (1997)
81. Wang, X.G., Yang, K., Kumar, J., Tripathy, S.K., Chittibabu, K.G., Li, L., Lindsay, G.: Heteroaromatic chromophore functionalized epoxy-based nonlinear optical polymers. Macromolecules **31**, 4126–4134 (1998)
82. Wang, X.G., Balasubramanian, S., Kumar, J., Tripathy, S.K., Li, L.: Azo chromophore-functionalized polyelectrolytes. 1. Synthesis, characterization, and photoprocessing. Chem. Mater. **10**, 1546–1553 (1998)
83. He, Y.N., Wang, X.G., Zhou, Q.X.: Epoxy-based azo polymers: synthesis, characterization and photoinduced surface-relief-gratings. Polymer **43**, 7325–7333 (2002)
84. Wang, X.L., Yin, J.J., Wang, X.G.: Self-structured surface patterns on epoxy-based azo polymer films induced by laser light irradiation. Macromolecules **44**, 6856–6867 (2011)
85. Wang, X.L., Yin, J.J., Wang, X.G.: Epoxy-based polymers functionalized with bisazo chromophores: synthesis, characterization and photoresponsive behavior. Polymer **52**, 3344–3356 (2011)
86. Zhou, Y.Q., Tang, B., Wang, X.G.: Photoinduced deformation behavior of a series of newly synthesized epoxy-based polymers bearing *push-pull* azo chromophores. Polymer **60**, 292–301 (2015)
87. Zhou, Y.Q., Wang, Q.G.: Photodeformable microspheres from amphiphilic azo polyurethane. Macromol. Chem. Phys. **216**, 2040–2047 (2015)
88. Wang, J.S., Matyjaszewski, K.: Controlled/"living" radical polymerization. Atom transfer radical polymerization in the presence of transition-metal complexes. J. Am. Chem. Soc. **117**, 5614–5615 (1995)
89. Matyjaszewski, K. (ed.): Controlled Radical Polymerization. American Chemical Society, Washington DC (1997)
90. Chiefari, J., Chong, Y.K., Ercole, F., Krstina, J., Jeffery, J., Le, T.P.T., Mayadunne, R.T.A., Meijs, G.F., Moad, C.L., Moad, G., Rizzardo, E., Thang, S.H.: Living free-radical polymerization by reversible addition-fragmentation chain transfer: the RAFT process. Macromolecules **31**, 5559–5562 (1998)
91. Matyjaszewski, K., Xia, J.H.: Atom transfer radical polymerization. Chem. Rev. **101**, 2921–2990 (2001)
92. Braunecker, W.A., Matyjaszewski, K.: Controlled/living radical polymerization: features, developments, and perspectives. Prog. Polym. Sci. **32**, 93–146 (2007)
93. Lowe, A.B., McCormick, C.L.: Reversible addition-fragmentation chain transfer (RAFT) radical polymerization and the synthesis of water-soluble (co)polymers under homogeneous conditions in organic and aqueous media. Prog. Polym. Sci. **32**, 283–351 (2007)
94. Han, Y.H., Dufour, B., Wu, W., Kowalewski, T., Matyjaszewski, K.: Synthesis and characterization of new liquid-crystalline block copolymers with *p*-cyanoazobene moieties and poly

(*n*-butyl acrylate) segments using atom-transfer radical polymerization. Macromolecules **37**, 9355–9365 (2004)
95. Angiolini, L., Benelli, T., Giorgini, L., Salatelli, E.: Optically active photochromic methacrylic polymers with controlled average molecular weight and defined end-groups by atom transfer radical polymerization. Polymer **46**, 2424–2432 (2005)
96. Wang, D.R., He, Y.N., Deng, W., Wang, X.G.: The photoinduced surface-relief-grating formation behavior of side-chain azo polymers with narrow MW distribution. Dyes Pigments **82**, 286–292 (2009)
97. Jeng, R.J., Chen, Y.M., Jain, A.K., Kumar, J., Tripathy, S.K.: Stable second-order nonlinear optical polyimide/inorganic composite. Chem. Mater. **4**, 1141–1144 (1992)
98. Jeng, R.J., Chen, Y.M., Jain, A.K., Kumar, J., Tripathy, S.K.: Second order optical nonlinearity on a modified sol-gel system at 100 °C. Chem. Mater. **4**, 972–975 (1992)
99. Xu, C.Z., Wu, B., Todorova, O., Dalton, L.R., Shi, Y.Q., Ranon, P.M., Steler, W.H.: Stabilization of the dipole alignment of poled nonlinear optical polymers by ultrastructure synthesis. Macromolecules **26**, 5303–5309 (1993)
100. Boogers, J.A.F., Klaase, P.T.A., de Vlieger, J.J., Alkema, D.P.W., Tinnemans, A.H.A.: Cross-linked polymer materials for nonlinear optics. 1. UV-cured acrylic monomers bearing azobenzene dyes. Macromolecules **27**, 197–204 (1994)
101. Marturunkakul, S., Chen, J.I., Li, L., Jeng, R.J., Kumar, J., Tripathy, S.K.: An interpenetrating polymer network as a stable second-order nonlinear optical material. Chem. Mater. **5**, 592–594 (1993)
102. Ranon, P.M., Shi, Y.Q., Steier, W.H., Xu, C.Z., Wu, B., Dalton, L.R.: Efficient poling and thermal crosslinking of randomly bonded main-chain polymers for stable second-order nonlinearities. Appl. Phys. Lett. **62**, 2605–2607 (1993)
103. Boogers, J.A.F., Klaase, P.T.A., de Vlieger, J.J., Tinnemans, A.H.A.: Cross-linked polymer materials for nonlinear optics. 2. Polyurethanes bearing azobenzene dyes. Macromolecules **27**, 205–209 (1994)
104. Tsutsumi, N., Yoshizaki, S., Sakaim, W., Kiyotsukuri, T.: Nonlinear optical polymers. 1. Novel network polyurethane with azobenzene dye in the main frame. Macromolecules **28**, 6437–6442 (1995)
105. Tsutsumi, N., Matsumoto, O., Sakai, W., Kiyotsukuri, T.: Nonlinear optical polymers. 2. Novel NLO linear polyurethane with dipole moments aligned transverse to the main backbone. Macromolecules **29**, 592–597 (1996)
106. Tomer, R., Florence, A.T.: Photo-responsive hydrogels for potential responsive release applications. Int. J. Pharm. **99**, R5–R8 (1993)
107. Shantha, K.L., Ravichandran, P., Rao, K.P.: Azo polymeric hydrogels for colon targeted drug delivery. Biomaterials **16**, 1313–1318 (1995)
108. Akala, E.O., Kopečková, P., Kopeček, J.: Novel pH-sensitive hydrogels with adjustable swelling kinetics. Biomaterials **19**, 1037–1047 (1998)
109. Liu, Z.L., Hu, H., Zhuo, B.X.: Konjac glucomannan-graft-acrylic acid hydrogels containing azo crosslinker for colon-specific delivery. J. Polym. Sci. A **42**, 4370–4378 (2004)
110. Wang, D., Dušek, K., Kopečkova, P., Dušková-Smrčková, M., Kopeček, J.: Novel aromatic azo-containing pH-sensitive hydrogels: synthesis and characterization. Macromolecules **35**, 7791–7803 (2002)
111. Mao, G.P., Wang, J.G., Clingman, S.R., Ober, C.K., Chen, J.T., Thomas, E.L.: Molecular design, synthesis, and characterization of liquid-crystal-coil diblock copolymers with azobenzene side groups. Macromolecules **30**, 2556–2567 (1997)
112. Yoshida, T., Doi, M., Kanaoka, S., Aoshima, S.: Polymer surface modification using diblock copolymers containing azobenzene. J. Polym. Sci. A **43**, 5704–5709 (2005)
113. Tian, Y.Q., Watanabe, K., Kong, X.X., Abe, J., Iyoda, T.: Synthesis, nanostructure, and functionality of amphiphilic liquid crystalline block copolymers with azobenzene moieties. Macromolecules **35**, 3739–3747 (2002)

114. Yang, J., Levy, D., Deng, W., Keller, P., Li, M.H.: Polymer vesicles formed by amphiphilic diblock copolymers containing a thermotropic liquid crystalline polymer block. Chem. Commun. **14**(34), 4345–4347 (2005)
115. He, X.H., Zhang, H.L., Yan, D.Y., Wang, X.Y.: Synthesis of side-chain liquid-crystalline homopolymers and triblock copolymers with *p*-methoxyazobenzene moieties and poly(ethylene glycol) as coil segments by atom transfer radical polymerization and their thermotropic phase behavior. J. Polym. Sci. A Polym. Chem. **41**, 2854–2864 (2003)
116. Kadota, S., Aoki, K., Nagano, S., Seki, T.: Photocontrolled microphase separation of block copolymers in two dimensions. J. Am. Chem. Soc. **127**, 8266–8267 (2005)
117. Yu, H.F., Shishido, A., Ikeda, T., Iyoda, T.: Novel amphiphilic diblock and triblock liquid-crystalline copolymers with well-defined structures prepared by atom transfer radical polymerization. Macromol. Rapid Commun. **26**, 1594–1598 (2005)
118. He, X.H., Sun, W.Q., Yan, D.Y., Xie, M.R., Zhang, Y.Q.: Synthesis and characterization of side-chain liquid crystalline ABC triblock copolymers with p-methoxyazobenzene moieties by atom transfer radical polymerization. J. Polym. Sci. A Polym. Chem. **46**, 4442–4450 (2008)
119. Wang, D.R., Ye, G., Wang, X.G.: Synthesis of aminoazobenzene-containing diblock copolymer and photoinduced deformation behavior of its micelle-like aggregates. Macromol. Rapid Commun. **28**, 2237–2243 (2007)
120. Wang, D.R., Ren, H.F., Wang, X.Q., Wang, X.G.: Amphiphilic diblock copolymers functionalized with strong *push-pull* azo chromophores: synthesis and multi-morphological aggregation. Macromolecules **41**, 9382–9388 (2008)
121. Cui, L., Zhao, Y., Yavrian, A., Galstian, T.: Synthesis of azobenzene-containing diblock copolymers using atom transfer radical polymerization and the photoalignment behavior. Macromolecules **36**, 8246–8252 (2003)
122. Cui, L., Dahmane, S., Tong, X., Zhu, L., Zhao, Y.: Using self-assembly to prepare multifunctional diblock copolymers containing azopyridine moiety. Macromolecules **38**, 2076–2084 (2005)
123. Yu, H.F., Shishido, A., Iyoda, T., Ikeda, T.: Novel wormlike nanostructures self-assembled in a well-defined liquid crystalline diblock copolymer with azobenzene moieties. Macromol. Rapid Commun. **28**, 927–931 (2007)
124. Cui, L., Tong, X., Yan, X.H., Liu, G.J., Zhao, Y.: Photoactive thermoplastic elastomers of azobenzene-containing triblock copolymers prepared through atom transfer radical polymerization. Macromolecules **37**, 7097–7104 (2004)
125. Deng, W., Albouy, P.A., Lacaze, E., Keller, P., Wang, X.G., Li, M.H.: Azobenzene-containing liquid crystal triblock copolymers: synthesis, characterization, and self-assembly behavior. Macromolecules **41**, 2459–2466 (2008)
126. Wang, G., Tong, X., Zhao, Y.: Preparation of azobenzene-containing amphiphilic diblock copolymers for light-responsive micellar aggregates. Macromolecules **37**, 8911–8917 (2004)
127. Wang, D.R., Liu, J.P., Ye, G., Wang, X.G.: Amphiphilic block copolymers bearing strong *push-pull* azo chromophores: synthesis, micelle formation and photoinduced shape deformation. Polymer **50**, 418–427 (2009)
128. Qi, B., Zhao, Y.: Fluorescence from an azobenzene-containing diblock copolymer micelle in solution. Langmuir **23**, 5746–5751 (2007)
129. Ravi, P., Sin, S.L., Gan, L.H., Gan, Y.Y., Tam, K.C., Xia, X.L., Hu, X.: New water soluble azobenzene-containing diblock copolymers: synthesis and aggregation behavior. Polymer **46**, 137–146 (2005)
130. Sin, S.L., Gan, L.H., Hu, X., Tam, K.C., Gan, Y.Y.: Photochemical and thermal isomerizations of azobenzene-containing amphiphilic diblock copolymers in aqueous micellar aggregates and in film. Macromolecules **38**, 3943–3948 (2005)
131. Su, W., Han, K., Luo, Y.H., Wang, Z., Li, Y.M., Zhang, Q.J.: Formation and photoresponsive properties of giant microvesicles assembled from azobenzene-containing amphiphilic diblock copolymers. Macromol. Chem. Phys. **208**, 955–963 (2007)

132. Su, W., Luo, Y.H., Yan, Q., Wu, S., Han, K., Zhang, Q.J., Gu, Y.Q., Li, Y.M.: Photoinduced fusion of micro-vesicles self-assembled from azobenzene-containing amphiphilic diblock copolymers. Macromol. Rapid Commun. **28**, 1251–1256 (2007)
133. Zhao, Y., Tremblay, L., Zhao, Y.: Doubly photoresponsive and water-soluble block copolymers: synthesis and thermosensitivity. J. Polym. Sci. A **48**, 4055–4066 (2010)
134. Zhang, Y.Y., Cheng, Z.P., Chen, X.R., Zhang, W., Wu, J.H., Zhu, J., Zhu, X.L.: Synthesis and photoresponsive behaviors of well-defined azobenzene-containing polymers via RAFT polymerization. Macromolecules **40**, 4809–4817 (2007)
135. Xu, J., Zhang, W., Zhou, N.C., Zhu, J., Cheng, Z.P., Xu, Y., Zhu, X.L.: Synthesis of azobenzene-containing polymers via RAFT polymerization and investigation on intense fluorescence from aggregates of azobenzene-containing amphiphilic diblock copolymers. J. Polym. Sci. A Polym. Chem. **46**, 5652–5662 (2008)
136. Zhao, Y., Qi, B., Tong, X., Zhao, Y.: Synthesis of double side-chain liquid crystalline block copolymers using RAFT polymerization and the orientational cooperative effect. Macromolecules **41**, 3823–3831 (2008)
137. Zhu, Y., Zhou, Y.Q., Chen, Z., Lin, R., Wang, X.G.: Photoresponsive diblock copolymers bearing strong push-pull azo chromophores and cholesteryl groups. Polymer **53**, 3566–3576 (2012)
138. Zhu, Y., Wang, X.G.: Photoresponsive diblock copolymers bearing strong push-pull azo chromophores and mesogenic biphenyl groups. Dyes Pigments **97**, 222–229 (2013)
139. Bai, S.Y., Zhao, Y.: Azobenzene-containing thermoplastic elastomers: coupling mechanical and optical effects. Macromolecules **34**, 9032–9038 (2001)
140. Bai, S.Y., Zhao, Y.: Azobenzene elastomers for mechanically tunable diffraction gratings. Macromolecules **35**, 9657–9664 (2002)
141. Lee, H.I., Pietrasik, J., Matyjaszewski, K.: Phototunable temperature-responsive molecular brushes prepared by ATRP. Macromolecules **39**, 3914–3920 (2006)
142. Vögtle, F., Gestermann, S., Hesse, R., Schwierz, H., Windisch, B.: Functional dendrimers. Prog. Polym. Sci. **25**, 987–1041 (2000)
143. Grayson, S.M., Fréchet, M.J.: Convergent dendrons and dendrimers: from synthesis to applications. Chem. Rev. **101**, 3819–3867 (2001)
144. Buhleier, E., Wehner, W., Vögtle, F.: Cascade-chain-like and nonskid-chain-like syntheses of molecular cavity topologies. Synthesis **155**, 158 (1978)
145. Tomalia, D.A., Baker, H., Dewald, J., Hall, M., Kallos, G., Martin, S., Roeck, J., Ryder, J., Smith, P.: A new class of polymers: starburst-dendritic macromolecules. Polym. J. **17**, 117–132 (1985)
146. Newkome, G.R., Yao, Z.Q., Baker, G.R., Gupta, V.K.: Cascade molecules: a new approach to micelles. A [27]-Arborol. J. Org. Chem. **50**, 2004–2006 (1985)
147. Hawker, C.J., Fréchet, J.M.J.: Preparation of polymers with controlled molecular architecture. A new convergent approach to dendritic macromolecules. J. Am. Chem. Soc. **112**, 7638–7647 (1990)
148. Kuckling, D., Wycisk, A.: Stimuli-responsive star polymers. J. Polym. Sci. A **51**, 2980–2994 (2013)
149. Deloncle, R., Caminade, A.M.: Stimuli-responsive dendritic structures: the case of light-driven azobenzene-containing dendrimers and dendrons. J. Photochem. Photobiol. C: Photochem. Rev. **11**, 25–45 (2010)
150. Mekelburger, H.B., Vögtle, F., Rissanen, K.: Repetitive-synthesis of bulky dendrimers – a reversibly photoactive dendrimer with Six azobenzene side chains. Chem. Ber. (German) **126**, 1161–1169 (1993)
151. Jiang, D.L., Aida, T.: Photoisomerization in dendrimers by harvesting of low-energy photons. Nature **388**, 454–456 (1997)
152. Junge, D.M., McGrath, D.V.: Photoresponsive dendrimers. Chem. Commun. 857–858 (1997)
153. Junge, D.M., McGrath, D.V.: Photoresponsive azobenzene-containing dendrimers with multiple discrete states. J. Am. Chem. Soc. **121**, 4912–4913 (1999)

154. Ghosh, S., Banthia, A.K.: Synthesis of photoresponsive polyamidoamine (PAMAM) dendritic architecture. Tetrahedron Lett. **42**, 501–503 (2001)
155. Ghosh, S., Banthia, A.K., Chen, Z.: Synthesis and photoresponsive study of azobenzene centered polyamidoamine dendrimers. Tetrahedron **61**, 2889–2896 (2005)
156. Archut, A., Vögtle, F., Cola, L.D., Azzellini, G.C., Balzani, V., Ramanujam, P.S., Berg, R.H.: Azobenzene-functionalized cascade molecules: photoswitchable supramolecular systems. Chem. Eur. J. **4**, 699–706 (1998)
157. Dirksen, A., Zuidema, E., Williams, R.M., Cola, L.D., Kauffmann, C., Vögtle, F., Roque, A., Pina, F.: Photoactivity and pH sensitivity of methyl orange functionalized poly (propyleneamine) dendrimers. Macromolecules **35**, 2743–2747 (2002)
158. Vögtle, F., Gorka, M., Hesse, R., Ceroni, P., Maestri, M., Balzani, V.: Photochemical and photophysical properties of poly(propylene amine) dendrimers with peripheral naphthalene and azobenzene groups. Photochem. Photobiol. Sci. **1**, 45–51 (2002)
159. Schenning, A.P.H.J., Elissen-Román, C., Weener, J.W., Baars, M.W.P.L., van der Gaast, S.J., Meijer, E.W.: Amphiphilic dendrimers as building blocks in supramolecular assemblies. J. Am. Chem. Soc. **120**, 8199–8208 (1998)
160. Alcalá, R., Giménez, R., Oriol, L., Piñol, M., Serrano, J.L., Villacampa, B., Viñuales, A.I.: Synthesis, characterization, and induction of stable anisotropy in liquid crystalline photoaddressable PPI dendrimers. Chem. Mater. **19**, 235–246 (2007)
161. del Barrio, J., Tejedor, R.M., Chinelatto, L.S., Sánchez, C., Piñol, M., Oriol, L.: Photocontrol of the supramolecular chirality imposed by stereocenters in liquid crystalline azodendrimers. Chem. Mater. **22**, 1714–1723 (2010)
162. Cheon, K.S., Kazmaier, P.M., Keum, S.R., Park, K.T., Buncel, E.: Azo-functionalized dendrimers. Can. J. Chem. **82**, 551–566 (2004)
163. Bobrovsky, A., Ponomarenko, S., Boiko, N., Shibaev, V., Rebrov, E., Muzafarov, A., Stumpe, J.: Photochemistry and photoorientational phenomena in carbosilane dendrimers with terminal azobenzene groups. Macromol. Chem. Phys. **203**, 1539–1546 (2002)
164. Yokoyama, S., Nakahama, T., Otomo, A., Mashiko, S.: Preparation and assembled structure of dipolar dendrons based electron donor/acceptor azobenzene branching. Chem. Lett. 1137–1138 (1997)
165. Yokoyama, S., Nakahama, T., Otomo, A., Mashiko, S.: Intermolecular coupling enhancement of the molecular hyperpolarizability in multichromophoric dipolar dendrons. J. Am. Chem. Soc. **122**, 3174–3184 (2000)
166. Wang, S.X., Wang, X.M., Li, L.J., Advincula, R.C.: Design, synthesis, and photochemical behavior of poly(benzyl ester) dendrimers with azobenzene groups throughout their architecture. J. Org. Chem. **69**, 9073–9084 (2004)
167. Shen, X.Q., Liu, H.W., Li, Y.S., Liu, S.Y.: Click-together azobenzene dendrons: synthesis and characterization. Macromolecules **41**, 2421–2425 (2008)
168. Sebastián, R.M., Blais, J.C., Caminade, A.M., Majoral, J.P.: Synthesis and photochemical behavior of phosphorus dendrimers containing azobenzene units within the braches and/or on the surface. Chem. Eur. J. **8**, 2172–2183 (2002)
169. Gao, C., Yan, D.: Hyperbranched polymers: from synthesis to applications. Prog. Polym. Sci. **29**, 183–275 (2004)
170. He, X.H., Yan, D.Y.: Branched azobenzene side-chain liquid-crystalline copolymers obtained by self-condensing ATR copolymerization. Macromol. Rapid Commun. **25**, 949–953 (2004)
171. Jin, M., Lu, R., Bao, C.Y., Xu, T.H., Zhao, Y.Y.: Synthesis and characterization of hyperbranched azobenzene-containing polymers via self-condensing atom transfer radical polymerization and copolymerization. Polymer **45**, 1125–1131 (2004)
172. He, Y.N., Wang, X.G., Zhou, Q.X.: Synthesis and characterization of a novel photoprocessible hyperbranched azo polymer. Synth. Met. **132**, 245–248 (2003)

173. Ding, L., Zhang, L.Y., Han, H.J., Huang, W., Song, C.M., Xie, M.R., Zhang, Y.Q.: Hyperbranched azo-polymers synthesized by acyclic diene metathesis polymerization of an AB_2 monomer. Macromolecules **42**, 5036–5042 (2009)
174. Che, P.C., He, Y.N., Zhang, Y., Wang, X.G.: Synthesizing hyperbranched azo polymer through azo-coupling reaction. Chem. Lett. **33**, 22–23 (2004)
175. Che, P.C., He, Y.N., Wang, X.G.: Hyperbranched azo-polymers synthesized by azo-coupling reaction of an AB_2 monomer and postpolymerization modification. Macromolecules **38**, 8657–8663 (2005)
176. Xie, J.D., Deng, X.X., Cao, Z.Q., Shen, Q.S., Zhang, W.Q., Shi, W.F.: Synthesis and second-order nonlinear optical properties of hyperbranched polymers containing pendant azobenzene chromophores. Polymer **48**, 5988–5993 (2007)
177. Chen, X.B., Zhang, Y.H., Liu, B.J., Zhang, J.J., Wang, H., Zhang, W.Y., Chen, Q.D., Pei, S. H., Jiang, Z.H.: Novel photoactive hyperbranched poly(aryl ether)s containing azobenzene chromophores for optical storage. J. Mater. Chem. **18**, 5019–5026 (2008)
178. Tsai, C.C., Chao, T.Y., Lin, H.L., Liu, Y.H., Chang, H.L., Liu, Y.L., Jeng, R.J.: The facile synthesis and optical nonlinearity of hyperbranched polyaspartimides with azobenzene dyes. Dyes Pigm. **82**, 31–39 (2009)
179. Chao, D.M., He, L.B., Berda, E.B., Wang, S.T., Jia, X.T., Wang, C.: Multifunctional hyperbranched polyamide: synthesis and properties. Polymer **54**, 3223–3229 (2013)
180. Li, Z., Qin, A.J., Lam, J.W.Y., Dong, Y.P., Dong, Y.Q., Ye, C., Williams, I.D., Tang, B.Z.: Facile synthesis, large optical nonlinearity, and excellent thermal stability of hyperbranched poly(aryleneethylnylene)s containing azobenzene chromophores. Macromolecules **39**, 1436–1442 (2006)
181. Li, Z.A., Wu, W.B., Ye, C., Qin, J.G., Li, Z.: New hyperbranched polyaryleneethynylene containing azobenzene chromophore moieties in the main chain: facile synthesis, large optical nonlinearity and high thermal stability. Polym. Chem. **1**, 78–81 (2010)
182. Li, Z.A., Wu, W.B., Ye, C., Qin, J.G., Li, Z.: New main-chain hyperbranched polymers: facile synthesis, structural control, and second-order nonlinear optical properties. Polymer **53**, 153–160 (2012)
183. Xie, J.D., Hu, L.H., Shi, W.F., Deng, X.X., Cao, Z.Q., Shen, Q.S.: Synthesis and nonlinear optical properties of hyperbranched polytriazole containing second-order nonlinear optical chromophore. J. Polym. Sci. B **46**, 1140–1148 (2008)
184. Pandey, S., Mishra, S.P., Kolli, B., Kanai, T., Samui, A.B.: Hyperbranched photoresponsive and liquid crystalline azo-siloxane polymers synthesized by click chemistry. J. Polym. Sci. A **50**, 2659–2668 (2012)
185. Wang, G.J., Wang, X.G.: A novel hyperbranched polyester functionalized with azo chromophores: synthesis and photoresponsive properties. Polym. Bull. **49**, 1–8 (2002)
186. Marcos, M., Alcalá, R., Barberá, J., Romero, P., Sánchez, C., Serrano, J.L.: Photosensitive ionic nematic liquid crystalline complexes based on dendrimers and hyperbranched polymers and a cyanoazobenzene carboxylic acid. Chem. Mater. **20**, 5209–5217 (2008)
187. Yu, B., Jiang, X.S., Wang, R., Yin, J.: Multistimuli responsive polymer nanoparticles on the basis of the amphiphilic azobenzene-contained hyperbranched poly(ether amine) (hPEA-AZO). Macromolecules **43**, 10457–10465 (2010)
188. Angiolini, L., Benelli, T., Giorgini, L., Salatelli, E.: Optically active photochromic polymers with three-arm star structure by atom transfer radical polymerization. Macromolecules **39**, 3731–3737 (2006)
189. Erdogan, T., Gungor, E., Durmaz, H., Hizal, G., Tunca, U.: Photoresponsive poly(methyl methacrylate)$_2$-(polysterene)$_2$ miktoarm star copolymer containing an azobenzene moiety at the core. J. Polym. Sci. A **44**, 1396–1403 (2006)
190. Liao, X.J., Zhang, H.L., Chen, J.F., Wang, X.Y.: Preparation and properties of azobenzene-containing amphiphilic miktoarm star polymers. Polym. Bull. **58**, 819–828 (2007)

191. He, X.H., Yan, D.Y., Mai, Y.Y.: Synthesis of novel multi-arm star azobenzene side-chain liquid crystalline copolymers with a hyperbranched core. Eur. Polym. J. **40**, 1759–1765 (2004)
192. del Barrio, J., Oriol, L., Alcalá, R., Sánchez, C.: Azobenzene-containing linear-dendritic diblock copolymers by click chemistry: synthesis, characterization, morphological study, and photoinduction of optical anisotropy. Macromolecules **42**, 5752–5760 (2009)
193. Shi, Z.H., Lu, H.J., Chen, Z.C., Cheng, R.S., Chen, D.Z.: Rational design, syntheses, characterization and solution behavior of amphiphilic azobenzene-containing linear-dendritic block copolymers. Polymer **53**, 359–369 (2012)
194. Reddinger, J.L., Reynolds, J.R.: Molecular engineering of π-conjugated polymers. In: Abe, A., Albertsson, A.C., Cantow, H.J., Dusek, K., Edwards, S., Hocker, H., Joanny, J.F., Kausch, H.H., Kobayashi, T., Lee, K.S., McGarth, J.E., Monnerie, L., Stupp, S.I., Suter, U.W., Thomas, E.L., Wegner, G., Young, R.J.A. (eds.) Advances in Polymer Science, vol. 145, pp. 57–122. Springer, Berlin (1999)
195. Lévesque, I., Leclerc, M.: Novel dual photochromism in pholythiophene derivatives. Macromolecules **30**, 4349–4352 (1997)
196. Zagórska, M., Kulszewicz-Bajer, I., Proń, A., Sukiennik, J., Raimond, P., Kajzar, F., Attias, A.J., Łapkowski, M.: Preparation and spectroscopic and spectroelectrochemical characterization of copolymers of 3-alkylthiophenes and thiophene functionalized with an azo chromophore. Macromolecules **31**, 9146–9153 (1998)
197. Zhao, X.Y., Hu, X., Yue, C.Y., Xia, X.L., Gan, L.H.: Synthesis, characterization and dual photochroic properties of azo-substituted polythiophene derivatives. Thin Solid Films **417**, 95–100 (2002)
198. Wang, G.J., Li, M., Chen, X.F., Wu, F., Tian, W.J., Shen, J.C.: Preparation and characterization of a novel poly(1,4-phenylenevinylene) derivative with an azobenzene side chain. Macromol. Rapid Commun. **20**, 591–594 (1999)
199. Wang, G.J., Li, M., Guo, C.W., Wu, F., Tian, W.J., Chen, X.F., Shen, J.C.: Soluble conjugated copolymers based on poly(1,4-phenylenevinylene). Polymer **41**, 2309–2312 (2000)
200. Matsui, T., Nagata, T., Ozaki, M., Fujii, A., Onoda, M., Teraguchi, M., Masuda, T., Yoshino, K.: Novel properties of conducting polymers containing azobenzene moieties in side chain. Synth. Met. **119**, 599–600 (2001)
201. Harbron, E.J., Vicente, D.A., Hoyt, M.T.: Fluorescence modulation via isomer-dependent energy transfer in an azobenzene-functionalized poly(phenylenevinylene) derivative. J. Phys. Chem. B **108**, 18789–18792 (2004)
202. Harbron, E.J., Vicente, D.A., Hadley, D.H., Imm, M.R.: Phototriggered fluorescence color changes in azobenzene-functionalized conjugated polymers. J. Phys. Chem. A **109**, 10845–10853 (2005)
203. Grimes, A.F., Call, S.E., Harbron, E.J., English, D.S.: Wavelength-resolved studies of Förster energy transfer in azobenzene-modified conjugated polymers: the competing roles of exciton migration and spectral resolution. J. Phys. Chem. C **111**, 14257–14265 (2007)
204. Teraguchi, M., Masuda, T.: Synthesis and properties of polyacetylenes having azobenzene pendant groups. Macromolecules **33**, 240–242 (2000)
205. Sukwattanasinitt, M., Wang, X.G., Li, L., Jiang, X.L., Kumar, J., Tripathy, S.K., Sandman, D. J.: Functionalizable self-assembling polydiacetylenes and their optical properties. Chem. Mater. **10**, 27–29 (1998)
206. Sukwattanasinitt, M., Lee, D.C., Kim, M., Wang, X.G., Li, L., Yang, K., Kumar, J., Tripathy, S.K., Sandman, D.J.: New processable, functionalizable polydiacetylenes. Macromolecules **32**, 7361–7369 (1999)
207. Huang, K., Qiu, H.J., Wan, M.,.X.: Synthesis of highly conducting polyaniline with photochromic azobenzene side groups. Macromolecules **35**, 8653–8655 (2002)

208. Izumi, A., Teraguchi, M., Nomura, R., Masuda, T.: Synthesis of poly(*p*-phenylene)-based photoresponsive conjugated polymers having azobenzene units in the main chain. Macromolecules **33**, 5347–5352 (2000)
209. Izumi, A., Teraguchi, M., Nomura, R., Masuda, T.: Synthesis of conjugated polymers with azobenzene moieties in the main chain. J. Polym. Sci. A **38**, 1057–1063 (2000)
210. Izumi, A., Nomura, R., Masuda, T.: Design and synthesis of stimuli-responsive conjugated polymers having azobenzene units in the main chain. Macromolecules **34**, 4342–4347 (2001)
211. Sogawa, H., Shiotsuki, M., Matsuoka, H., Sanda, F.: Synthesis, chiroptical properties, and photoresponsiveness of optically active poly(*m*-phenyleneethynylene)s containing azobenzene moieties. Macromolecules **44**, 3338–3345 (2011)
212. Sogawa, H., Shiotsuki, M., Sanda, F.: Synthesis and photoresponse of helically folded poly (phenyleneethynylene)s bearing azobenzene moieties in the main chain. Macromolecules **46**, 4378–4387 (2013)
213. Apaydin, D.H., Akpinar, H., Sendur, M., Toppare, L.: Electrochromism in multichromic conjugated polymers: Thiophene and azobenzene derivatives on the main chain. J. Electroanal. Chem. **665**, 52–57 (2012)
214. Zhang, W., Yoshida, K., Fujiki, M., Zhu, X.L.: Unpolarized-light-driven amplified chiroptical modulation between chiral aggregation and achiral disaggregation of an azobenzene-*alt*-fluorene copolymer in limonene. Macromolecules **44**, 5105–5111 (2011)
215. Grebel-Koehler, D., Liu, D.J., Feyter, S., Enkelmann, V., Weil, T., Engels, C., Samyn, C., Müllen, K., Schryver, F.C.: Synthesis and photomodulation of rigid polyphenylene dendrimers with an azobenzene core. Macromolecules **36**, 578–590 (2003)
216. Willner, I., Rubin, S.: Control of the structure and functions of biomaterials by light. Angew. Chem. Int. Ed. Engl. **35**, 367–385 (1996)
217. Ercole, F., Davisa, T.P., Evans, R.A.: Photo-responsive systems and biomaterials: photochromic polymers, light-triggered self-assembly, surface modification, fluorescence modulation and beyond. Polym. Chem. **1**, 37–54 (2010)
218. Pieroni, O., Fissi, A., Popova, G.: Photochromic polypeptides. Prog. Polym. Sci. **23**, 81–123 (1998)
219. Feringa, B.L., van Delden, R.A., Koumura, N., Geertsema, E.M.: Chiroptical molecular switches. Chem. Rev. **100**, 1789–1816 (2000)
220. Pieroni, O., Fissi, A., Angelini, N., Lenci, F.: Photoresponsive polypeptides. Acc. Chem. Res. **34**, 9–17 (2001)
221. Goodman, M., Kossoy, A.: Conformational aspects of polypeptide structure. XIX. Azoaromatic side-chain effects. J. Am. Chem. Soc. **88**, 5010–5015 (1966)
222. Goodman, M., Falxa, M.L.: Conformational aspects of polypeptide structure. XXIII. Photoisomerization of azoaromatic polypeptides. J. Am. Chem. Soc. **89**, 3863–3867 (1967)
223. Ueno, A., Anzai, J.I., Osa, T., Kadoma, Y.: Light-induced conformational changes of polypeptides. Random copolymers of γ-benzyl-L-glutamate with m- and p-phenylazobenzyl-L-aspartates. Bull. Chem. Soc. Jpn **50**, 2995–2999 (1977)
224. Ueno, A., Takahashi, K., Anzai, J.I., Osa, T.: Conformational changes of azoaromatic polyaspartate induced by solvent and/or light. Macromolecules **13**, 459–460 (1980)
225. Ueno, A., Takahashi, K., Anzai, J.I., Osa, T.: Photocontrol of polypeptide helix sense by *cis-trans* isomerism of side-chain azobenzene moieties. J. Am. Chem. Soc. **103**, 6410–6415 (1981)
226. Sisido, M., Ishikawa, Y., Itoh, K., Tazuke, S.: Helically arranged azobenzene chromophores along a polypeptide chain. 1. Synthesis and circular dichroism. Macromolecules **24**, 3993–3998 (1991)
227. Sisido, M., Ishikawa, Y., Harada, M., Itoh, K.: Helically arranged azobenzene chromophores along a polypeptide chain. 2. Prediction of conformations and calculation of theoretical circular dichroism. Macromolecules **24**, 3999–4003 (1991)
228. Houben, J.L., Pieroni, O., Fissi, A., Ciardelli, F.: Optical acitivity and photochromism of the azo chromophore bound to poly(L-glutamic acid). Biopolymers **17**, 799–804 (1978)

229. Pieroni, O., Houben, J.L., Fissi, A., Costanfino, P.: Reversible conformational changes induced by light in poly(L-glutamic acid) with photochromic side chains. J. Am. Chem. Soc. **102**, 5913–5915 (1980)
230. Pieroni, O., Fissi, A., Houben, J.L., Ciardelli, F.: Photoinduced aggregation changes in photochromic polypeptides. J. Am. Chem. Soc. **107**, 2990–2991 (1985)
231. Ciardelli, F., Pieroni, O., Fissic, A.: Photocontrol of the solubility of arobenzene-containing poly(L-glutamic acid. J. Chem. Soc. Chem. Commun. 264–265 (1986)
232. Sato, M., Kinoshita, T., Takizawa, A., Tsujita, Y.: Photoinduced conformational transition of polypeptides containing azobenzenesulfonate in the side chains. Macromolecules **21**, 1612–1616 (1988)
233. Higuchi, M., Takizawa, A., Kinoshita, T., Tsujita, Y.: Photocontrol of vesicular adhesion and functions by a photoresponsive polypeptide. Macromolecules **20**, 2888–2892 (1987)
234. Yamamoto, H., Nishida, A.: Light-induced reversible conformational change in poly (L-lysine) with photochromic side chains. Macromolecules **19**, 943–944 (1986)
235. Yamamoto, H.: Synthesis and reversible photochromism of azo aromatic poly(L-lysine). Macromolecules **19**, 2472–2476 (1986)
236. Fissi, A., Pieroni, O., Balestreri, E., Amato, C.: Photoresponsive polypeptides. Photomodulation of the macromolecular structure in poly(N^{ε}((phenylazophenyl)sulfonyl)-L-lysine). Macromolecules **29**, 4680–4685 (1996)
237. Berg, R.H., Hvilsted, S., Ramanujam, P.S.: Peptide oligomers for holographic data storage. Nature **383**, 505–508 (1996)

Chapter 4
Photoinduced Orientation and Anisotropy

Abstract This chapter is devoted to the photoinduced orientation and anisotropy of azo polymers, which are intensively investigated as a way for holographic recording and data storage based on the polarization gratings. The orientation induced by linearly polarized light endows the polymer films with dichroism and birefringence. Moreover, out-of-plane orientation and chirality can also be induced by irradiation with unpolarized light and circularly polarized light. The main types of the photoanisotropic materials include liquid crystalline azo polymers, amorphous azo polymers, amphiphilic azo polymers, azo polyelectrolytes, and azo block copolymers. The preparation, structure–property relationship, mechanism and models, and possible application in holography are presented in this chapter.

Keywords Polarized light • Dichroism • Birefringence • Photoinduced orientation • Mechanism and models • Photoanisotropic materials • Optical grating • Holography

Orientation of anisotropic organic molecules, such as elongated and disklike molecules, has aroused tremendous research enthusiasm in the past decades. Those intensively explored materials, which are closely linked to the molecular orientation, include liquid crystals (LCs), LC polymers and the second-order nonlinear optical (NLO) polymers, and others. By incorporating anisotropic building blocks (mesogenic units) into polymeric systems, many types of LC polymers have been developed [1]. The second-order NLO polymers are functionalized with anisotropic chromophores. The second-order susceptibility can be obtained only when chromophores are oriented in a non-centrosymmetric way. Distinct from spontaneous orientation of LC polymers, the chromophore orientation in a NLO polymer is typically induced by electric field poling at a temperature above its glass transition temperature [2]. Azobenzene moieties are a typical type of anisotropic building blocks, which have been widely used to fabricate LC polymers (LCPs) and NLO polymers [3–6].

This chapter will discuss another very important orientation behavior, which is unique to azo polymers. Distinguished from the spontaneous orientation of LCs and electric field poling-induced orientation of NLO polymers, this type of molecular orientation is induced by light irradiation. According to Dumont, the light-driven orientation can be classified into *photoinduced anisotropy* (*PIA*), *photoassisted*

X. Wang, *Azo Polymers*, Soft and Biological Matter,
DOI 10.1007/978-3-662-53424-3_4

electric poling (*PAEP*), and *all-optical poling* (*AOP*) [7, 8]. PAEP and AOP are mainly applied as innovative poling methods to obtain the second-order NLO polymer films. PAEP uses optical pumping to increase the mobility of molecules, which allows chromophores to be oriented by the electric field at room temperature. For AOP, an azo polymer is coherently pumped by light at the fundamental frequency and the second harmonic of a laser beam, which means resonant at double frequency to induce NLO chromophore orientation. As PIA is much more intensively investigated than the others, this chapter will only discuss this very interesting function of azo polymers in detail. For PAEP and AOP, we refer the interested reader to the original articles [9–11] and reviews [4, 12, 13].

4.1 Background and Terminology

Photoinduced anisotropy (*PIA*) is also known as *Weigert effect*, which was first reported by Weigert [14]. According to his observation, light-sensitive layers of photographic papers became dichroic after exposure to the linearly polarized light. Since then, "Weigert effect" has been used by many researchers to refer to the birefringence and dichroism induced by a polarized light excitation. In the 1930s, optical anisotropy in azo-dye-colored gelatin films was observed to be caused by irradiation with linearly polarized violet light [15 and references there]. Intensive studies in later years show that PIA of azo polymers is a result of molecular orientation driven by *trans–cis* photoisomerization of azobenzene moieties. *Photoinduced orientation* is a term frequently used to describe this type of structure variations of azo polymers occurring at the molecular level, which is also termed *photoinduced reorientation* by many authors, especially for LC azo polymers.

Photoinduced orientation is an interesting research frontier persistently stimulated by its mysterious physical and chemical mechanisms. On the other hand, research enthusiasm for PIA of azo polymers has been aroused not only by the scientific curiosity but also by many potential applications related to this unique function. Azo polymers have been considered as new high-efficiency PIA materials for optical storage and polarization holography. One specific area for these applications is recording, detecting, manipulating, and reconstructing of polarized light by using polarization grating and holography [16].

In the 1970s, Kakichashvili carried out the pioneering study on polarization grating and holography for recording and reconstructing of polarized light [17, 18]. Later in the 1980s, it was reported by Todorov et al. that polyvinyl alcohol (PVA) containing an azo dye (methyl orange), where the azo dye was incorporated as a dissolved guest in a polymer host to form the guest–host system, is a new high-efficient organic material with reversible photoinduced birefringence [19, 20]. Some applications of the polarization holographic recording were also explored for the guest–host system containing azo dye [21].

The development from the guest–host system to azo polymers is a significant breakthrough in this research area. The polymers developed in the initial stage are side-chain azo polymers, which can be divided into liquid crystal (LC) azo polymers [22–24] and amorphous polymers [25, 26]. Since then, photoinduced orientation and its optical applications have been widely explored for various azo polymer systems. Many different types of azo polymers in various condensation states have been developed for this purpose, which has become one of the major branches of azo polymer research.

4.2 Photoinduced Orientation

For azo polymers, photoinduced orientation is a key concept to understand the anisotropy induced by light irradiation and to distinguish it from other types of the Weigert effect. PIA observed in azo polymers reflects the light-driven orientation of azo chromophores and adjacent segments at the molecular level. It means that the orientation of azo chromophores and adjacent segments can be manipulated by irradiation with polarized lights. Azo polymers can show a variety of orientation and reorientation behavior as responses to different forms of the polarized light. The polarized light is not only a way to introduce the orientation but also a powerful tool to detect the orientation and PIA effects. Some background knowledge about the polarized light seems to be required to start the discussion.

4.2.1 Polarized Light

In classical physics, light is synonymous with electromagnetic radiation in a proper wavelength range. Based on the Maxwell's equations, light is an electromagnetic wave propagating with harmonic oscillation. The transverse wave can be described by three vectors $\boldsymbol{E}$, $\boldsymbol{B}$, and $\boldsymbol{k}$, which represent the electric field, magnetic field, and wave vector. During the wave traveling process, $\boldsymbol{E}$ and $\boldsymbol{B}$ oscillate in a plane perpendicular to $\boldsymbol{k}$, which specifies the propagation direction of the wave. In most cases, $\boldsymbol{E}$ is used to denote the polarization properties as represented by the equation in the complex notation:

$$\boldsymbol{E} = \frac{1}{2}\boldsymbol{E}_0\, e^{i(\boldsymbol{k}\cdot\boldsymbol{r}-\omega t)} + c.c. \tag{4.1}$$

where $\boldsymbol{r}$ and t are position and time of the electric field. By this definition, polarized light is the wave exhibiting preference for $\boldsymbol{E}$ vibration patterns, such as transverse direction or handedness, which are specified as linear polarization or circular polarization. Equation 4.1 can be rewritten with the quantity $\phi = \boldsymbol{k}\cdot\boldsymbol{r} - \omega t$, where ϕ is the phase angle at position $\boldsymbol{r}$ and time t:

$$\boldsymbol{E} = \frac{1}{2}\,\boldsymbol{E}_0\, e^{i\varnothing} \ + c.c. \tag{4.2}$$

By using this expression, a polarized beam of a monochromatic beam can be represented by a combination of two components in the right-handed Cartesian coordinates (x, y, z):

$$\boldsymbol{E} = \frac{1}{2}\left(\boldsymbol{E}_{0x}\, e^{i\varnothing_x} + \boldsymbol{E}_{0y}\, e^{i\varnothing_y}\right) + c.c \tag{4.3}$$

where x and y are the vibration directions and z is the propagation direction. By this definition, the different forms of polarized light can be described by $\boldsymbol{E}_{0x}$, $\boldsymbol{E}_{0y}$, and the difference $\gamma = \phi_y - \phi_x$. When $\gamma = 0$, the light beam is linearly polarized; when $\gamma = 90°$ or $-90°$, the beam is right-circularly polarized or left-circularly polarized. In the general case, when the x and y components have different amplitudes, and γ has other values, the expression represents the elliptically polarized light. In a sectional drawing, a linearly polarized beam is represented by a short line to denote the vibrational direction. Right circular and left circular polarizations are represented by circles with clockwise and anticlockwise senses. Polarized light can be more accurately described by the Poincaré sphere, the Stokes vector, and the Jones vector [27].

4.2.2 Dichroism and Birefringence

Dichroism and *birefringence* are two most important properties related to polarized light. Dichroism and birefringence substances and materials have been widely used to produce, analyze, and manipulate polarized light [27]. These two terms refer to the anisotropic light absorption and refraction properties, respectively. *Dichroism* is defined as a property of substances whose light absorption extents depend on the polarization direction of the incident beam. When used quantitatively, dichroism usually denotes the difference between the principal absorbancy coefficients. Absorption indicatrix is a general way to describe the absorption properties of an absorbing material. Substances with this property can be used to fabricate dichroic polarizers [27].

Birefringence is usually defined by considering the refractoanisotropic effect of substances. When a light ray strikes a cube cut from the refractoanisotropic body, two refracted rays are produced, and each of them has its own ray direction, wave front, and wave normal. When an unpolarized beam obliquely strikes the cube, two orthogonally polarized beams will be produced. Depending on their depolarization forms, the body can be classified into linearly, circularly, and elliptically birefringent. The refractive index of a refractoanisotropic body depends on the $\boldsymbol{E}$ vibration direction of the rays. A given refraction indicatrix is typically used to describe the refractive indices of a refractoanisotropic body. Generally, the birefringence of the refractoanisotropic body is described by three numbers, i.e., which are the differences between the principal refractive indices [27]. Birefringence substances have been used to fabricate the optical retarders and other devices.

For azo polymers, the photoinduced orientation can cause significant dichroism and birefringence, which are the macroscopic effects caused by microscopic structure variations. A piece of thin film on proper substrate is usually used for dichroism and birefringence studies. In this case, the definition of dichroism and birefringence is relatively simple. The dichroic ratio, which is given as ratio of the absorbance of the characteristic bands in UV–vis or IR ranges, is typically used to characterize the molecular orientation. The birefringence (Δn) is defined by the difference between the major and minor refractive indices of the films. The dichroism and birefringence induced by light irradiation are most important PIA effects of azo polymers, which can be used to characterize the molecular orientation and for different real applications.

4.2.3 Orientation Induced by Linearly Polarized Light

When illuminated with linearly polarized light, the azo chromophores tend to orient in a way perpendicular to the polarization direction of the incident light. The chromophore orientation is caused by *trans–cis–trans* isomerization cycles under light absorption [28, 29]. Due to this preferential orientation, the light irradiation will cause the macroscopic dichroism and birefringence, which are termed *photoinduced dichroism and birefringence*. For azo polymers, actinic light with different polarization forms can induce various types of dichroism and birefringence, which record the polarization forms of the light. The substances with the photoinduced dichroism and birefringence as optical elements can be used to manipulate polarized light for many interesting applications. It is a common practice to induce the orientation in a piece of thin films by irradiation with linearly polarized light in normal incidence. The photoinduced orientation can be both in the film plane and out of the film plane, where the interaction of the azo chromophores with the pumping beam is minimized. Other orientation can also be achieved by controlling the light polarization and the angle of incidence. The photoinduced orientations in azo polymer films can be characterized by spectral and other methods, such as spectroscopic dichroism analysis, birefringence measurement, and Raman spectroscopy.

The photoinduced orientation can be characterized by measuring dichroism of specific absorption bands using UV–vis and infrared spectroscopy. The dichroic measurements are performed by irradiating the thin films of azo polymers with a linearly polarized light at a proper wavelength. In a real-time manner or immediately after irradiation, absorption spectra are recorded with the polarized probe light. The polarization of the probe light can be parallel and perpendicular to the actinic light polarization as well as other predetermined direction. Typically, the absorbance parallel ($A_{\parallel}$) and perpendicular ($A_{\perp}$) to the polarization of actinic light is measured and used to give dichroic ratios, $D = A_{\parallel}/A_{\perp}$.

For UV–vis spectroscopic study, the absorbance of the π–π* transition band is normally used to obtain the dichroic ratio, which reflects the orientation extent of the azo chromophores [28]. It is assumed that the transition dipole moment of an azo chromophore is coincident with its elongated direction. Considering the orientation both in-plane and out-of-plane, the second-order orientation order parameter can be given by the following equation [30]:

$$P_2(\cos\theta) = \frac{A_{\parallel} - A_{\perp}}{A_{\parallel} + 2A_{\perp}} \tag{4.4}$$

The orientation degrees of different groups and segments can be obtained by infrared (IR) spectroscopy by analyzing the dichroic ratios of the characteristic vibration bands [31]. Figure 4.1 shows the geometry of the setup for the measurement. For the IR spectroscopic method, to exactly specify the orientation of a group or segment requires some knowledge about its orientation relative to the transition dipole moment of a given vibration mode. In practice, some groups with easy assignment of their transition dipole moment direction, such as C=O and CN groups, are used for the purpose. The angular dependence of absorbance (A) for the different bands, both from UV–vis and IR spectroscopy, is frequently used to extract more detailed information of the orientation in the molecular level.

Orientation of azo polymers can also be characterized by Raman spectroscopy through the polarization measurement [32, 33]. Figure 4.2 shows the 90° geometry in Raman scattering to obtain the information. An advantage of Raman spectroscopy is that both the second-order and fourth-order orientation order parameters can be obtained by the measurements. By combination of infrared dichroism and polarized Raman measurements, it is possible to afford valuable information

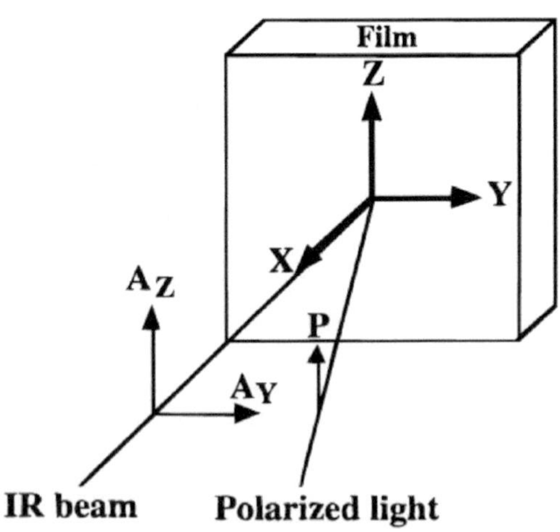

Fig. 4.1 Geometry of incoming infrared and laser beams at the sample and definition of the X, Y, Z coordinate system. The angle between the polarized laser and the infrared beam is approximately 20° (Reprinted with permission from Ref. [31]. Copyright (1998) American Chemical Society)

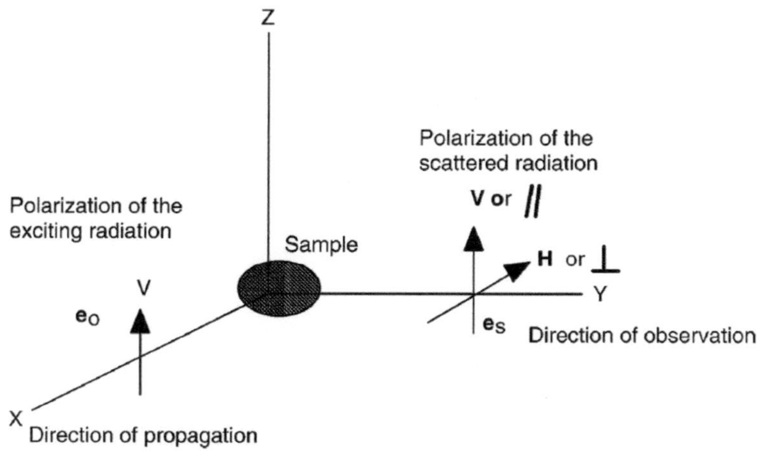

Fig. 4.2 The 90° geometry in Raman scattering to obtain the orientation information (Reprinted with permission from Ref. [33]. Copyright (2004) American Chemical Society)

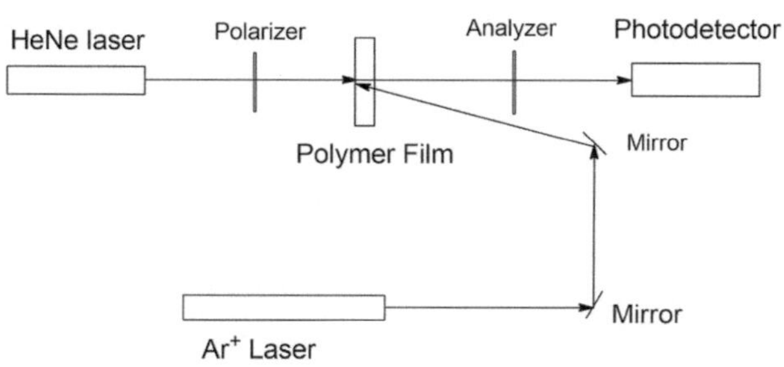

Fig. 4.3 Setup to characterize the photoinduced orientation through the birefringence measurements

about a full picture of the molecular orientation distribution function in an anisotropic polymer film.

Another method widely used to characterize the photoinduced orientation is through the birefringence measurements [26, 34], which is more closely related to the holographic applications of azo polymers. The optically induced anisotropy is measured by placing a piece of the thin film between crossed linear polarizer and analyzer (Fig. 4.3). Orientation in the polymer film is induced by the actinic light, such as a linearly polarized Ar^{+} laser beam, where its polarization is typically fixed at an angle of 45° with respect to transmission axis of the polarizer. A low-power He–Ne laser (632.8 nm) is typically used as the light probe to measure the birefringence. The power of the transmitted laser beam through this optical setup is recorded and plotted versus writing time to show the dynamic variation of the birefringence. The probe beam intensity and the birefringence (Δn) have the following relationship:

$$I = I_0 \sin^2\left(\frac{\pi}{\lambda}\Delta n\, d\right) \tag{4.5}$$

where I_0 and I are the probe light intensity before and after the optical setup, λ is the wavelength of the probe light at vacuum, and d is the film thickness [34].

4.2.4 Out-of-Plane Orientation Induced by Unpolarized Light

Reorientation in a specific out-of-plane direction can be realized by controlling the light irradiation condition. For a typical method, films of liquid crystal polymers with dominant azobenzenes in the *cis*-state are first prepared by UV light irradiation. Then, when the films are exposed to unpolarized light at 436 nm, the azobenzene moieties tend to align along the propagation direction of the irradiation light [35–37]. The information about the out-of-plane orientation is obtained by measuring absorption spectra of the polymer films with a linearly polarized probe light at different incident angles. The induced anisotropy is usually large, with an order parameter about 0.36, and highly stable [36]. This approach extends the photoinduced reorientation from biaxial manipulation to three-dimensional control.

More details about three-dimensional orientation have been obtained by monitoring the process by spectral analysis together with conoscopic observation [38]. Experiment results showed that strong H-aggregation is induced during the reorientation of the azo chromophores. The tilt orientation of the azobenzene moieties can be more easily achieved by irradiation with light at a high temperature even above the glass transition temperature. The alkylene spacer between the polymeric backbone and side group shows the effect on the three-dimensional orientation, where the homeotropical orientation (perpendicular to the substrate) is enhanced as the spacer length increases. Subsequent anneal after the light irradiation shows the effect to enhance the orientation [39].

4.2.5 Chirality Induced by Circularly Polarized Light

Chiral orientation can be induced by irradiation with circularly polarized light. Very large circular anisotropy, both circular birefringence and circular dichroism, can be induced in films of side-chain LC azo polyesters by illumination with circularly polarized light [40]. In this case, no chiral group is required for the azo polymer films to show the chirality. As revealed by circular dichroism spectroscopy, chirality is induced by exposing films of an achiral LC azo polymer to circularly polarized light [41]. By using the experimental setup shown in Fig. 4.4, circularly polarized light with opposite handedness can produce enantiomeric structures in the films. The creation of the chiral suprastructure by irradiation

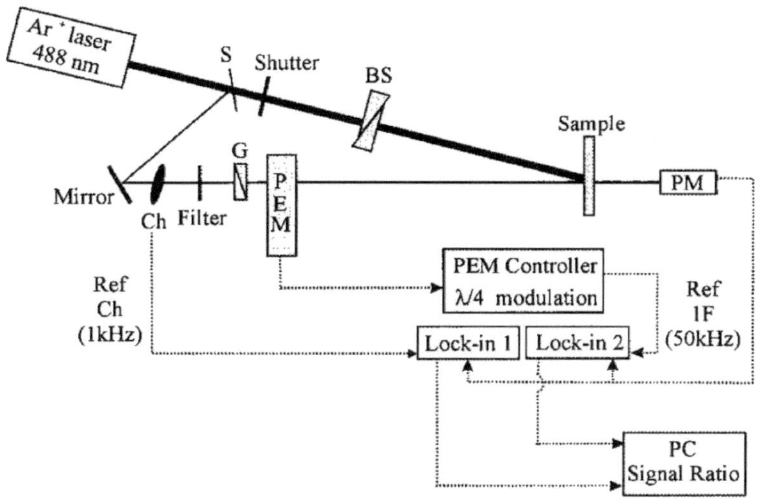

Fig. 4.4 Experimental setup for dynamical recording of chirality switching: *S* beam splitter, *BS* Babinet–Soleil compensator, *PM* photomultiplier, *G* Glan–Taylor polarizer, *Ch* chopper, *PEM* photoelastic modulator (Reprinted with permission from Ref. [41]. Copyright (2000) American Chemical Society)

with circularly polarized light involves the effects of the azobenzene group, the polymer backbone, and the LC alignment. A reversible chiroptical switch between chiral supramolecular structures in the polymer films is demonstrated by adjusting the handedness of the light. For achiral LC polymers containing 7 or 15 mol % azo chromophores, circular dichroism is observed not only for the azobenzene moieties but also for the nonphotoactive cyanobiphenyl mesogens due to a cooperative motion [42].

Besides the LC polymers, chirality can also be induced for guest–host systems and amorphous polymers. When a nonchiral azo-dye-doped polymer film is exposed to a helicoidal standing wave, light-induced macroscopic chirality is achieved [43]. Polarization properties of obtained helicoidal Bragg gratings are similar to cholesteric liquid crystals. Reversible chiral structure in amorphous polymer films is induced by irradiation with circularly polarized light and used for light polarization control [44]. The photoinduced structure strongly affects the polarization of the light propagating in it, which converses from circular to linear polarization and changes the optical activity. The direction of the optical rotation (clockwise or anticlockwise) can be controlled by means of the handedness of the circularly polarized light.

The photoinduced chiral properties have been compared for amorphous and LC azo polymers [45]. In amorphous azo polymers, light induces a macroscopic chiral structure whose pitch depends on the ellipticity of the actinic light. In LC azo polymers, circularly polarized light induces the formation of many microscopic spirals, whose optical axes possess random orientation. This property makes the material analogue to the classical optically active media.

4.3 Mechanism and Models

As discussed above, azobenzene and its derivatives, both in the guest–host systems and covalently tethered on polymer chains, can be efficiently oriented by polarized light in different forms. It has been shown by some early experiments that the chromophores tend to stabilize at positions with their long axis perpendicular to the polarization direction of the light [19, 26, 28], which has also been confirmed by many later investigations. It is generally agreed that the photoinduced orientation is caused by the *trans–cis* isomerization. To understand the correlation between the isomerization and orientation is an exploration of a multidisciplinary nature, which has been actively performed for several decades.

To understand the process, microscopic insight into the mechanism and key factors is necessary. Some puzzling issues needed to be addressed include: how is the chromophore rotation induced by the photoisomerization, what is the effect of polymeric chains on the rotation, and to what extent the rotation is restrained by the cage effect related to the small free volume around the azo chromophores in the solid state? In a pioneering study, Dumont et al. have systematically investigated the orientation mechanism of the photoisomerizable molecules induced by resonant pumping beam. The PIA orientation is attributed to angular hole burning (AHB) by polarized light and angular redistribution (AR) during the photoisomerization, which are closely related to the lifetime of the *cis* isomer and the relaxation (spontaneous or photoinduced) to the stable *trans* isomeric [7, 8, 46].

According to the quantum chemistry, the probability to be excited to a specific orbit is given by the square of the corresponding coefficient $\alpha_k(t)$ in Eq. (1.5). Both $[\alpha_k(t)]^2$ and the oscillator strength f are proportional to the square of the transition dipole moment, which is approximately parallel to the long axis of azo chromophores. Therefore, the excitation of an azo chromophore is dependent on the angle (θ) between its long axis and the light polarization. A linearly polarized light beam tends to excite an azo chromophore with a probability proportional to $\cos^2\theta$. The irradiation with the polarized light results in anisotropic depletion of the *trans* isomer, which is termed *angular hole burning* (*AHB*). The photoinduced orientation is caused the rotation of azo chromophores in the *trans–cis–trans* photoisomerization cycles. Although the rotation at each pumping cycle is random, the resultant effect is to accumulate the chromophores in the direction with the weakest excitation, which is termed *angular redistribution* (*AR*).

The AHB and AR processes are modeled by rate equations representing the correlations between the variables [7]. The angular distributions are given by $n_T(\Omega)$ and $n_C(\Omega)$, where $\Omega = \{\varphi, \theta, \xi\}$ are three Euler angles and T and C in the subscripts denote the *trans* and *cis* isomers. The equations include the probability of AHB ($P_r(\Omega)$), AR (R_{TC} and R_{CT}), and diffusion as shown below:

$$\frac{dn_T(\Omega)}{dt} = -\Phi_{TC}P_r(\Omega)n_T(\Omega) + \frac{1}{\tau_c}\int R_{CT}(\Omega' \to \Omega)n_C(\Omega')d\Omega' + \left(\frac{dn_T(\Omega)}{dt}\right)_{Diff} \tag{4.6}$$

$$\frac{dn_C(\Omega)}{dt} = \Phi_{TC}\int R_{TC}(\Omega' \to \Omega)P_r(\Omega')n_T(\Omega')d\Omega' - \frac{1}{\tau_C}n_C(\Omega) + \left(\frac{dn_C(\Omega)}{dt}\right)_{Diff} \tag{4.7}$$

In these expressions, *trans*-to-*cis* transition is supposed to undergo a short living on the excited levels and Φ_{TC} is the quantum yield of the isomerization. If the angular distribution is independent on the azimuthal angle φ and chromophore has an axis of symmetry, $n_T(\Omega)$ and $n_C(\Omega)$ can be simplified to $n_T(\theta)$ and $n_C(\theta)$ expanded in Legendre polynomials. By using a proper thermal diffusion model, above equations can be solved quantitatively. A modified model from the same group can be seen in their later publication [46]. A quantitative agreement with the experimental results obtained from DR1-PMMA film has been given there. This approach is also used to model photoisomerization and photoinduced orientation of azobenzene molecules in a viscous environment [47]. The experimental behavior is well explained by the model and quantitative calculations.

Considering the cooperative orientation, such as in liquid crystalline (LC) polymers, a different model has been proposed by Palto et al. [48, 49]. This model considers that the molecules in a domain are aligned along one predominant direction, which is termed the director of the domain in liquid crystal terminology. The model quantitatively describes the domain director rotation caused by molecule motion in an anisotropic molecular field. Polarized light excitation of a molecular absorption band causes different rates of collisions (friction) between excited and ground state molecules and a substrate. If an angular distribution function of domain molecules has a nonzero width, the different friction coefficients will result in the domain director rotation. The domain director rotation is a result of the conservation of the angular momentum for the whole system, which includes the substrate. This model has been further developed by Perdersen et al. by combining an excitation scheme based on photoinduced *trans–cis* isomerization to explain the experimentally observed photoinduced anisotropy in LC azobenzene side-chain polyesters [50]. This mechanism including the multi-domain picture is able to account for the long-term anisotropic orientation stability of LC polymers.

Besides above models and theories, some other mechanisms have also been proposed. A mechanism considering the collective motion has been proposed after studying the photoreorientation of substituted aminoazobenzene in Langmuir–Blodgett films [51]. The reorientation is modeled as a collective process caused by the vibronic coupling of excited states through a radiationless deactivation channel. The electronic excitation energy is thermalized and distributed across neighboring molecules of the excited chromophore. It causes the excited chromophore to rotate an infinitesimal amount within the steric potential of its neighbors.

This model can describe the light intensity dependence observed in the experiments and obtain the magnitude of a characteristic intensity threshold.

A totally different model without consideration of the molecular orientation has also been proposed [15]. In this model, optical anisotropy in azo-dye-colored films is attributed to the generation of anisotropic grains. This model is more closely related to the explanation of the general Weigert effect rather than mechanism of photoinduced orientation.

4.4 Azo Polymers as Photoanisotropic Materials

The guest–host polymeric system, such as methyl orange/polyvinyl alcohol (PVA), is among the earliest polymer-related materials used to explore the photoanisotropic (i.e., photoinduced anisotropy, PIA) effect for polarization holography applications [19, 20]. When exposed to polarized light, azo polymers become optically anisotropic, which can record the type and direction of the polarization of the actinic light. To pursue the best PIA materials, many azo polymers and related materials have been developed in the past decades. Their performances to record and reconstruct both the amplitude and the polarization of the light field have been extensively investigated. As so many publications are available, I have no intention to discuss all the contents here. Only some representative examples from some typical catalogues are presented below.

4.4.1 Liquid Crystalline Azo Polymer

In the 1980s, the pioneering study on polymeric liquid crystal for applications in optical storage was carried out by Eich and Wendorff [22, 23]. Figure 4.5 shows the chemical structure of the liquid crystalline (LC) polymer used in the investigation. In this study, 7 μm thick preoriented films were used in the optical storage. Both simple mask exposure and holographic storage techniques were investigated. A high resolving power (3000 lines/mm) and large diffraction efficiencies (50 %)

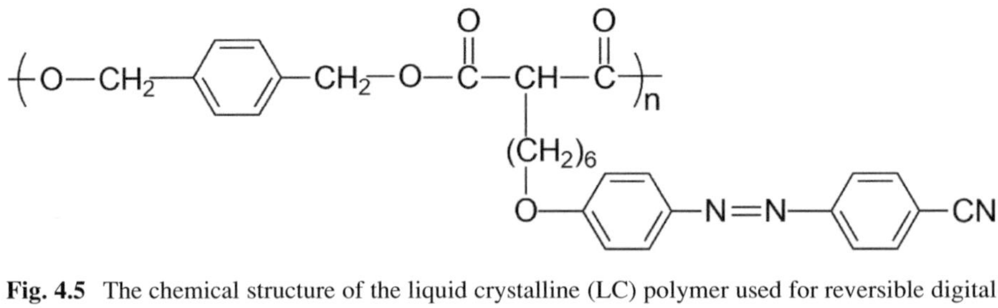

Fig. 4.5 The chemical structure of the liquid crystalline (LC) polymer used for reversible digital and holographic optical storage

were achieved for the LC polymers [52]. Following this study, a series of investigations has been carried out to understand the PIA effect triggered by the light irradiation for the LC polymer systems. Spectroscopic investigation proves that molecular reorientation occurs upon irradiation with the polarized light, where the preferential orientation direction is perpendicular to the vibration direction of the light [28, 53].

Since then, various LC azo polymers have been synthesized and investigated for the PIA effect, material optimization, and reversible optical storage applications. The polymers can be classified into addition polymers and condensation polymers depending on the synthetic methods, which are separately discussed below.

Most LC azo polymers classified into addition polymers are obtained from radical polymerization of methacrylate-/acrylate-based monomers. One of the most commonly adopted types is the LC copolymers containing both azobenzene and nonphotoactive mesogens in the side chain [54, 55]. Owing to a cooperative motion existing between the photoactive and nonphotoactive mesogens, the photoinduced orientation effect is amplified and enhanced [24]. The photoinduced orientation of the azobenzene side groups results in a reorientation of non-photochromic, rodlike side groups by a cooperative mechanism even in the glassy state [56]. For liquid crystalline terpolymers containing azobenzene and dye moieties, the light-induced orientational order generated in the glassy state is significantly amplified by the subsequent annealing of the irradiated films at temperatures within the mesophase of the co- and terpolymers [57].

A systematic study has been carried out on methacrylate-based LC copolymer (MACB-ABn) with a low content of azo units [58]. One advantage of the molecular design is that relative thick films can be adopted for the LC copolymer owing to the relatively low absorbance and deep penetration of the writing light. The orientation in the polydomain films occurs after irradiation with linearly polarized light at 366 nm. Figure 4.6 shows the angular-dependent absorbance of the azobenzene and cyanobiphenyl moieties measured by UV–vis and FTIR spectroscopy. In the nematic state, alignment can take place even for a low intensity (0.7 mW/cm^2) of the irradiation light. As a general rule, photoinduced orientation can be induced in the copolymer with low content of azo units, while phase transition occurs only for copolymers with higher contents of the azo units [59].

The photoinduced birefringence is closely related to the structural factors of LC azo polymers. It is caused by the high rate of *cis–trans* isomerization and the mobility of mesogens. The photoinduced orientation is dependant on the flexible spacer between the azobenzene moieties and polymer main chain [60]. Orientation is more difficult to be induced for the copolymer having a short spacer although the spacer length does not obviously affect the *trans–cis* isomerization. The restriction of the polymer main chain on azobenzene moieties is the main factor to affect the alignment efficiency of the azo mesogens and polymer liquid crystals. Polymer liquid crystal containing strong donor–acceptor pairs in the azobenzene moiety shows high alignment efficiency [61]. LC copolymers containing tolane-type

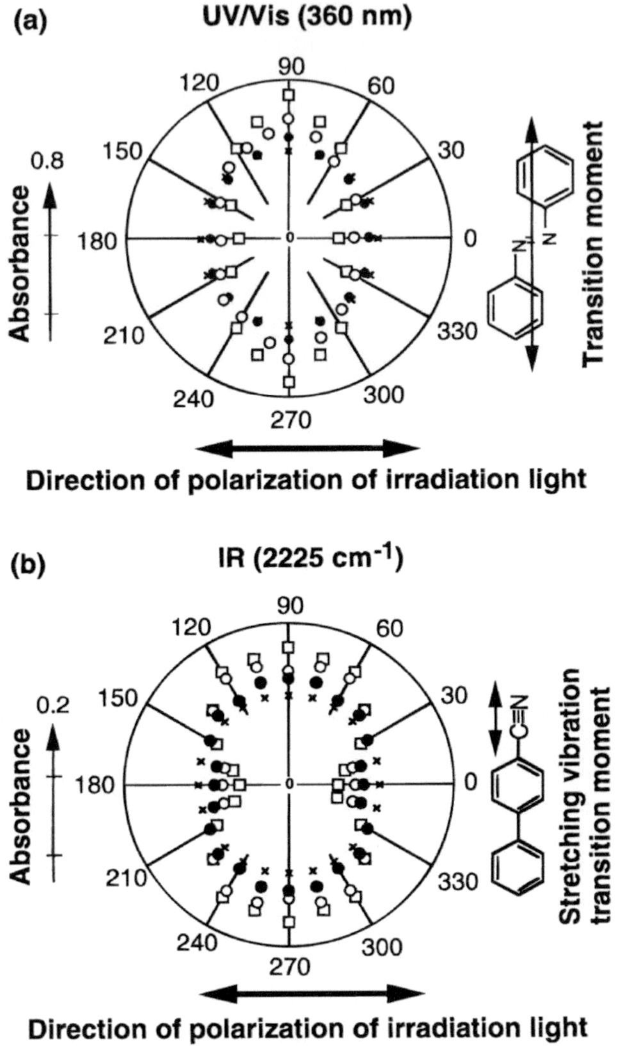

Fig. 4.6 Angular-dependent absorbance of azobenzene (**a**) and cyanobiphenyl (**b**) moieties at 360 nm and 2225 cm^{-1}, respectively. The measurement was performed after the film was irradiated with 366 nm linearly polarized light (2.8 mW/cm^2) at 95 °C for various times: (×) 0 min, (•) 3 min, (o) 5 min, and (□) 30 min (Reprinted with permission from Ref. [58]. Copyright (1998) American Chemical Society)

mesogenic units show high birefringence efficiency and are highly efficient PIA materials [62].

As discussed in Chap. 3, condensation LC azo polymers are synthesized by step polymerization. A series of side-chain LC polyesters has been synthesized and studied for optical information storage [63, 64]. The azo polymers were synthesized by a base-catalyzed transesterification of mesogenic 1,3-propanediols and diphenyl diesters. Optical storage properties of thin unoriented polyester films were investigated by measuring polarization anisotropy and holography. The polarization holography shows a resolution of over 5000 lines/mm and diffraction efficiencies about 40 % with long-term stability. The recording can be completely erased by heating the films to about 80 °C and the films can be rewritten many times. Many times of writing, reading, and erasing cycles can be optically performed in the side-chain liquid crystalline azo polyester [65].

The PIA effect and its correlation with polymer structures have been studied for the LC polyesters by the polarized FTIR spectroscopy [66]. The orientation of chromophores and other segments was analyzed on the basis of dichroic ratios of characteristic absorption bands. Figure 4.7 shows the chemical structure of the azo polymers and a typical polar plot of the integrated absorbance area for one of the polymers. For the polyesters with long methylene main-chain segments, such as tetradecanedioates and dodecanedioates, not only the light-sensitive azo chromophore but also the main-chain methylene segment and the side-chain flexible spacer are preferentially oriented. But, the nonphotoactive groups can only be oriented to a smaller extent. The extent of orientation perpendicular to the laser light polarization increases with increasing spacer length. On the other hand, for the shorter adipate main chain, only the chromophore and the side-chain spacer are oriented under the same conditions.

Recently, cyclic azobenzene-containing side-chain liquid crystalline polymers (SCLCPs) have been synthesized by using "click" cyclization of the linear polymeric precursor with alkyne and azide groups at chain ends [67]. The coupling between mesogenic side groups and chain backbone can be affected by the main-chain cyclic structure in SCLCPs. This additional topological effect shows some photoinduced birefringence behavior distinct from its linear analogue.

4.4.2 Amorphous Azo Polymer

Although LC azo polymers have been proved to be highly efficient for PIA applications, liquid crystallinity is not a necessary and exclusive condition for a polymer to exhibit this effect. Photoinduced orientation can also be achieved for amorphous azo polymers with high T_g [25, 26]. The distinctive characteristics of LC and amorphous azo polymers for photo-storage applications have been discussed by Natansohn et al. [4, 26], which is briefly discussed below.

For an LC azo polymer, the overall orientation can be achieved only when writing is performed above or near T_g of the polymer to enter the LC state [28]. A long spacer between the mesogenic units and main chain is usually required to decrease T_g of the polymer. Moreover, erasing is usually realized by heating above T_g of the polymer as the PIA is highly stable against the optical erasure. On the other hand, for an amorphous side-chain azo polymer, PIA can be induced at room temperature. A polymer with very short spacers and relatively high T_g can be used, which results in a significant increase in the stability of the written material. Also, erasing can be done by optical means, not necessarily by heating the sample above a certain temperature. Moreover, the synthetic procedure for an amorphous azo polymer is relatively simpler and cheaper. The amorphous azo polymers can

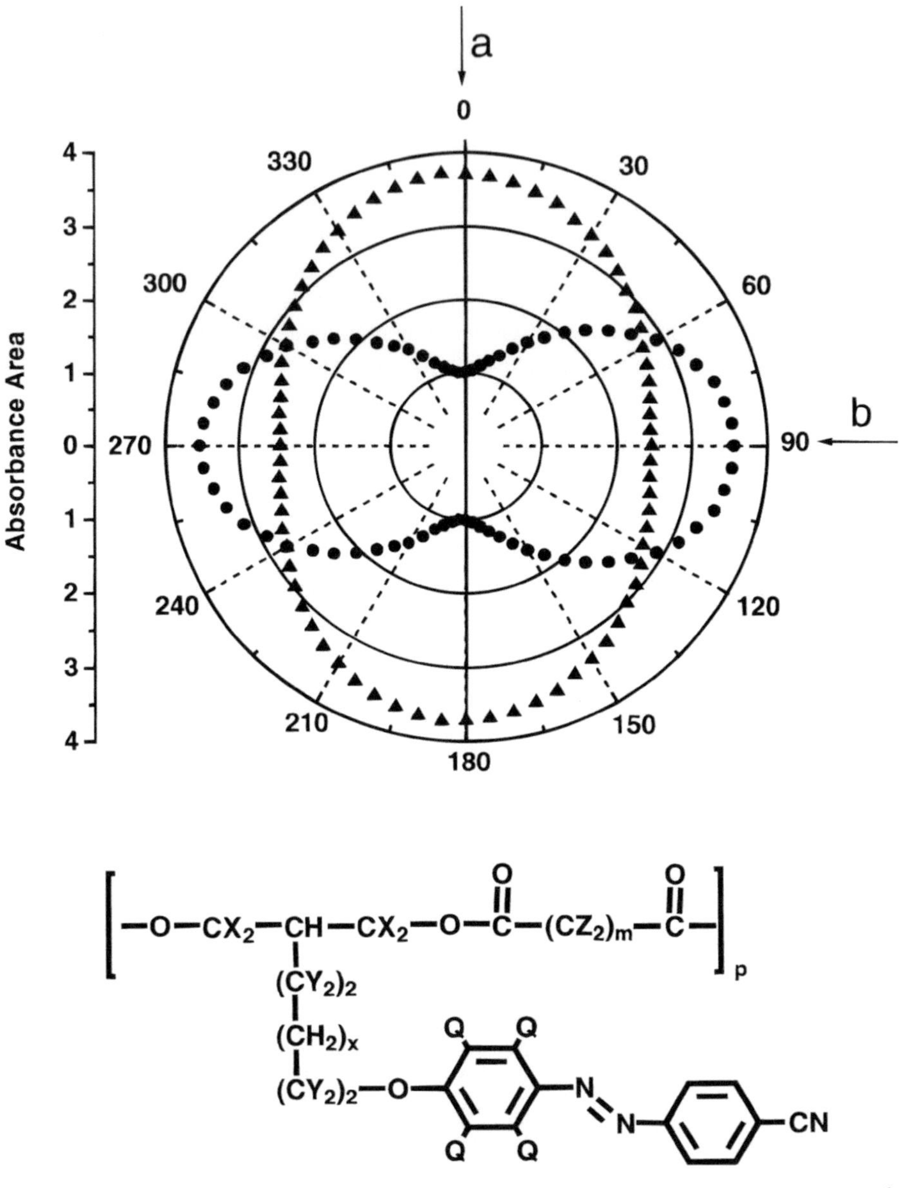

Fig. 4.7 Polar plot of the integrated absorbance area of the aromatic $\nu(\mathrm{C{=}C})_{\mathrm{ring}}$ (•, 1601 cm^{-1}) and aliphatic $\nu(\mathrm{CD_2})$ (♦, 2091 cm^{-1}) stretching vibration of **P6,12Z** ($x=2$, $m=12$, X=H, Y=H, Z=D, Q=H) after Ar^{+} laser irradiation (488 nm, 300 $\mathrm{mW/cm^2}$, 1000 s): (←, *a*) $\alpha=0°$, 180°; (←, *b*) $\alpha=90°$, 270° (Reprinted with permission from Ref. [66]. Copyright (1998) American Chemical Society)

also be synthesized by addition polymerization and condensation polymerization to produce addition polymers and condensation polymers accordingly.

Majority of the additional polymers are obtained from radical polymerization of methacrylate-/acrylate-based monomers as discussed in Chap. 3. The amorphous azo polymers for optical storage have been thoroughly investigated by Natansohn, Rochon, and their coworkers. Reversible optical storage was first demonstrated by using an amorphous azo polymer pDR1A (disperse-red 1-type azo dye) containing 4-amino-4′-nitroazobenzene chromophores [26, 68]. The chemical structure of pDR1A has been given in Fig. 3.2. The photoinduced orientation behavior of the polymer, characterized by birefringence measurement, is shown in Fig. 4.8. A linearly polarized Ar^+ laser at 514.5 nm was used as the writing beam, and the same light source was used as the erasing laser beam, but its polarization was circular instead of planar. The writing–relaxation–erasing behavior shown in the figure is typical for amorphous azo polymers. Later studies have been carried out on various azo polymers and copolymers in this category to fully understand the structure–property correlations.

Azo chromophores in the polymers play a critically important role in affecting the PIA properties. The orientability of the azo dipoles is correlated with the intramolecular dipolar interaction affects [69]. A triad sequence distribution model has been used to explain this observation. The effect of the electron-withdrawing nitro group has been investigated by comparing an analogue without this group [70]. The polymer containing 4-aminoazobenzene moieties (pMEA)

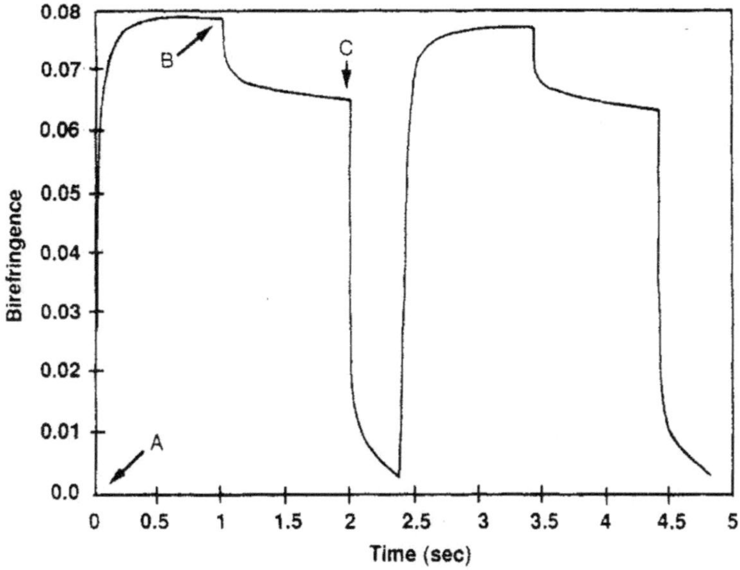

Fig. 4.8 Writing–relaxation–erasing curve for pDR1A. A = writing laser is turned on; B = writing laser is turned off; C = erasing laser is turned on (Reprinted with permission from Ref. [68]. Copyright (1994) American Chemical Society)

shows a fairly slow *cis–trans* thermal isomerization rate. The lower birefringence level of pMEA is attributed to a low absorptivity for the actinic light in the visible region. With the increasing content of the *push–pull*-type azo chromophores, the copolymers containing these two-type azo chromophores show an increase in both the maximum photoinduced birefringence and the remaining fraction after relaxation [71]. On the other hand, the methacrylate-based polymer containing bulkier azo chromophores exhibits a slower writing rate [72].

Similar to LC azo polymers, the cooperative motion or molecular addressing can be observed for amorphous polymers. Cooperative orientation was first observed for a semicrystalline azo polymer pDR19T (disperse-red 19-type azo dye) with main-chain rigid groups, which contains the azobenzene moieties in the side chains [68]. Later, cooperative motion has also been observed for amorphous azo copolymers. An amorphous azo copolymer containing non-photochromic 4-nitrophenyl benzoate (BEM) side-chain groups shows high photoinduced birefringence with low azo contents [73]. Infrared spectroscopic investigation showed that cooperative orientation of both azobenzene and phenyl benzoate groups occurs when the copolymer is irradiated with polarized light [74]. For copolymers with similar rigid non-photochromic groups, this cooperative motion is significantly reduced for less polar groups. Based on above observation, the cooperative motion below T_g for amorphous azo polymers is attributed to the polarity interaction of the side-chain groups involved in orientation process [75].

Amorphous azo polymers synthesized by step polymerization, which belong to condensation polymers, have also been investigated for the PIA properties. Depending on the locations of the azobenzene moieties, it can also be classified into side-chain and main-chain azo polymers. Among them, side-chain condensation polymers are much more widely explored for this purpose than their main-chain counterparts. Side-chain azo polymers include polyester [25, 68, 76, 77], polyethers [78], polyurethanes [79, 80], polyimides and polyamides [81, 82], and epoxy-based polymers [83, 84]. In several cases, the PIA properties of side-chain condensation polymers have been compared with the addition polymers in a side-by-side way [68, 76, 77]. Main-chain polymers as PIA materials are relatively rare, which include polyesters [85, 86] and polyurethanes [87, 88].

4.4.3 Amphiphilic Azo Polymer

Amphiphilic azo polymers are azo polymers containing both hydrophobic and hydrophilic groups/segments. As a unique characteristic of these polymers, Langmuir–Blodgett (LB) multilayers can be assembled through a layer-by-layer dipping approach. The dipping process results in a tilted arrangement of the amphiphilic moieties to form thin films with well-controlled structures and thickness. Amphiphilic azo polymers based on 2-hydroxyethyl polyacrylates and "hairy rod" polyglutamates are two types of the polymers first used to demonstrate the photoinduced orientation in LB multilayers [89]. Irradiation with linearly polarized

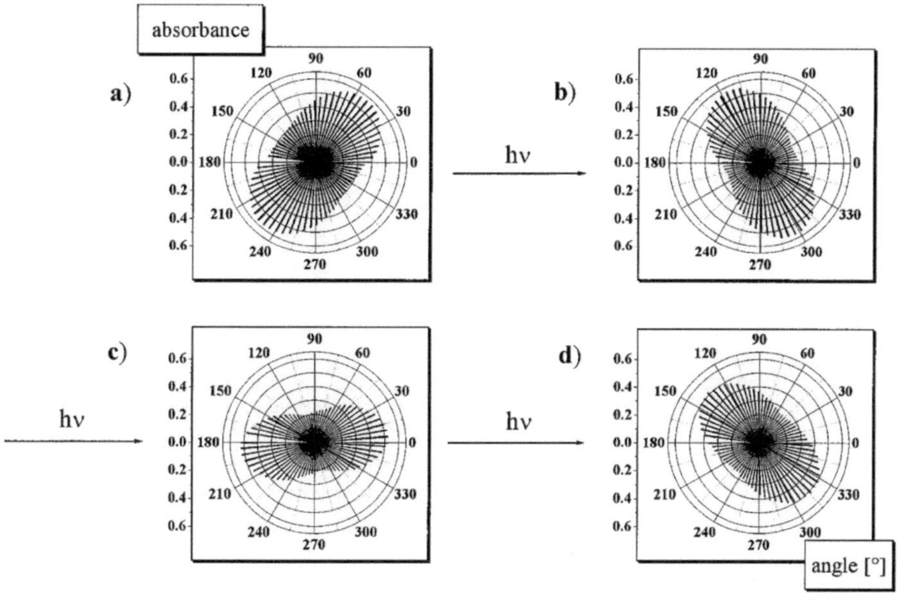

Fig. 4.9 Multistep photoreorientation started after the UV/visible irradiation cycle: change of the angular-dependent absorbance of a LB multilayer at 335 nm upon polarized irradiation ($\lambda_{ex} = 436$ nm, 2 mW/cm^2, $P = 1.64$ J/cm^2) with different directions of polarization (**a**) 135°, (**b**) 30°, (**c**) 95°, and (**d**) 45° at room temperature (Reprinted with permission from Ref. [90]. Copyright (1997) American Chemical Society)

UV and visible light can induce a reversible reorientation of the azobenzene side groups, while the layered structure maintains integrity in the LB films [90]. As shown in Fig. 4.9, the dichroism of 0.34 is photogenerated after UV irradiation (step a). In step b, the direction of the optical axis is changed upon irradiation with a different angle of the *E* vector of the incident light. A complete reorientation (steps c and d) can be induced by this method. In this way, the optical anisotropy of the LB films can be induced in a reversible and continuous way.

Mixed LB films composed of disperse-red 19 isophorone polyurethane (DR19-IPPU) and cadmium stearate (CdSt) have been fabricated to demonstrate the photoinduced birefringence [91]. Figure 4.10 shows the chemical structure of DR19-IPPU and writing–relaxation–erasing sequence characterized by birefringence measurement. For this system, the amplitude of induced birefringence decreases with the number of layers in the composite LB films, while the film thickness shows no effect on the time required to achieve maximum birefringence. Mixed LB films have also been fabricated a methacrylate-based azo polymer (HPDR13) and CdSt [92]. Similarly, the birefringence induced by a linearly polarized laser light decreases with the number of layers. On the other hand, the birefringence increases with the weight percentage of HPDR13 in the mixed LB film. Distance-dependent cooperative interactions between azobenzene chromophores show influence on the time to achieve the birefringence and the remaining

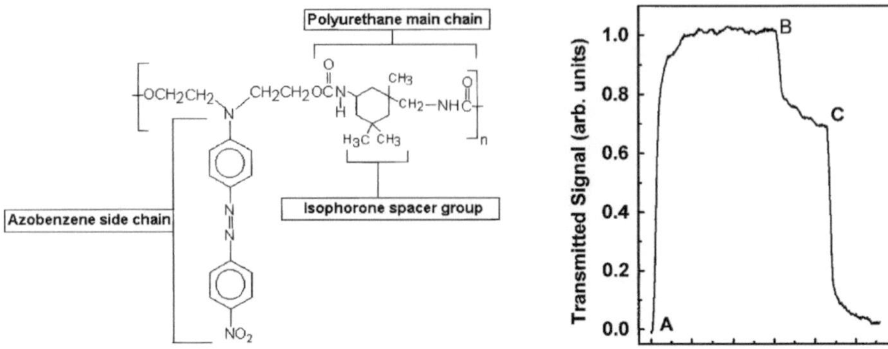

Fig. 4.10 The chemical structure of DR19-IPPU and the writing–relaxation–erasing sequence of the LB film (61 layers) of DR19-IPPU and cadmium stearate, where the transmitted signal is normalized (Reprinted with permission from Ref. [91]. Copyright (1999) American Chemical Society)

value after the writing beam is switched off [93]. Recently, mixed LB films have also been prepared by using block copolymers and CdSt for photoinduced birefringence investigation [94].

4.4.4 Azo Polyelectrolyte and Supramolecular Assembling

Electrostatic layer-by-layer system and other supramolecular assembled systems have been intensively investigated for various applications. In these systems, electrostatic and other non-covalent bonding interactions are exploited to build up materials through more controllable and feasible manners [95–97]. Different azo polymers and thin films have been prepared by this type of methods for investigating PIA properties. This type of approaches shows some advantages in feasibility and controllability, especially for preparing multilayers and thin films, which distinguish them from the spin-coating and LB methods.

A typical method to prepare this type of PIA materials is through layer-by-layer deposition methodology [95]. Oppositely charged polyelectrolytes or other substances can be used to fabricate multilayers through electrostatic sequential adsorption (Fig. 4.11). Thin films containing azo chromophores have been fabricated using Congo red (CR) and poly(dimethyldiallyl ammonium chloride) (PDAC) by electrostatic sequential adsorption [98]. The CR/PDAC film shows a large birefringence $\Delta n \sim 0.1$ without any relaxation after switching off the light. The electrostatic layer-by-layer films have also been fabricated by alternate deposition of a methacrylic acid-based azo copolymer (MA-co-DR13) and PDAC [99], chitosan and the azo dye (Ponceau-S) [100], azo polymer, and dendrimer [101].

Besides the layer-by-layer approach obtaining solid thin films for PIA applications, azobenzene-containing materials can be first prepared by supramolecular

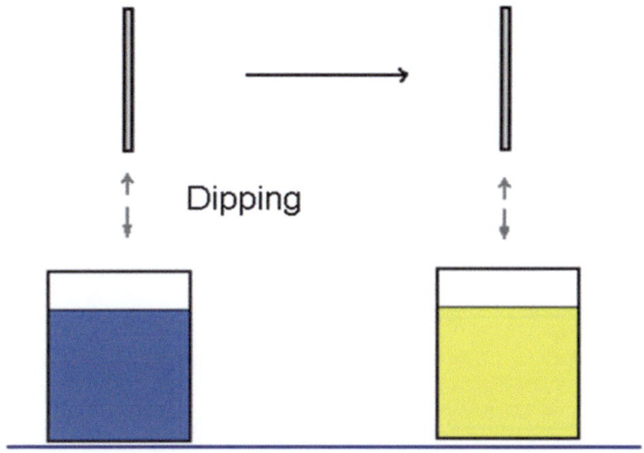

Fig. 4.11 Multilayer films are prepared by alternately dipping in solutions of oppositely charged polyelectrolytes

assembly through ionic interaction, hydrogen bonding, and other methods. Then, the thin films for PIA applications are prepared by solution-casting or spin-coating instead of the sequential deposition. The azobenzene-containing materials have been prepared by the ionic interactions between anionic azo compounds and polycations [102]. It is easy to produce the materials that are thermally very stable. By exploiting self-complementary quadruple hydrogen bonds, main-chain supramolecular azo polymers have been fabricated [103]. Figure 4.12 shows the synthetic route of the azobenzene-containing compound and the proposed geometry of self-complementary quadruple hydrogen bonds. The supramolecular polymer film shows significant photoinduced dichroism. Thin films have also been prepared from supramolecular 4-nitro-4′-hydroxyazobenzene and poly(4-vinylpyridine) complexes [104]. Hydrogen bonding between the phenol and pyridine moieties binds the chromophore to each repeat unit of the polymer. The supramolecular approach yields a superior optical performance compared to guest–host polymers and is comparable with the properties of azo polymers functionalized through covalent linkages.

4.4.5 Azo Block Copolymer

Block polymers with microphase-separated properties have been developed as a new type of PIA materials [105, 106]. One way to obtain the block copolymers starts from a sequential living anionic polymerization of butadiene and styrene [105]. The poly(1,2-butadiene) segment is then hydroborated and converted by a polymer-analogous reaction to incorporate the azo chromophores. PIA films are composed of the photoaddressable microphase containing *p*-methoxy-substituted azobenzene side-chain groups dispersed in polystyrene matrix. To obtain thick films suitable for volume holographic storage, some of the azo chromophores in the azo block are replaced with mesogenic side groups [106]. The diblock

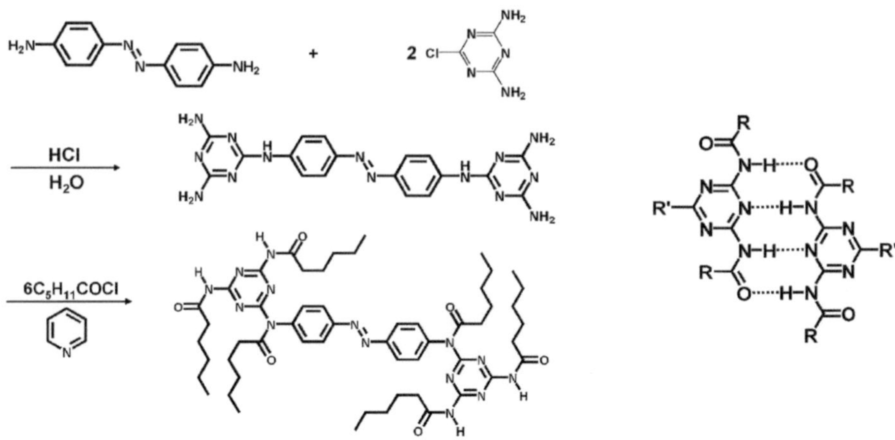

Fig. 4.12 The synthetic route of the azobenzene-containing compound and the proposed geometry of self-complementary quadruple hydrogen bonds (Reprinted with permission from Ref. [103]. Copyright (2007) American Chemical Society)

copolymer consists of a photoaddressable mesogen-containing microphase dispersed in polystyrene matrix. This approach is very efficient to reduce the optical density and avoid possible surface modulation during the holographic recording, which are the requirements for high-density optical storage.

In order to perform angular multiplexing of a large number of holograms at the same spatial position, holographic films with thicknesses in the millimeter range have been prepared. The films are fabricated by blending an azo block copolymer composed of a polystyrene block with polystyrene (PS) [107] or by blending an azo block copolymer composed of a poly(methyl methacrylate) block with poly(methyl methacrylate) (PMMA) [108]. Both PS and PMMA are commercially available polymers with good optical transparency. In this approach, the optical density can be adjusted by blending a different amount of the block copolymer with the homopolymers. Multiplexed volume holographic gratings can be written, read, and erased by purely optical means. Figure 4.13 shows the structure of the block copolymer and pictures of an injection-molded sample with the inscribed holograms between crossed polarizers. The sample consists of 77.5 wt% polystyrene as the matrix and the dispersed phase is formed by the block copolymer.

Diblock copolymers composed of two side-chain liquid crystalline blocks (SCLCP1-block-SCLCP2) have been synthesized by the reversible addition–fragmentation chain transfer (RAFT) polymerization [109]. The structure of the block copolymers can be seen in Fig. 3.14 (Chap. 3) together with its synthesis. It is interesting to observe that the cooperative photoaddressing can be effective for the phase-separated nonphotoactive microphase due to the interaction of the two mesogens via the interface. The photoinduced orientation of azobenzene groups on one block can bring biphenyl groups on the other block to orient in the same direction inside their respective lamellar domains.

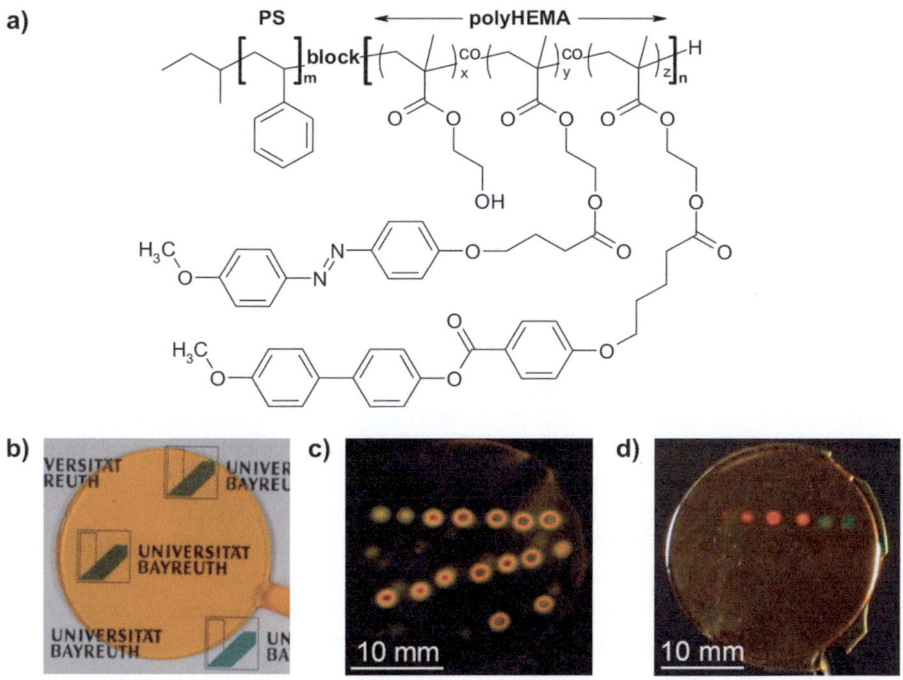

Fig. 4.13 Structure of the block copolymer and pictures of an injection-molded sample. (**a**) Structure of the polystyrene-*block*-poly(hydroxyethyl methacrylate)(PS-*b*-polyHEMA) with photoaddressable azo chromophores and mesogenic side chains in the HEMA block forming the minority phase ($m = 528$, $n = 40$). The HEMA block contains azobenzene and mesogenic side chains in a ratio of 46:54 and residual unconverted HEMA moieties ($x = 0.24$, $y = 0.35$, $z = 0.41$). (**b**) Picture of an injection-molded sample with a diameter of 25 mm and a thickness of 1.1 mm. (**c**) Sample with several inscribed holograms between crossed polarizers. (**d**) Sample showing the first diffraction order of some of these holograms under illumination with white light (Reprinted with permission from Ref. [107]. Copyright (2007) John Wiley and Sons)

RAFT polymerization has been used to synthesize diblock copolymers composed of the blocks bearing strong *push–pull* azo chromophores and mesogenic cholesteryl groups [110] and diblock copolymers composed of blocks bearing strong *push–pull* azo chromophores and mesogenic biphenyl groups [111]. The function of the copolymers as PIA materials has been investigated by measuring the photoinduced dichroism and birefringence. The microphase separation shows no obvious restraint on the photoinduced orientation of the azo chromophores. On the other hand, surface modulation during the optical writing can be avoided due to the microphase separation, which is required for volume holographic recording.

4.5 Optical Grating and Holography

The photoinduced dichroism and birefringence effects have been widely explored for the applications in optical gratings and holography. As demonstrated by Todorov et al. in 1985, such effects can be efficiently used for the applications of the polarization holographic recording [21]. Polarization grating and holography can be considered as optical diffractive devices specifically used to deal with polarized light. Optical gratings and holography in general are topics of many articles and monographs. Only basic knowledge necessary for the following discussion is presented here. For more details, we refer the readers to some monographs [16, 112, 113].

Gratings are basic optical elements used to record, detect, manipulate, and reconstruct light. Gratings have been fabricated from different dielectric and metallic materials to obtain the structures with spatially modulated optical properties, such as refractive index and coefficient of absorption. By using photosensitive materials, gratings have been produced by photographic processes for many years. Besides the well-known silver halide photographic emulsions, photoresists and photochromic materials have been used for grating fabrication.

A simple mask exposure technique can be used for the grating fabrication. Nowadays, two-beam interference technique is more widely used to produce gratings through exposure with spatially modulated light field. Holography, originated from Gabor's demonstration of holographic imaging [114], is now widely used as a light recording and display technology. The essence of holography is the idea of recording both the phase and the amplitude of the light waves from an object. It is realized by recording the interference pattern produced by the light waves from the object and a reference beam. Azo polymers can be used to prepare polarization gratings and holograms owing to the unique PIA characteristics.

The two-beam interference is a typical way for the production of laser-induced gratings [112]. If two coherent beams with equal intensity intersect at an angle, the phase relation will vary from one position to another in the region of intersection. In a basic setup, a laser beam is split into two beams that intersect at an angle θ at the sample to create the interference pattern. The beams are specified by wave vectors $\boldsymbol{k}_1$ and $\boldsymbol{k}_2$, electric field amplitude $\boldsymbol{E}_{01}$ and $\boldsymbol{E}_{02}$, and intensities I_1 and I_2. Figure 4.14 shows the wave vectors and the coordinate to define them. Therefore, the grating vector $\boldsymbol{q}$ of the interference pattern is given as

$$\boldsymbol{q} = \pm(\boldsymbol{k}_1 - \boldsymbol{k}_2) \tag{4.8}$$

$$q = |\boldsymbol{q}| = \frac{2\pi}{\lambda} \cdot \left(2\sin\left(\frac{\theta}{2}\right)\right) \tag{4.9}$$

The spatial period Λ can be expressed by the following equation in terms of q, the wavelength λ of light, and angle θ:

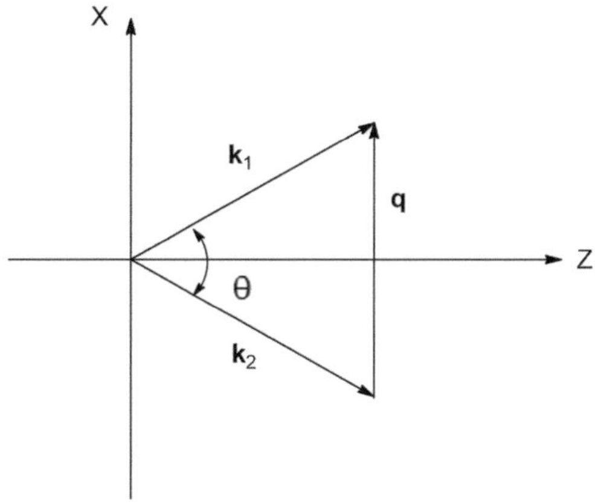

Fig. 4.14 The wave vectors of the two interfering beams and the coordinate to define them

$$\Lambda = \frac{2\pi}{q} = \frac{\lambda}{2\sin(\theta/2)} \tag{4.10}$$

In the interference region, the electric field amplitude $\boldsymbol{A}$, total time-dependent field $\boldsymbol{E}(\boldsymbol{r}, t)$, and intensity I are given as

$$\boldsymbol{A} = \boldsymbol{E}_{01}e^{ik_x x} + \boldsymbol{E}_{02}e^{-ik_x x} \tag{4.11}$$

$$\boldsymbol{E}(r,t) = \frac{\boldsymbol{A}}{2}e^{i(k_z z - \omega t)} + c.c. \tag{4.12}$$

$$I = I_1 + 2\Delta I \cos 2k_x x + I_2 \tag{4.13}$$

where ΔI is the intensity modulation amplitude and $c.c.$ denote the complex conjugate [112].

In the general case, both the amplitude and polarization of the resultant field are spatially modulated for non-collinear beams. For grating recording, the interference field is produced in the plane perpendicular to the bisectrix of the two writing beams. Superposition of two equal-intensity beams with different polarizations will produce the interference field modulated in both intensity and polarization.

A unique feature of azo polymers is that both the intensity and polarization can be recorded through the photoinduced molecular orientation and the PIA effects. Figure 4.15 shows a typical setup to write the birefringence gratings in azo polymer films. Holograms are produced by recording an interference pattern between an object beam and a reference beam. Polarization holography employs two beams with different polarizations for recording information. In this case, both intensity and polarization state of the superimposed beams are recorded on a suitable azo polymer medium.

Several experimental conditions have been typically used for recording the interference patterns. The setup of s, s polarizations, where polarizations of both beams are perpendicular to the plane of incidence, is a simplest experimental

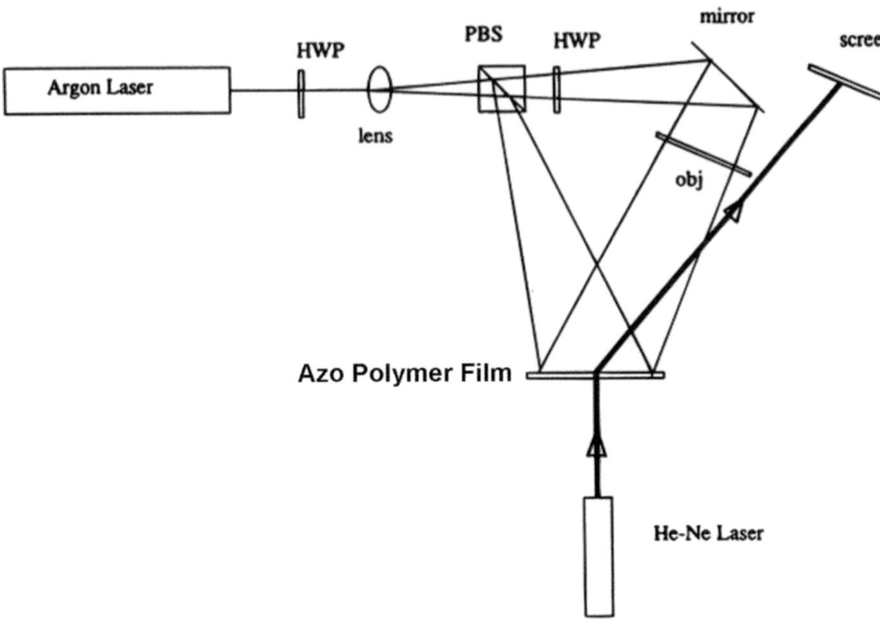

Fig. 4.15 Setup for conventional two-beam polarization holography. *Hwp* half-wave plate, *PBS* polarizing beam splitter (Reprinted with permission from Ref. [64]. Copyright (1995) American Chemical Society)

condition. The polarization is not modulated in the recording plane, but the intensity is fully modulated, which can be described by equation

$$I = 2I_1\ (1 + \cos qx) \tag{4.14}$$

The p, p polarizations, where polarizations of both beams are in the plane of incidence, result in a polarization varying between linear and elliptic depending on the relative phase qx. For the case of s, p mixed polarizations, polarization of the superimposed field changes similar to the p, p case, but the intensity is kept the same without modulation. For opposite circular polarizations, the intensity modulation is proportional to $\sin^2\theta$. It becomes small for $\theta \rightarrow 0$, while the resulted polarization tends to be linearly polarized and rotate with the grating period.

The recorded interference patterns can be treated as a diffractive element. The reconstruction of the object information is accomplished by reconstruction of the original object beam. The information is obtained in the form of diffraction orders of a conjugate reference beam. The thin polarization holograms can be quantitatively analyzed by using the Poincaré sphere, the Stokes vector, and the Jones matrix [16]. For transmission volume holograms, the Bragg condition needs to be considered with respect to the angle of incidence of the reconstructing wave. The writing and readout of thick volume polarization holograms in photoanisotropic organic materials have been studied by Huang and Wagner [115].

Compared with conventional holography, polarization holographic storage shows several advantages, such as high diffraction efficiency, ability to record both intensity and polarization information, and easy one-step recording process. Besides the optical storage devices, it is possible to use the polarization gratings for other optical elements and applications, such as spectral Stokesmeter, bifocal holographic lens, polarization optical phase conjugation, and joint transform correlation [16, 21]. These applications are promising by developing various azo polymers and exploiting their PIA effects. For more details, we refer the interested reader to a comprehensive monograph published in 2009 [16].

References

1. Platé, N. (ed.): Liquid crystal polymers I. Springer, Berlin/Heidelberg (1984)
2. Burland, D.M., Miller, R.D., Walsh, C.A.: Second-order nonlinearity in poled-polymer systems. Chem. Rev. **94**, 31–75 (1994)
3. Xie, S., Natansohn, A., Rochon, P.: Recent developments in aromatic azo polymers research. Chem. Mater. **5**, 403–411 (1993)
4. Natansohn, A., Rochon, P.: Photoinduced motions in azo-containing polymers. Chem. Rev. **102**, 4139–4175 (2002)
5. Shibaev, V., Bobrovsky, A., Boiko, N.: Photoactive liquid crystal polymer systems with light-controlled structure and optical properties. Prog. Polym. Sci. **28**, 729–836 (2003)
6. Yesodha, S.K., Pillai, C.K.S., Tsutsumi, N.: Stable polymeric materials for nonlinear optics: a review based on azobenzene systems. Prog. Polym. Sci. **29**, 45–74 (2004)
7. Dumont, M.: Photoinduced orientational order in dye-doped amorphous polymeric films. Mol. Cryst. Liq. Cry. **282**, 437–450 (1996)
8. Fischer, M., El Osman, A., Blanche, P.A., Dumont, M.: Photoinduce dichroism as a tool for understanding orientational mobility of photoisomerizable dyes in amorphous matrices. Syn. Metals. **115**, 139–144 (2000)
9. Sekkat, Z., Dumont, M.: Photoassisted poling of azo dye doped polymeric films at room temperature. Appl. Phys. B **54**, 486–489 (1992)
10. Charra, F., Kajzar, F., Nunzi, J.M., Raimond, P., Idiart, E.: Light-induced second-harmonic generation in azo-dye polymers. Opt. Lett. **18**, 941–943 (1993)
11. Chalupczak, W., Fiorini, C., Charra, F., Nunzi, J.M., Raimond, P.: Efficient all-optical poling of an azo-dye copolymer using a low power laser. Opt. Commun. **126**, 103–107 (1996)
12. Delaire, J.A., Nakatani, K.: Linear and nonlinear optical properties of photochromic molecules and materials. Chem. Rev. **100**, 1817–1845 (2000)
13. Ikeda, T.: Photomodulation of liquid crystal orientations for photonic applications. J. Mater. Chem. **13**, 2037–2057 (2003)
14. Weigert, F.: A new effect of the rays in light-sensitive layers. Verhandlungen der Deutschen Physikalischen Gesellschaft (German) **21**, 479–491 (1919)
15. Ebralidze, T.D., Mumladze, A.N.: Light-induced anisotropy in azo-dye-colored materials. Appl. Opt. **29**, 446–447 (1990)
16. Nikolova, L., Ramanujam, P.S.: Polarization holography. Cambridge University Press, New York (2009)
17. Kakichashvili, S.D.: Polarization recording of holograms. Optika I Spektroskopiya (Russian) **33**, 324 (1972)
18. Kakichashvili, S.D.: Polarizational (anisotripic-vectorial) holographic recording on real photo-anisotropic materials. Optika I Spektroskopiya (Russian) **42**, 390–394 (1977)

19. Todorov, T., Nikolova, L., Tomova, N.: Polarization holography. 1: a new high-efficiency organic material with reversible photoinduced birefringence. Appl. Opt. **23**, 4309–4312 (1984)
20. Todorov, T., Nikolova, L., Tomova, N.: Polarization holography. 2: polarization holographic gratings in photoanisotropic materials with and without intrinsic birefringence. Appl. Opt. **23**, 4588–4591 (1984)
21. Todorov, T., Nikolova, L., Stoyanova, K., Tomova, N.: Polarization holography. 3: some applications of polarization holographic recording. Appl. Opt. **24**, 785–788 (1985)
22. Eich, M., Wendorff, J.H., Reck, B., Ringsdorf, H.: Reversible digital and holographic optical storage in polymeric liquid crystals. Makromol. Chem. Rapid Commun. **8**, 59–63 (1987)
23. Eich, M., Wendorff, J.H.: Erasable holograms in polymeric liquid crystals. Makromol. Chem. Rapid Commun. **8**, 467–471 (1987)
24. Stumpe, J., Muller, L., Kreysig, D., Hauck, G., Koswig, H.D., Ruhmann, R., Rubner, J.: Photoreaction in mesogenic media 5. Photoinduced optical anisotropy of liquid-crystalline side-chain polymers with azochromophores by linearly polarized light of low intensity. Makromol. Chem. Rapid Commun. **12**, 81–87 (1991)
25. Shi, Y.Q., Steier, W.H., Yu, L.P., Chen, M., Dalton, L.R.: Large photoinduced birefringence in an optically nonlinear polyester polymer. Appl. Phys. Lett. **59**, 2935–2937 (1991)
26. Natansohn, A., Rochon, P., Gosselin, J., Xie, S.: Azo polymers for reversible optical storage. 1. Poly[4′-[[2-(acryloyloxy)ethyl]ethylamino]-4-nitroazobenzene]. Macromolecules **25**, 2268–2273 (1992)
27. Shurcliff, W.A.: Polarized light, production and use. Harvard University Press, Cambridge, MA (1962)
28. Anderle, K., Birenheide, R., Werner, M.J., Wendorff, J.H.: Molecular addressing? Studies on light-induced reorientation in liquid-crystalline side chain polymers. Liq. Cryst. **9**, 691–699 (1991)
29. Jones, C., Day, S.: Shedding light on alignment. Nature **351**, 15–15 (1991)
30. Buffeteau, T., Labarthet, F.L., Pézolet, M., Sourisseau, C.: Photoinduced orientation of azobenzene chromophores in amorphous polymers as studied by real-time visible and FTIR spectroscopies. Macromolecules **31**, 7312–7320 (1998)
31. Buffeteau, T., Pézolet, M.: Photoinduced orientation in azopolymers studied by infrared spectroscopy: cooperative and biaxial orientation in semicrystalline polymers. Macromolecules **31**, 2631–2635 (1998)
32. Labarthet, F.L., Buffeteau, T., Sourisseau, C.: Molecular orientations in azopolymer holographic diffraction gratings as studied by Raman confocal microspectroscopy. J. Phys. Chem. B **102**, 5754–5765 (1998)
33. Sourisseau, C.: Polarization measurements in macro- and micro-Raman spectroscopies: molecular orientations in thin films and azo-dye containing polymer systems. Chem. Rev. **104**, 3851–3891 (2004)
34. Holme, N.C.R., Ramanujam, P.S., Hvilsted, S.: Photoinduced anisotropy measurements in liquid-crystalline azobenzene side-chain polyesters. Appl. Opt. **35**, 4622–4627 (1996)
35. Wu, Y.L., Ikeda, T., Zhang, Q.J.: Three-dimensional manipulation of an azo polymer liquid crystal with unpolarized light. Adv. Mater. **11**, 300–302 (1999)
36. Wu, Y.L., Mamiya, J., Kanazawa, A., Shiono, T., Ikeda, T., Zhang, Q.J.: Photoinduced alignment of polymer liquid crystals containing azobenzene moieties in the side chain. 6. Biaxiality and three-dimensional reorientation. Macromolecules **32**, 8829–8835 (1999)
37. Ichimura, K., Han, M.N., Morino, S.Y.: Photochemistry determined by light propagation. Part 1 three-dimensional photomanipulation of self-organized azobenzenes in liquid-crystalline polymers. Chem. Lett. **28**, 85–86 (1999)
38. Han, M.N., Ichimura, K.: Tilt orientation of *p*-methoxyazobenzene side chains in liquid crystalline polymer films by irradiation with nonpolarized light. Macromolecules **34**, 82–89 (2001)

39. Han, M.N., Ichimura, K.: In-plane and tilt reorientation of *p*-methoxyazobenzene side chains tethered to liquid crystalline polymethacrylates by irradiation with 365 nm light. Macromolecules **34**, 90–98 (2001)
40. Nikolova, L., Todorov, T., Ivanov, M., Andruzzi, F., Hvilsted, S., Ramanujam, P.S.: Photoinduced circular anisotropy in side-chain azobenzene polyesters. Opt. Mater. **8**, 255–258 (1997)
41. Iftime, G., Labarthet, F.L., Natansohn, A., Rochon, P.: Control of chirality of an azobenzene liquid crystalline polymer with circularly polarized light. J. Am. Chem. Soc. **122**, 12646–12650 (2000)
42. Wu, Y.L., Natansohn, A., Rochon, P.: Photoinduced chirality in thin films of achiral polymer liquid crystals containing azobenzene chromophores. Macromolecules **37**, 6801–6805 (2004)
43. Birabassov, B., Galstian, T.V.: Light-induced macroscopic chirality in azo-dye-doped polymers. J. Opt. Soc. Am. B **18**, 1469–1473 (2001)
44. Cipparrone, G., Pagliusi, P., Provenzano, C., Shibaev, V.P.: Reversible photoinduced chiral structure in amorphous polymer for light polarization control. Macromolecules **41**, 5992–5996 (2008)
45. Nedelchev, L., Nikolova, L., Todorov, T., Petrova, T., Tomova, N., Dragostinova, V., Ramanujam, P.S., Hvilsted, S.: Light propagation through photoinduced chiral structures in azobenzene-containing polymers. J. Opt. A, Pure Appl. Opt. **3**, 304–310 (2001)
46. Dumont, M., El Osman, A.: On spontaneous and photoinduced orientational mobility of dye molecules in polymers. Chem. Phys. **245**, 437–462 (1999)
47. Sekkat, Z., Wood, J., Knoll, W.: Reorientation mechanism of azobenzenes within the *trans* → *cis* photoisomerization. J. Phys. Chem. **99**, 17226–17234 (1995)
48. Palto, S.P., Blinov, L.M., Yudin, S.G., Grewer, G., Schönhoff, M., Lösche, M.: Photoinduced optical anisotropy in organic molecular films controlled by an electric field. Phys. Chem. Lett. **202**, 308–314 (1993)
49. Palto, S.P., Durand, G.: Friction model of photoinduced reorientation of optical-axis in photo-oriented Langmuir-Blodgett films. J. Physique II **5**, 963–978 (1995)
50. Pedersen, T.G., Johansen, P.M., Holme, N.C.R., Ramanujam, P.S., Hvilsted, S.: Theoretical model of photoinduced anisotropy in liquid-crystalline azobenzene side-chain polyesters. J. Opt. Soc. Am. B **15**, 1120–1129 (1998)
51. Schönhoff, M., Mertesdorf, M., Lösche, M.: Mechanism of photoreorientation of azobenzene dyes in molecular films. J. Phys. Chem. **100**, 7558–7565 (1996)
52. Eich, M., Wendorff, J.: Laser-induced gratings and spectroscopy in monodomains of liquid-crystalline polymers. J. Opt. Soc. Am. B **7**, 1428–1436 (1990)
53. Anderle, K., Birenheide, R., Eich, M., Wendorff, J.H.: Laser-induced reorientation of the optical axis in liquid-crystalline side chain polymers. Makromol. Chem. Rapid Commun. **10**, 477–483 (1989)
54. Wiesner, U., Reynolds, N., Boeffel, C., Spiess, H.W.: Photoinduced reorientation in liquid crystalline polymers below the glass transition temperature studied by time-dependent infrared spectroscopy. Makromo. Chem. Rapid Commun. **12**, 457–464 (1991)
55. Wiesner, U., Reynolds, N., Boeffel, C., Spiess, H.W.: An infrared spectroscopic study of photo-induced reorientation in dye containing liquid-crystalline polymers. Liq. Cryst. **11**, 251–267 (1992)
56. Stumpe, J., Läsker, L., Fischer, T., Rutloh, M., Kostromin, S., Ruhmann, R.: Photoorientation in amorphous and aligned films of photochromic liquid crystalline polymers. Thin Solid Films **284–285**, 252–256 (1996)
57. Rosenhauer, R., Fischer, T., Stumpe, J., Giménez, R., Piňol, M., Serrano, J.L., Viňuales, A., Broer, D.: Light-induced orientation of liquid crystalline terpolymers containing azobenzene and dye moieties. Macromolecules **38**, 2213–2222 (2005)
58. Wu, Y.L., Demachi, Y., Tsutsumi, O., Kanazawa, A., Shiono, T., Ikeda, T.: Photoinduced alignment of polymer liquid crystals containing azobenzene moieties in the side chain. 1. Effect of light intensity on alignment behavior. Macromolecules **31**, 349–354 (1998)

59. Wu, Y.L., Zhang, Q.J., Kanazawa, A., Shiono, T., Ikeda, T., Nagase, Y.: Photoinduced alignment of polymer liquid crystals containing azobenzene moieties in the side chain. 5. Effect of the azo contents on alignment behavior and enhanced response. Macromolecules **32**, 3951–3956 (1999)
60. Wu, Y.L., Demachi, Y., Tsutsumi, O., Kanazawa, A., Shiono, T., Ikeda, T.: Photoinduced alignment of polymer liquid crystals containing azobenzene moieties in the side chain. 2. Effect of spacer length of the azobenzene unit on alignment behavior. Macromolecules **31**, 1104–1108 (1998)
61. Wu, Y.L., Demachi, Y., Tsutsumi, O., Kanazawa, A., Shiono, T., Ikeda, T.: Photoinduced alignment of polymer liquid crystals containing azobenzene moieties in the side chain. 3. Effect of structure of photochromic moieties on alignment behavior. Macromolecules **31**, 4457–4463 (1998)
62. Yoneyama, S., Yamamoto, T., Tsutsumi, O., Kanazawa, A., Shiono, T., Ikeda, T.: High-performance material for holographic gratings by means of a photoresponsive polymer liquid crystal containing a tolane moiety with high birefringence. Macromolecules **35**, 8751–8758 (2002)
63. Hvilsted, S., Andruzzi, F., Ramanujam, P.S.: Side-chain liquid-crystalline polyesters for optical information storage. Opt. Lett. **17**, 1234–1236 (1992)
64. Hvilsted, S., Andruzzi, F., Kulinna, C., Siesler, H.W., Ramanujam, P.S.: Novel side-chain liquid crystalline polyester architecture for reversible optical storage. Macromolecules **28**, 2172–2183 (1995)
65. Holme, N.C.R., Ramanujam, P.S., Hvilsted, S.: 10,000 optical write, read, and erase cycles in an azobenzene sidechain liquid-crystalline polyester. Opt. Lett. **21**, 902–904 (1996)
66. Kulinna, C., Hvilsted, S., Hendann, C., Siesler, H.W., Ramanujam, P.S.: Selectively deuterated liquid crystalline cyanoazobenzene side-chain polyesters. 3. Investigations of laser-induced segmental mobility by Fourier transform infrared spectroscopy. Macromolecules **31**, 2141–2151 (1998)
67. Han, D.H., Tong, X., Zhao, Y., Galstian, T., Zhao, Y.: Cyclic azobenzene-containing side-chain liquid crystalline polymers: synthesis and topological effect on mesophase transition, order, and photoinduced birefringence. Macromolecules **43**, 3664–3671 (2010)
68. Natansohn, A., Rochon, P., Pézolet, M., Audet, P., Brown, D., To, S.: Azo polymers for reversible optical storage. 4. Cooperative motion of rigid groups in semicrystalline polymers. Macromolecules **27**, 2580–2585 (1994)
69. Brown, D., Natansohn, A., Rochon, P.: Azo polymers for reversible optical storage. 5. Orientation and dipolar interactions of azobenzene side groups in copolymers and blends containing methyl methacrylate structural units. Macromolecules **28**, 6116–6123 (1995)
70. Natansohn, A., Rochon, P., Ho, M.S., Barrett, C.: Azo polymers for reversible optical storage. 6. Poly[4-[2-(methacryloyloxy)ethyl]azobenzene. Macromolecules **28**, 4179–4183 (1995)
71. Ho, M.S., Natansohn, A., Rochon, P.: Azo polymers for reversible optical storage. 9. Copolymers containing two types of azobenzene side groups. Macromolecules **29**, 44–49 (1996)
72. Ho, M.S., Natansohn, A., Rochon, P.: Azo polymers for reversible optical storage. 7. The effect of the size of the photochromic groups. Macromolecules **28**, 6124–6127 (1995)
73. Meng, X., Natansohn, A., Barrett, C., Rochon, P.: Azo polymers for reversible optical storage. 10. Cooperative motion of polar side groups in amorphous polymers. Macromolecules **29**, 946–952 (1996)
74. Buffeteau, T., Natansohn, A., Rochon, P., Pézolet, M.: Study of cooperative side group motions in amorphous polymers by time dependent infrared spectroscopy. Macromolecules **29**, 8783–8790 (1996)
75. Natansohn, A., Rochon, P., Meng, X.S., Barrett, C., Buffeteau, T., Bonenfant, S., Pézolet, M.: Molecular addressing? Selective photoinduced cooperative motion of polar ester groups in copolymers containing azobenzene groups. Macromolecules **31**, 1155–1161 (1998)

76. Han, M.N., Kidowaki, M., Ichimura, K., Ramanujam, P.S., Hvilsted, S.: Influence of structures of polymer backbones on cooperative photoreorientation behavior of *p*-cyanoazobenzene side chains. Macromolecules **34**, 4256–4262 (2001)
77. Nedelchev, L., Matharu, A.S., Hvilsted, S., Ramanujam, P.S.: Photoinduced anisotropy in a family of amorphous azobenzene polyesters for optical storage. Appl. Opt. **42**, 5918–5927 (2003)
78. Cha, S.W., Choi, D.H., Jin, J.I.: Unusually fast optically induced birefringence in polyoxetanes bearing 4-(N, N-diphenyl)amino-4′-nitroazobenzene chromophores. Adv. Funct. Mater. **11**, 355–360 (2001)
79. Tawa, K., Kamada, K., Kiyohara, K., Ohta, K., Yasumatsu, D., Sekkat, Z., Kawata, S.: Photoinduced reorientation of azo dyes bonded to polyurethane studied by polarized FT-IR spectroscopy. Macromolecules **34**, 8232–8238 (2001)
80. Hasegawa, M., Ikawa, T., Tsuchimori, T., Watanabe, O.: Photochemically induced birefringence in polyurethanes containing donor–acceptor azobenzenes as photoresponsive moieties. J. Appl. Polym. Sci. **86**, 17–22 (2002)
81. Chen, J.P., Labarthet, F.L., Natansohn, A., Rochon, P.: Highly stable optically induced birefringence and holographic surface gratings on a new azocarbazole-based polyimide. Macromolecules **32**, 8572–8579 (1999)
82. Sava, I., Bruma, M., Köpnick, T., Sapich, B., Wagner, J., Schulz, B., Stumpe, J.: Photoinduced optical anisotropy in new polymers containing azobenzene side groups. High Perform. Polym. **19**, 296–310 (2007)
83. Fernández, R., Mondragon, I., Galante, M.J., Oyanguren, P.A.: Bonding and molecular environment effects on photoorientation in epoxy-based polymers having azobenzene units. Eur. Polym. J. **45**, 788–794 (2009)
84. Wang, Y.W., He, Y.N., Wang, X.G.: Photoinduced orientation and cooperative motion of three epoxy-based azo polymers. Polym. Bull. **68**, 1731–1746 (2012)
85. Xu, Z.S., Drnoyan, V., Natansohn, A., Rochon, P.: Novel polyesters with amino-sulfone azobenzene chromophores in the main chain. J. Polym. Sci. A **38**, 2245–2253 (2000)
86. Sapich, B., Vix, A.B.E., Rabe, J.P., Stumpe, J.: Photoinduced self-organization and photoorientation of a LC main-chain polyester containing azobenzene moieties. Macromolecules **38**, 10480–10486 (2005)
87. Wu, Y.L., Natansohn, A., Rochon, P.: Photoinduced birefringence and surface relief gratings in novel polyurethanes with azobenzene groups in the main chain. Macromolecules **34**, 7822–7828 (2001)
88. Wu, Y.L., Natansohn, A., Rochon, P.: Photoinduced birefringence and surface relief gratings in polyurethane elastomers with azobenzene chromophore in the hard segment. Macromolecules **37**, 6090–6095 (2004)
89. Stumpe, J., Geue, T., Fischer, T., Menzel, H.: Photo-orientation in LB multilayers of amphotropic polymers. Thin Solid Films **284–285**, 606–611 (1996)
90. Geue, T., Ziegler, A., Stumpe, J.: Light-induced orientation phenomena in Langmuir-Blodgett multilayers. Macromolecules **30**, 5729–5738 (1997)
91. Dhanabalan, A., Santos, D., Jr, D.S., Mendonca, C.R., Misoguti, L., Balogh, D.T., Giacometti, J.A., Zilio, S.C., Oliveira Jr., O.N.: Optical storage in mixed Langmuir-Blodgett (LB) films of disperse red-19 isophorone polyurethane and cadmium stearate. Langmuir **15**, 4560–4564 (1999)
92. Mendonca, C.R., Dhanabalan, A., Balogh, D.T., Misoguti, L., Santos, D., Jr, D.S., Pereira-da-Silva, M.A., Giacometti, J.A., Zilio, S.C., Oliveira Jr., O.N.: Optically induced birefringence and surface relief gratings in composite Langmuir-Blodgett (LB) films of poly[4′-[[2-(methacryloyloxy)ethyl]ethylamino]-2-chloro-4-nitroazobenzene] (HPDR13) and cadmium stearate. Macromolecules **32**, 1493–1499 (1999)
93. Dhanabalan, A., Mendonca, C.R., Balogh, D.T., Misoguti, L., Constantino, C.J.L., Giacometti, J.A., Zilio, S.C., Oliveira Jr., O.N.: Storage studies of Langmuir-Blodgett (LB) films of

methacrylate copolymers derivatized with disperse red-13. Macromolecules **32**, 5277–5284 (1999)
94. Pavinatto, F.J., Barletta, J.Y., Sanfelice, R.C., Cardoso, M.R., Balogh, D.T., Mendonca, C.R., Oliveira Jr., O.N.: Synthesis of azopolymers with controlled structure and photoinduced birefringence in their LB film. Polymer **50**, 491–498 (2009)
95. Decher, G., Hong, J.D., Schmitt, J.: Buildup of ultrathin multilayer films by a self-assembly process. 3. Consecutively alternating adsorption of anionic and cationic polyelectrolytes on charged surfaces. Thin Solid Films **210**, 831–835 (1992)
96. Decher, G.: Fuzzy nanoassemblies: toward layered polymeric multicomposites. Science **277**, 1232–1237 (1997)
97. Whitesides, G.M., Grzybowski, B.: Self-assembly at all scales. Science **295**, 2418–2421 (2002)
98. Bian, S.P., He, J.A., Li, L., Kumar, J., Tripathy, S.K.: Large photoinduced birefringence in azo dye/polyion films assembled by electrostatic sequential adsorption. Adv. Mater. **12**, 1202–1205 (2000)
99. Zucolottoa, V., Mendonca, C.R., Santos, D., Jr, D.S., Balogh, D.T., Zilio, S.C., Oliveira Jr., O.N., Constantino, C.J.L., Aroca, R.F.: The influence of electrostatic and H-bonding interactions on the optical storage of layer-by-layer films of an azopolymer. Polymer **43**, 4645–4650 (2002)
100. Santos, D., Jr, D.S., Bassi, A., Rodrigues Jr., J.J., Misoguti, L., Oliveira Jr., O.N., Mendonca, C.R.: Light-induced storage in layer-by-layer films of chitosan and an azo dye. Biomacromolecules **4**, 1502–1505 (2003)
101. Santos, D., Jr, D.S., Cardoso, M.R., Leite, F.L., Aroca, R.F., Mattoso, L.H.C., Oliveira Jr., O. N., Mendonca, C.R.: The role of azopolymer/dendrimer layer-by-layer film architecture in photoinduced birefringence and the formation of surface-relief -gratings. Langmuir **22**, 6177–6180 (2006)
102. Kulikovska, O., Goldenberg, L.M., Stumpe, J.: Supramolecular azobenzene-based materials for optical generation of microstructures. Chem. Mater. **19**, 3343–3348 (2007)
103. Gao, J., He, Y.N., Xu, H.P., Song, B., Zhang, X., Wang, Z.Q., Wang, X.G.: Azobenzene-containing supramolecular polymer films for laser-induced surface relief gratings. Chem. Mater. **19**, 14–17 (2007)
104. Priimagi, A., Vapaavuori, J., Rodriguez, F.J., Faul, C.F.J., Heino, M.T., Ikkala, O., Kauranen, M., Kaivola, M.: Hydrogen-bonded polymer-azobenzene complexes: enhanced photoinduced birefringence with high temporal stability through interplay of intermolecular interactions. Chem. Mater. **20**, 6358–6363 (2008)
105. Frenz, C., Fuchs, A., Schmidt, H.W., Theissen, U., Haarer, D.: Diblock copolymers with azobenzene side-groups and polystyrene matrix: synthesis, characterization and photoaddressing. Macromol. Chem. Phys. **205**, 1246–1258 (2004)
106. Häckel, M., Kador, L., Kropp, D., Frenz, C., Schmidt, H.W.: Holographic gratings in diblock copolymers with azobenzene and mesogenic side groups in the photoaddressable dispersed phase. Adv. Funct. Mater. **15**, 1722–1727 (2005)
107. Häckel, M., Kador, L., Kropp, D., Schmidt, H.W.: Polymer blends with azobenzene-containing block copolymers as stable rewritable volume holographic media. Adv. Mater. **19**, 227–231 (2007)
108. Breiner, T., Kreger, K., Hagen, R., Häckel, M., Kador, L., Müller, A.H.E., Kramer, E.J., Schmidt, H.W.: Blends of poly(methacrylate) block copolymers with photoaddressable segments. Macromolecules **40**, 2100–2108 (2007)
109. Zhao, Y., Qi, B., Tong, X., Zhao, Y.: Synthesis of double side-chain liquid crystalline block copolymers using RAFT polymerization and the orientational cooperative effect. Macromolecules **41**, 3823–3831 (2008)
110. Zhu, Y., Zhou, Y.Q., Chen, Z., Lin, R., Wang, X.G.: Photoresponsive diblock copolymers bearing strong push-pull azo chromophores and cholesteryl groups. Polymer **53**, 3566–3576 (2012)

111. Zhu, Y., Wang, X.G.: Photoresponsive diblock copolymers bearing strong push-pull azo chromophores and mesogenic biphenyl groups. Dyes Pigm. **97**, 222–229 (2013)
112. Eichler, H.J., Günter, P., Pohl, D.W.: Laser-induced dynamic gratings. Springer, Berlin (1986)
113. Collier, R.J., Burckhardt, C.B., Lin, L.H.: Optical holography. Academic, New York (1971)
114. Gabor, D.: A new microscopic principle. Nature **161**, 777–778 (1948)
115. Huang, T., Wagner, K.H.: Holographic diffraction in photoanisotropic organic materials. J. Opt. Soc. Am. A **10**, 306–315 (1993)

Chapter 5
Photoinduced Mass Transport

Abstract This chapter discusses another fascinating function of azo polymers, photoinduced mass transport, in detail. This photoinduced effect is unprecedented in polymers and other materials, which has been intensively studied to understand the mechanism and also for possible applications in many areas such as surface modification and patterning. The original discovery and many following investigations from materials to functions are presented in a step-by-step way. The basic concepts and background are introduced at the beginning. Then, material design and preparation, optical setup, structure–property relationship, mechanism, and models are discussed in sequence. Meanwhile, some new observations and insights are discussed to document the recent progresses in this area.

Keywords Surface-relief-grating • Optical setup • Characterization • Molecular structure • Parameters • Model and theory • Mechanical property • Real-time imaging • Spontaneous patterning

Photoinduced mass transport is one of the most fascinating properties of azo polymers. This effect started to draw wide attention as soon as the scientific community found it from two reports on the pioneering work [1, 2]. As reported in the articles, surface-relief-grating (SRG) can be inscribed on the azo polymer films upon exposure to interference patterns of laser beams. The grating formation was soon recognized to be caused by *the mass transport* in nano-/micrometer scale, which shows several unusual and puzzling characteristics. Upon irradiation with interfering laser beams at modest light intensity, sinusoidal surface patterns can be developed on the azo polymer films at temperature well below the glass transition temperature (T_g) of the polymers. The large surface modulation, typically several hundreds of nanometers, is achieved by one-step irradiation. The gratings can be erased by heating the samples above their T_gs without degradation or charring of the polymer films. These unique features denying some simple explanations have inspired many researchers to carry out intensive investigations to understand the mechanism. The intensive investigations have been carried out in recent years to understand the mechanism, to elucidate the structure–property relationship, and to obtain materials with better performance.

The SRG formation caused by the mass transport has drawn wide attention not only because of the puzzling nature of this physical effect but also owing to

X. Wang, *Azo Polymers*, Soft and Biological Matter,
DOI 10.1007/978-3-662-53424-3_5

potential applications as a new approach to fabricate various optical elements, functional surfaces, and devices. The gratings with different amplitudes and periods can be obtained by adjusting the optical parameters such as the irradiation energy, intersection angle of the writing beams, and wavelength of the light. This method is capable of recording orthogonal gratings by exposure to two sets of gratings that are orthogonal to each other. A variety of complicated surface patterns can be inscribed on the azo polymer surfaces [3, 4]. The scope of the possible application has gone far beyond the holography, although it was one of the main goals in the initial developing stage.

Several excellent review articles have been published to document the developments at different stages and aspects. We refer the interested reader to some of the articles [3–8]. Although the authors in this research area use different terms to define this effect, such as mass transport, mass migration and polymer chain migration, and photofluidic motion, among others, they generally agreed that nano-/microscale transport of polymer chains is achieved by light irradiation based on their own experiments and observations. This chapter will discuss this fascinating property of azo polymers starting from the SRG formation to a variety of recent developments under the topic of photoinduced mass transport.

5.1 SRG Formation and Characterization

The terms "surface gratings" and *surface-relief-gratings* (*SRG*) were first used by the discoverers of this unique photoresponsive effect [1, 2]. Since then, these terms have been widely used by the researchers in this area. Micron-scale and reversible surface modulations are some unique features that distinguish SRG formation from other types of photoinduced surface effects. SRG inscription can be considered as a specific type of ways to record a hologram. Atomic force microscope (AFM) can supply highly visible images of the topographical profile and morphology of the surface patterns. Figure 5.1 shows AFM images of the SRGs published in a recent article. The optical setup and materials for inscribing SRGs are discussed in this section.

5.1.1 Materials

The solid films with smooth surfaces are required to inscribe SRGs, which are conventionally prepared by spin coating and vacuum drying. A suitable solution is obtained by dissolving an azo polymer in its good solvent. The solution is then filtered through a membrane and spin-coated onto the substrates with a proper spinning speed. The spin-coated films are dried under vacuum at a proper temperature, which depends on the boiling point of the solvent. The film thickness, typically in a range from submicron- to micron-scale, can be controlled by adjusting the solution concentration and spinning speed. Other methods that can produce films with smooth surface can also be used for this purpose.

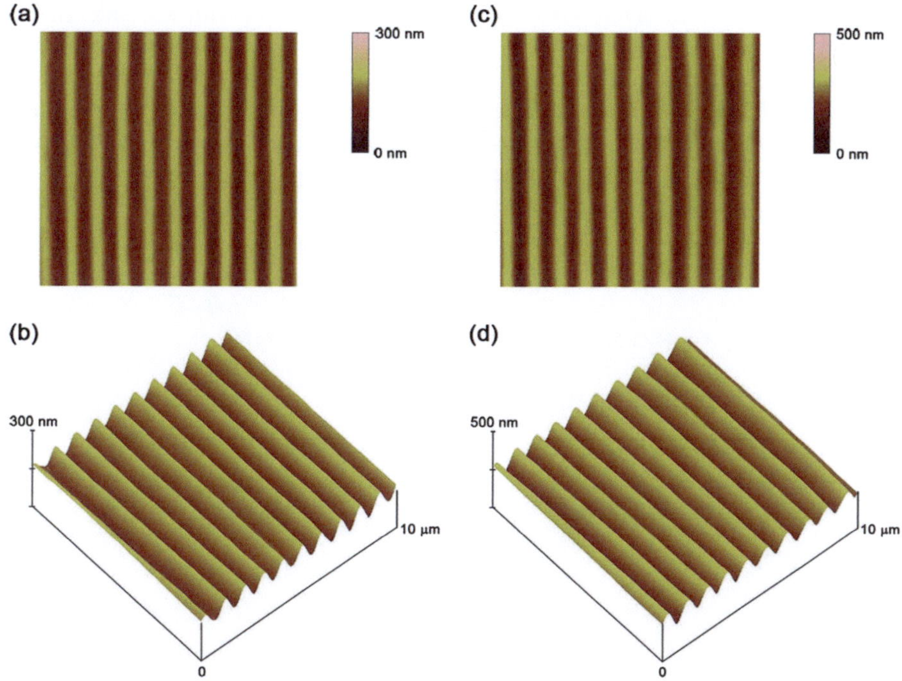

Fig. 5.1 Typical AFM images (10 μm × 10 μm) of the surface-relief-grating formed on the film of BP35-BAZ-35CN (see Fig. 5.2 for the polymer structure) with irradiation of incident laser beam at 488 and 532 nm: (**a**) 488 nm, 2D view; (**b**) 488 nm, 3D view; (**c**) 532 nm, 2D view; (**d**) 532 nm, 3D view; the thickness of the films was 500 nm. The intensity of the incident laser light was 200 mW/cm^2 and the irradiation time was 1200 s (Reprinted with permission from Ref. [9]. Copyright (2011) Elsevier)

Surface-relief-grating (SRG) formation has been observed for many types of azo polymers. The original investigations were carried out by using amorphous azo polymers bearing *push–pull* azo chromophores. The polymers include a variety of acrylate-/methacrylate-based polymers [1, 3] and epoxy-based azo polymers [2, 4, 10]. The chemical structure and syntheses of these azo polymers can be seen in Chap. 3. These polymers can be readily dissolved in polar organic solvents, such as tetrahydrofuran (THF) and *N*,*N*-dimethylformamide (DMF). The films with smooth surfaces and optical homogeneity can then be obtained by spin coating with the solutions. Besides the above two types of polymers, many other types of azo polymers have also been used for SRG inscription and investigations. Theoretically, almost every type of the azo polymers discussed in Chaps. 3 and 4 can be used for the SRG inscription. However, most reports indicate that amorphous azo polymers containing the *push–pull* azo chromophores can form SRGs in a more efficient way [3, 4, 7].

Other types of the azo polymers and related materials, which have also been used for SRG inscription and investigations, include side-chain liquid crystalline polyesters [11, 12], main-chain polyureas and polyurethanes [13–15], polyelectrolyte and electrostatic multilayers [16–19], conjugated polymers with azo side chains [20, 21], azobenzene-modified polypeptide [22], hyperbranched azo polymers and dendrimers [23, 24], liquid crystalline copolymers [25, 26], and supramolecular polymers [27, 28]. Besides typical azo chromophores, bisazo chromophores have also be used as functional groups for SRG inscription [9]. The chemical structure and synthetic route of the azo polymers are given in Fig. 5.2.

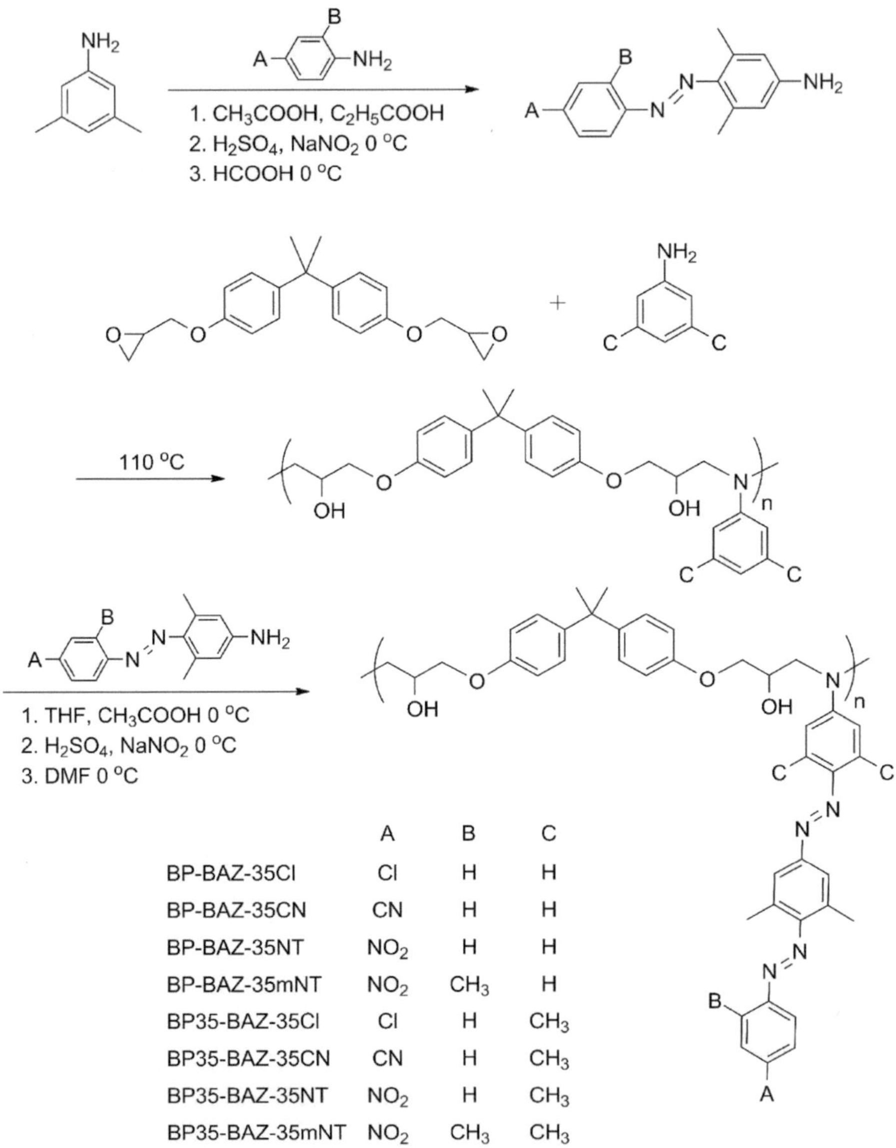

	A	B	C
BP-BAZ-35Cl	Cl	H	H
BP-BAZ-35CN	CN	H	H
BP-BAZ-35NT	NO_2	H	H
BP-BAZ-35mNT	NO_2	CH_3	H
BP35-BAZ-35Cl	Cl	H	CH_3
BP35-BAZ-35CN	CN	H	CH_3
BP35-BAZ-35NT	NO_2	H	CH_3
BP35-BAZ-35mNT	NO_2	CH_3	CH_3

Fig. 5.2 The synthetic route and chemical structure of the aminoazobenzene derivatives and bisazo polymers (Reprinted with permission from Ref. [9]. Copyright (2011) Elsevier)

Although the grating formation efficiency is closely related with the chromophoric and polymeric structures, SRGs can be inscribed on films of many types of azo polymers upon irradiation with proper light waves. In addition to the spin coating method, films suitable for inscribing SRGs can be obtained by other film fabrication methods depending on the processing properties of the materials. Even electrostatic layer-by-layer adsorption can be used to prepare the films [17–19]. It means that the SRG formation is one of the fundamental characteristic of azo polymers and other azobenzene-containing materials.

5.1.2 *Optical Setup for SRG Inscription*

The necessary experimental setup to inscribe SRGs includes a light source and interferometric apparatus. To match the absorption wavelength of the azo polymers, an Ar^+ laser (488 or 514 nm) and a diode-pumped frequency-doubled solid-state laser (532 nm) are frequently used as the light source. As discussed below, the polarization state of the interfering laser beams is an important factor for efficiently inscribing SRGs. Therefore, polarization states of the interfering laser beams need to be precisely controlled. Two types of optical setup, Lloyd mirror and two-arm interferometer (Fig. 5.3), have been typically used to write SRGs [29].

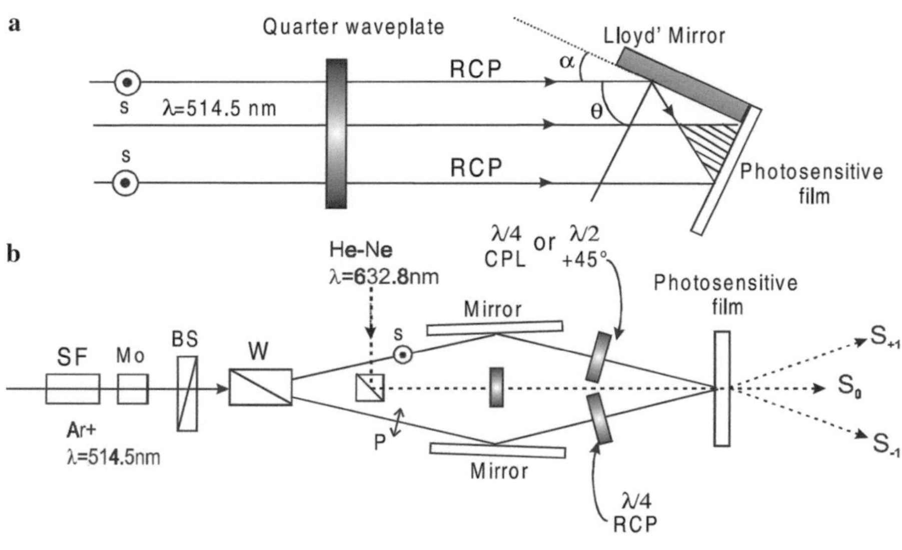

Fig. 5.3 Optical setup used for grating inscription. (**a**) Lloyd mirror setup. (**b**) Two-arm interferometer setup. The terms SF, Mo, BS, and W stand for spatial filter, microscope objective, Babinet–Soleil compensator, and Wollaston prism, respectively; the symbols λ/2 and λ/4 mean half- and quarter-wave plates, respectively (Reprinted with permission from Ref. [29]. Copyright (2002) Springer)

The Lloyd mirror setup has been used by the original investigators in this area [1, 2]. The Lloyd setup uses a mirror placed at right angles to a thin polymer film to write the gratings. A laser beam with proper wavelength is spatially filtered, expanded, and collimated. A half of the collimated laser beam is incident on the film directly, and the other portion of the beam is reflected from the mirror onto the film to form the interference pattern. The most obvious advantage of this simple setup is the robust against the perturbations, which is very important for recording SRGs over a long time period. On the other hand, the polarization state for the half beam reflected from the mirror might be modified. At a large angle of incidence, an initially right circularly polarized (RCP) beam becomes elliptically polarized with an azimuth of nearly $+45^{\circ}$ [29]. As the SRG formation and growth efficiency are dependent on the polarization state, to account for the polarization influence is not always straightforward and the interpretation becomes ambiguous in some cases.

The two-arm interferometer setup was also adopted in the early stage of the SRG study [11] and has been widely used in the later studies. In the two-arm interferometer setup, a laser beam is expanded, spatially filtered, and split into two by a polarization beam splitter. The two emerging beams, whose polarization states can be separately adjusted by half- and quarter-wave plates, are thus recombined onto the sample with a suitable incidence angle. The two-arm interferometer setup is typically applied in optical laboratories for recording holography. For SRG investigations, one of the advantages of the setup is that the polarization state of each beam can be precisely and separately altered, and the interpretation is relatively simple. However, if a long-time exposure is needed to write SGRs, the setup might suffer from the deteriorated quality due to the small vibration and other perturbations.

Besides using a continuous wave (cw) laser as the light source in most investigations, mode-locked lasers with the pulse of very short duration have also been used to inscribe SRGs [30]. In this way, a frequency-doubled yttrium–aluminum–garnet (YAG) laser operating at 532 nm is typically used as the light source, which has Q-switched pulses of 5–7 ns duration at a 20 Hz rate. With the energy of 10 mJ and a peak output of 1.6 MW, SRG is obtained just after exposure of the film of the cyanoazobenzene side-chain polyester to one 5 ns pulse. However, for using a nanosecond pulse to write SRG, effects with thermal origin might play an important role in inducing the gratings [31–34].

Moreover, in addition to the typical two-beam interference technique with the excitation wavelength corresponding to the absorption band, the biphotonic holographic recording method has been used to inscribe SRGs [35, 36]. SRGs are recorded by irradiating the films with blue incoherent light and simultaneous illumination with the red light pattern generated by the interference of two properly polarized beams from a He–Ne laser. SRGs can be recorded with simultaneous illumination of blue light and the intensity-modulated light pattern of the red light. However, the surface modulation generated by this method is obviously smaller compared with conventional method discussed above.

5.1.3 SRG Characterization

SRG formation can be characterized by different methods, which characterize the grating parameters (grating amplitude and period), surface profile, and dynamic process for the grating formation. Recently, more sophisticated methods have been developed to investigate the mechanism of the surface modulation. Two routine methods to characterize the topographic morphology and the SRG formation process are atomic force microscopy (AFM) and diffraction efficiency measurement.

Atomic force microscopy (AFM) is most widely used to investigate the SRGs, which can give the information about the grating parameters and surface profile in a straightforward manner. The first unambiguous evidence of the SRG formation was supplied by the AFM observation [1, 2]. On the other hand, as the AFM observation is usually carried out in an ex situ manner, it cannot monitor the dynamic process of the grating formation in most conventional cases. Recently, a new setup combining an atomic force microscope and two-beam interferometry has been developed [37]. With this approach, it is possible to detect the SRG topography change in situ.

The dynamic process of grating formation is usually monitored with the diffraction efficiency measurement. Typically, using a low-power He–Ne laser at 633 nm, the power of the diffracted beam is recorded during the writing process. To obtain the diffraction efficiency, the time-dependent intensities of the transmitted zero order and the diffracted first +1 order beams are detected simultaneously using two identical photodiodes. The diffraction efficiency is then calculated by their ratio and plotted versus the irradiation time. Diffraction by gratings has been considered as a classical problem in the general scope of light–matter interaction. Although the diffraction efficiency can be readily detected by a typical optical setup, the interpretation of the results is not always straightforward.

Most theoretical analyses deal with a plane wave incident upon a planar grating [38]. The terminology, thin and thick gratings, has been widely used to classify the diffraction behavior [39]. A thin grating is a grating that produces Raman–Nath diffraction. In this case, multiple grating diffraction orders can be observed, and the diffraction efficiencies η_i are described by

$$\eta_i = J_i^2\left(\frac{2\pi\Delta nd}{\lambda}\right) \tag{5.1}$$

where i is the integer representing the diffraction order, J_i is an integer-order ordinary Bessel function of the first kind, Δn is the refractive index modulation amplitude, and d is the grating thickness [38]. A thick grating is described as a grating that produces Bragg regime diffraction. It can be efficiently probed only if the Bragg condition

$$\boldsymbol{k}_m - \boldsymbol{k}_c = m\boldsymbol{q}\,, \qquad m = 1,\ 2,\ 3 \tag{5.2}$$

is satisfied, where $\boldsymbol{k}_{\mathrm{m}}$ and $\boldsymbol{k}_{\mathrm{c}}$ are the wave vectors of the diffracted wave and probe wave and $\boldsymbol{q}$ is the grating vector. A thin grating can be defined by $d/\Lambda < 10$, where d and Λ are the grating thickness and period. For more sophisticated analysis of the diffraction efficiency for SRG with arbitrary surface profiles, we refer the interested reader to the literature [40, 41]. Although the diffraction efficiency detected in a real-time manner can supply valuable information related to the grating growth process, the optical parameters of gratings cannot be unambiguously determined by the measurement. Therefore, it is usually necessary to monitor the grating growth with atomic force microscopy at the same time.

5.2 Factors Affecting SRG Formation

The SRG formation behavior is significantly affected by the polymer structures and optical parameters. The SRG formation can be characterized by the grating growth rate and saturated amplitude, which are directly related to the mass transport driven by the light. In many cases, the diffraction efficiency measurement is used to obtain the information in a real-time manner. The influences of different factors on SRG formation have been intensively investigated to understand the nature of this light-driven effect and to obtain materials with better properties. However, not all of these factors have been strictly compared in a side-by-side way, only some comparable results are discussed in this section.

5.2.1 Influences of Molecular Structures

It was observed in the early stage of the SRG investigations that the azo chromophores are necessary to produce the SRGs with large surface modulation [10, 42]. Polymers with different degrees of functionalization (DF) have designed and synthesized to understand the effect of DF on the SRG formation [43, 44]. Figure 5.4 shows relationship between the surface modulation amplitude and degree of functionalization. The effective SRG inscription rate increases with the loading density of azo chromophores. It indicates that the SGR formation is driven by light through the interaction with azo chromophores. On the other hand, a critical value of the DF is also observed for two series of azo polymers. Above the critical values, the SRG growth rate is independent of DF [43], and the surface relief even decreases with the DF [45]. The saturation of the inscription rate with the DF increase has been attributed to the decrease of the light penetration depth for films with the high chromophore density [43].

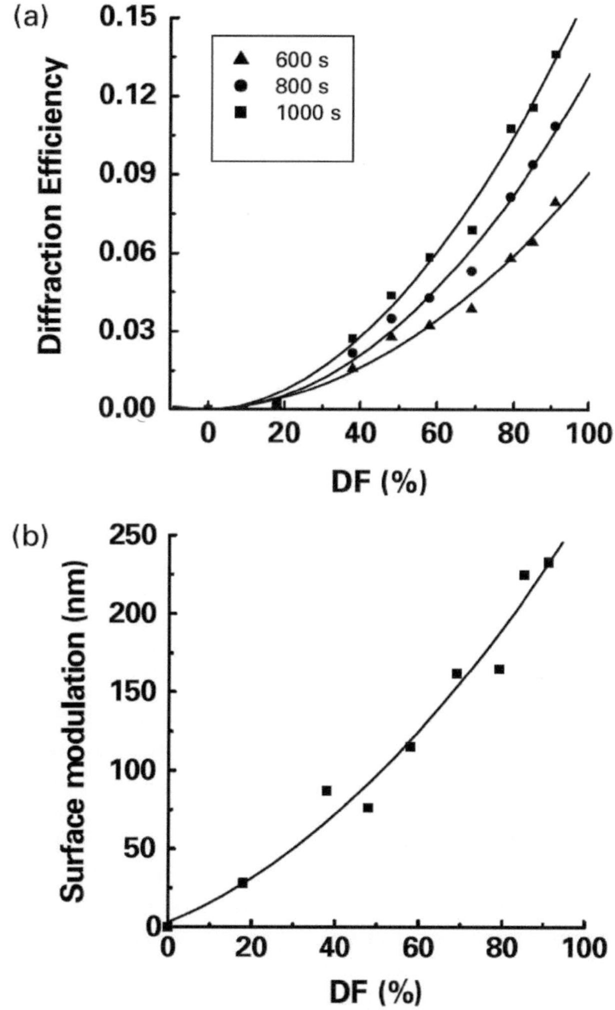

Fig. 5.4 Relationship between surface modulation amplitude and degree of functionalization for BP-AZ-CA (see Fig. 5.6 for the polymer structure): (**a**) measured from the diffraction efficiency experiments at 600, 800, and 1000 s; (**b**) measured from AFM images after 1000 s irradiation (Reprinted with permission from Ref. [44]. Copyright (2002) Elsevier)

The SRG formation behavior of three acrylate-/methacrylate-based azo polymers, PMEA, PDR1A, and PDR13A, has been investigated [42]. The main dissimilarity of the azo polymers is the bulkiness of the azo chromophores. The chemical structure and efficiencies of the SRG formation of the polymers are shown in Fig. 5.5. The sizes of the azo chromophores with these substituents were estimated by using molecular mechanics software to give the van der Waals occupied volumes of the isomerized azobenzene of 65 Å^3 for PDR13A, 34 Å^3 for PDR1A, and 23 Å^3 for PMEA. This observation suggests that the increase in free volume required for the photoisomerization is in the order, PDR13A (with the chloro- and nitro-substituted benzene ring) over PDR1A (with the nitrobenzene ring) over

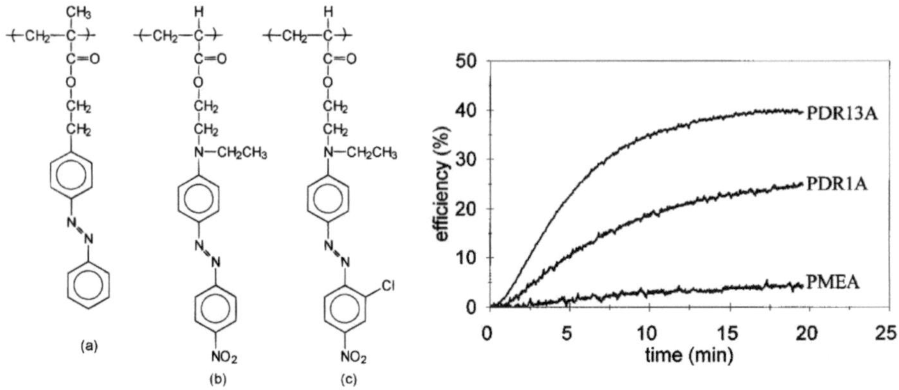

Fig. 5.5 Diffraction efficiency vs time for polymers PDR13A, PDR1A, and PMEA (Reprinted with permission from Ref. [42]. Copyright (1996) American Chemical Society)

PMEA (with no ring substituents). It is the same order as the efficiencies of the SRG formation of the polymers.

Investigation has been conducted on the correlation of the SRG formation behavior and the electron-withdrawing groups on the azo chromophores. As discussed in Chap. 2, the electron-withdrawing group has a significant effect on the absorption spectrum of the azo chromophores. Although most *push–pull*-type chromophores possess the amino group as the electron-donating group, the $\pi - \pi^*$ absorption band position can be adjusted in a wide wavelength range by using different electron-withdrawing groups. The post-polymerization azo-coupling scheme, discussed in Chap. 3, can be used to introduce the azo chromophores bearing different types of the electron-withdrawing groups. The degree of polymerization (DP) and the structures in other part of the polymers are the same by using the same batch of the precursor polymers. It has been clearly demonstrated that the electron-withdrawing group has a significant effect on the SRG formation. This effect of the electron-withdrawing group has been observed for different types of azo polymers [9, 23, 44, 46, 47].

A typical case is given here for the discussion, which is a comparison between three epoxy-based azo polymers, BP-AZ-CA, BP-AZ-CN, and BP-AZ-NT, bearing the carboxyl, cyano, and nitro as the electron-withdrawing groups on the azobenzene moieties [44]. Upon irradiation with interfering Ar^+ laser beams, the polymers containing nitro and carboxy as the electron-withdrawing groups exhibit the slowest and fastest rates for SRG formation among the three polymers (Fig. 5.6). Surprisingly, although the λ_{max} of BP-AZ-NT (482 nm) is close to the wavelength of the Ar^+ laser (488 nm), the SRG inscribing rate is obviously slower than the rates of BP-AZ-CN and BP-AZ-CA, which have $\lambda_{max} = 449$ and 431 nm, respectively. This observation contradicts the simple inference that irradiation at λ_{max} could maximize the inscription rates. DSC analysis shows that BP-AZ-CA containing carboxyl as the electron-withdrawing group has a considerably higher T_g (160 °C) compared with those of the other two polymers containing nitro or cyano

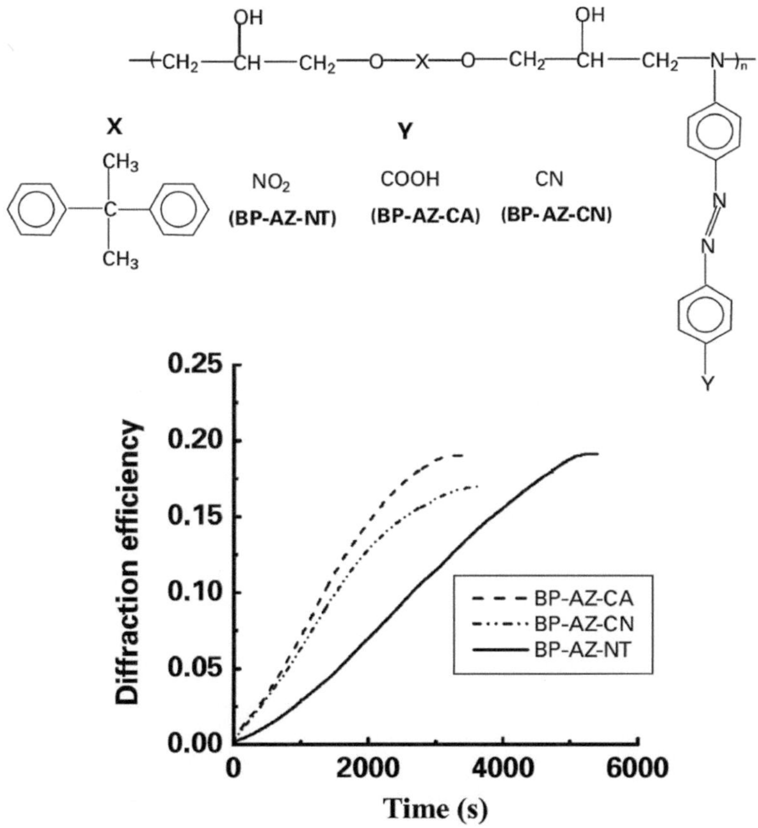

Fig. 5.6 Diffraction efficiency as a function of the irradiation time for azo polymers BP-AZ-CA, BP-AZ-CN, and BP-AZ-NT (Reprinted with permission from Ref. [44]. Copyright (2002) Elsevier)

as the electron-withdrawing groups (138 and 132 °C). It indicates that SRG formation rate is not directly correlated with the T_gs of the polymers.

Other building blocks of the molecular structures also show influences on the SRG formation. However, as SRG formation has been observed for azo polymers with diversified structures, it is difficult to directly compare the properties among them. Only those investigations, which have purposely designed to compare a specific structural factor, are justified for the comparison. Some of the studies show that the azo polymers with a more flexible backbone show a higher rate for the SRG formation [44, 47]. It should be related to the mobility of the polymer chain upon the light irradiation. For side-chain azo block copolymers, the higher SRG formation rate is observed for the polymer with short flexible spacer between the backbone and azo chromophore [48].

5.2.2 *Roles of Optical and Grating Parameters*

Optical parameters of the writing laser beam have important effects on the SRG formation, which include the wavelength, intensity, polarization, spot size, and intensity distribution. If short laser pulses are used to inscribe the SRGs, the instantaneous intensity and duration are parameters needed to be considered. As SRGs are inscribed by two-beam interference, the intersection angle is an important parameter to determine the grating periods. As discussed in Chap. 4, superposition of two beams with different polarizations can modulate both the intensity and polarization of the interference field, which show a significant effect on the SRG formation.

The effect of the excitation wavelength has been studied by using lasers with different wavelengths [9, 46, 49]. The excitation wavelength shows an important influence on the SRG inscription rate, which is related to the spectral location of the absorption bands of azo chromophores. This kind of studies can supply a complementary picture to the study on the influence of the electron-withdrawing groups discussed above. Generally, when the excitation wavelength is located between λ_{max}s and the absorption band tails at the low-energy side, the azo polymer films show the higher SRG growth rate compared with irradiation by light wave with a higher frequency. On the other hand, when the excitation wavelength approaches the absorbance band tail at the long wavelength side, the SRG formation rate is significantly reduced. This observation can be rationalized by considering the relative efficiency of the excitation at $\pi - \pi^*$ and $n - \pi^*$ bands to cause the *trans–cis* isomerization [49]. As discussed in Chap. 2, the *trans* → *cis* quantum yield is always larger for the low-energy ($n - \pi^*$) excitation compared to that obtained from the high-energy ($\pi - \pi^*$) excitation. For the pseudo-stilbene-type chromophores, the low-intensity $n - \pi^*$ band is buried beneath the intense $\pi - \pi^*$ band at the longer wavelength side, which is closer to the more efficient excitation wavelength.

Light intensity is an important parameter to determine the inscribing rate and maximum relief amplitude. In the medium light intensity range (below 100 mW/cm^{-2}), both the inscribing rate and saturated relief amplitude increase with the light intensity increase [2, 42]. The grating growth rate, which is defined in terms of the slope of the curves of the diffraction efficiency versus irradiation time, shows a linear increase with the light intensity [50]. Due to the possible side reaction and decomposition, the intensity and the exposure time have some limitation, which depends on the polymer structures and other conditions. The development of the grating has been followed by AFM for 20 h by repeatedly scanning the same area of the film [11]. The grating structure disappears after a few hours and surface roughness significantly increases in the process.

The amplitude of the surface modulation is related to the grating periods, which can be readily adjusted by changing the angle between the two writing beams. If the Lloyd mirror setup is used, grating period (Λ) is determined by the equation:

$$\Lambda = \frac{\lambda}{2\sin\theta} \tag{5.3}$$

where λ is the wavelength of the writing beam and θ is the angle between the beam propagation axis and the mirror plane, which is referred to as α in Fig. 5.3. The grating period corresponding to the highest value of the surface modulation was reported to be in a proper range, such as 800 nm [10] and 940 nm ($\theta = 15°$) [42].

The SRG formation is strongly dependent on the polarization of the writing beams [10, 12, 42]. The grating formation rate and efficiency are closely correlated with the polarization forms. The interferometric field distribution with the different polarizations of the interfering beams has been adopted for SRG inscription to understand this effect [51, 52]. According to the studies, the existence of the spatial variations of both intensity and direction of the resultant electric field vector in the films is an essential condition to efficiently induce surface modulations. The superposition of beams with the suitable polarizations such as *p*-/*p*-polarized and the left/right circularly polarized (LCP/RCP) is typically used to inscribe the surface-relief-gratings.

5.3 Correlation with Polarization Grating

As discussed in Chap. 4, irradiation with polarized light causes the orientation of azo chromophores and adjacent segments. Therefore, upon exposure to the interference patterns of the polarized light, the photoinduced dichroism and birefringence will result in the bulk polarization grating in the azo polymer films. As the spatial variation of both intensity and polarization direction is a necessary condition to efficiently inscribe SRGs, photoinduced anisotropy will inevitably occur during the SRG inscription process. Exposure to the light interference pattern can cause two dominant processes among others induced by the irradiation. The first process is the light-induced orientation of azobenzene moieties perpendicular to the polarization plane of the incident light, which is accompanied by the correlated alignment of adjacent groups of the polymer chains. The second process is the mass transport along the grating vector to form SRG. These two processes result in the generation of the volume polarization grating coincident with the SRG formation.

The diffraction efficiency measurements have been used to identify these two processes, i.e., the formation of the birefringence grating and SRG [12, 53, 54]. By polarization analyses of the (± 1) first-order diffracted beams, the contributions to the diffraction efficiency are separated into a birefringence part and a surface relief part. In the analyses, the Jones' matrix method is used. As the probe beam passes first through the SRG and then through the polarization grating, the total Jones' matrix of the system is given as

$$T = T_{\text{anis}} T_{\text{SRG}} = e^{i\varphi_0} \left\{ T_0 + \sum_{k=\pm 1}^{\pm\infty} e^{\pm ki\delta} T_{\pm k} \right\} \tag{5.4}$$

where T_{anis} and T_{SRG} are the Jones matrices of the polarization grating and SRG and T_0 and T_{k} are the transmission Jones' matrices of the zero-order transmitted beam and of the kth-order diffracted beams. The diffracted outcome is computed by multiplying the Jones' matrix by the Jones vector corresponding to the incident beams with different polarizations. The analyses show that during the growth of the gratings, both contributions show very different time scales for their dynamical responses. The polarization grating formation is completed in a much shorter time period (several seconds) compared with that to inscribe SRG with a large amplitude (hundreds to thousands second). The viscoelastic flow of the polymer involved in the SRG formation can partly erase the birefringence grating formed in the early stage [53].

The time variations of the phase shifts between the polarization grating and SRG can be generated during the inscribing process (Fig. 5.7). A nonstationary phase shift $\Delta\varphi(t)$ between the bulk polarization grating $\Delta n(t)$ and SRG $\Delta d(t)$ has been studied to understand diffraction efficiency as a function of time during the recording process [55]. The origin of the nonstationary phase shift is attributed to a slow shift of interference pattern due to delicate symmetry breaking under illumination conditions. Two most possible cases are the difference in beam intensities and deviation of exact symmetrical beam incidence angles on the sample. Without the change of the amplitudes of refractive index and relief gratings, different dynamic responses can be observed when the $\Delta\varphi(t)$ changes from 0 to π (Fig. 5.8).

Later, the complex diffraction efficiency variation during SRG formation has been attributed to the three-grating coupling [56]. Two of gratings originate from the refractive index changes in the bulk of material and the third one from the surface relief modulation. This result indicates that extra care must be taken to properly interpret the diffraction efficiency results. On the other hand, the diffraction efficiency from SRG is much stronger and dominant for the grating with a

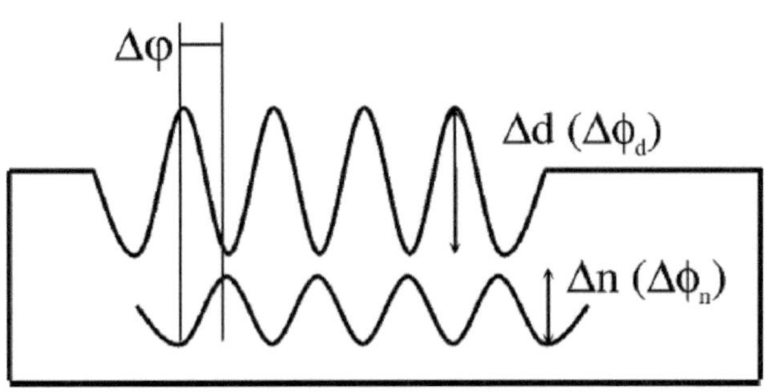

Fig. 5.7 Schematic illustration of the sinusoidal bulk refractive index Δn, the surface-relief-gratings Δd, and related phase parameters, $\Delta\varphi_{\text{n}}$ and $\Delta\varphi_{\text{d}}$. The phase shift $\Delta\varphi$ between the gratings is also indicated (Reprinted with permission from Ref. [55]. Copyright (2007) American Chemical Society)

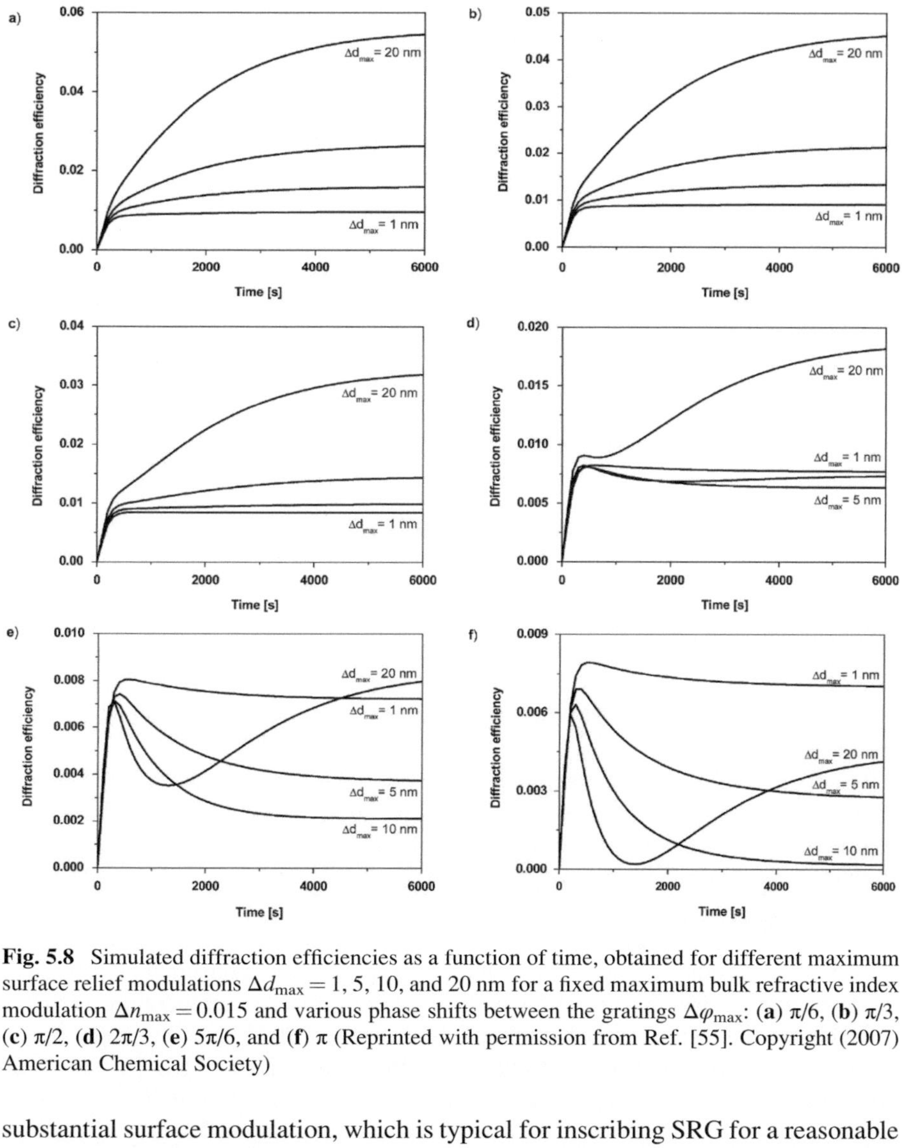

Fig. 5.8 Simulated diffraction efficiencies as a function of time, obtained for different maximum surface relief modulations $\Delta d_{\max} = 1, 5, 10$, and 20 nm for a fixed maximum bulk refractive index modulation $\Delta n_{\max} = 0.015$ and various phase shifts between the gratings $\Delta\varphi_{\max}$: (**a**) π/6, (**b**) π/3, (**c**) π/2, (**d**) 2π/3, (**e**) 5π/6, and (**f**) π (Reprinted with permission from Ref. [55]. Copyright (2007) American Chemical Society)

substantial surface modulation, which is typical for inscribing SRG for a reasonable exposure time. Therefore, the diffraction efficiency observed for most SRG inscribing processes can be considered as the contribution mainly from the SRG.

5.4 Models and Theories

Since the SRG formation was first observed more than 20 years ago, the puzzling issues related to its mechanism have inspired intensive investigations through different approaches. Especially, two most important issues among others have aroused tremendous research enthusiasm. The first is about the driving forces that cause the mass transport and SRG formation. Although the light and matter interaction has been consistently investigated for many years, it can hardly find an available model that is appropriate to explain this unprecedented observation. The second is how polymer chains can be moved in a micron-scale below the glass transition temperature (T_g) of bulk polymers. It is well documented in polymer textbooks that the segments of a polymer chain can undergo a migration in considerable scale only when temperature is above T_g of the polymer. The SRG formation, as an unusual phenomenon, seems to be against both instinct and conventional belief. This section will discuss some most well-known models and theories as the framework for further explorations.

5.4.1 Thermal Gradient Model

A preliminary explanation proposed at the early stage investigation is that the SRG formation is a result of mass diffusion of the polymer chains caused by the temperature gradient [1]. According to the explanation, supposed the rate of diffusion is proportional to the temperature gradient, the resulting profile should be proportional to the second derivative of the light intensity. It can temporarily explain the observations on most azo polymers, the light irradiation results in a sinusoidal height profile 180° out-of-phase with the interfering patterns of the excitation beams.

However, later studies proved that this model is not consistent with the results obtained from the investigations using the low-power laser irradiation. The large amplitude SRGs can only be formed on polymer films that contain a substantial amount of photoisomerizable azo chromophores [10, 42]. To understand the role of thermal effects in the SRG formation process, a cellular automaton simulation has been employed to model heat flow in thin films undergoing laser irradiation [57]. As shown by the simulation and experiments, for a typical light intensity of 50 mW/cm^2, the temperature rises on the order of 5 K in the film. The temperature gradient between the light maxima and minima is stabilized at 10^{-4} K within 2 μs. These results indicate that thermal effects play a negligible role during SRG formation for films of any thickness. This point is further confirmed by monitoring SRG formation on substrates with different thermal conductivity, which shows that grating inscription is insensitive to film temperature.

On the other hand, for irradiation with pulse laser, the controversy still exists over the role of the thermal gradient during the SRG formation. The above simulation suggests that the high-intensity pulsed irradiation can only lead to destructive effects and sample ablation, but not to reversible optical mass transport. On the contrary, shown by other studies, relief formation in pulsed holographic experiments could arise from thermal effects [31–34]. Instead of the mechanism considering the diffusion directly caused by temperature gradient, the inhomogeneous force density is attributed to the permittivity changes caused by the temperature variation [32]. However, no matter the scale of the temperature gradient, the thermal effect model seems to be inconsistent with the result that the SRG formation is strongly dependent on the polarization of the writing beams [10, 12, 42]. Therefore, to explain the SRG formation inscribed by the polarized low-power laser irradiation, other models and theories need to be considered.

5.4.2 Isomerization-Driven Free Volume Expansion Model

This model considers the free volume expansion as the driving force to cause the SRG formation [42, 50]. For SRG inscription, the interference of the two laser beams generates fringe regions with different light intensities, where the high *trans–cis–trans* isomerization corresponds to the phase addition regions bordered by regions of low isomerization. The free volume expansion is caused by the isomerization to the *cis* geometry and the dissipation of the large amount of thermal energy of the *cis* isomer. The free volume expansion has two effects necessary for the SRG formation. First, the expanded free volume is required for polymer segments and chains to migrate. Second, the volume change will cause the pressure increase in the polymer film. The pressure gradients cause the viscoelastic flow from these high-pressure areas to lower-pressure areas, which lead to the formation of the regularly spaced sinusoidal surface-relief-gratings. The pressure experienced in bulk is supposed to be proportional to the relative volume change as

$$P = B\Delta V \tag{5.5}$$

where B is experimentally determined to equal to 2×10^9 Pa.

Based on Newtonian fluid dynamic relations, this model has been formulated to quantitatively describe laser-induced mass transport in thin films of polymers containing isomerizable azo chromophores [50]. The Navier–Stokes equations for laminar flow of a viscous fluid are developed to relate velocity components to pressure gradients in the polymer film under the defined boundary layer conditions. The polymer is assumed to behave as an incompressible viscous fluid whose motion under stress or pressure follows the Navier–Stokes equations. By selecting Cartesian coordinate with x-axis parallel to the grating vector and y-axis perpendicular to substrate (Fig. 5.9), the reduced Navier–Stokes equation is given:

$$\frac{\partial^2 v_x}{\partial y^2} = \frac{1}{\eta}\frac{\partial P(x)}{\partial x} \tag{5.5}$$

where v_x is the velocity component along x, η is the kinematic viscosity, and $P(x)$ is the pressure. Considering the mass conservation, the change in the surface of film ($y = h$) with time t upon irradiation can be given as

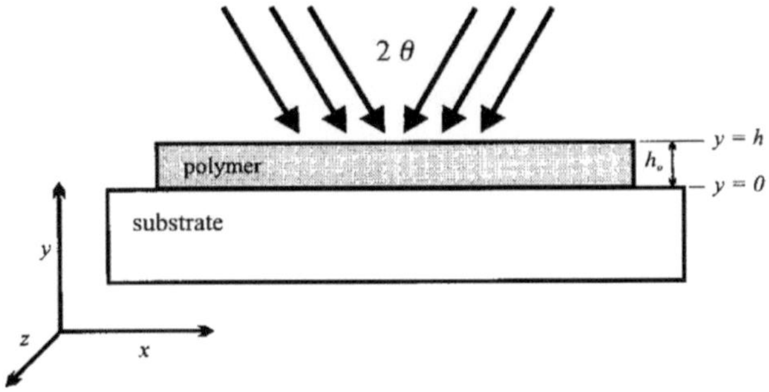

Fig. 5.9 Description of the grating inscription geometry and coordinate (Reprinted from Ref. [50] with the permission of AIP Publishing)

$$\frac{\partial h}{\partial t} = \frac{1}{3}\frac{h^3}{\eta}\frac{\partial^2 P(x)}{\partial x^2} \tag{5.6}$$

Solving this differential equation gives an expression for the time-dependent film thickness with the initial film thickness (h_0) as a parameter:

$$\frac{1}{h^2} = \frac{1}{h_0^2} - \frac{2}{3\eta}\frac{\partial^2 P(x)}{\partial x^2}t \tag{5.7}$$

The above model includes two related parts covering different scopes. The general laminar flow part based on Navier–Stokes equation is general, which is applicable to the formation of surface gratings through a variety of mechanisms. On the other hand, the specific mechanism, considering an isomerization-driven free volume expansion to produce internal pressure gradients, corresponds to a specific model developed to describe polymer flow resulting from laser-induced isomerization of the bulky azo chromophores.

Equation 5.7 gives an expression connecting the time evolution of the surface gratings to various preset or experimentally determined conditions and parameters, such as the optical setups of the SRG inscription and polymer properties. The correlations established by the model can be tested by experiments. The main predictions of the theory include the rate of grating inscription linearly increases with the intensity of the inscription laser, varies inversely with the molecular weight of the polymer below the limit of entanglement, and scales with the third power of the initial thickness of the film. Predictions from the model have been tested against the results of experiments, which show good agreement in most cases. As shown in Fig. 5.10, the cubic dependence appears to be followed for thin films. However, as the critical absorption limit is approached, the absorbance of the material becomes appreciable, and the thin film assumption is no longer valid as there is no longer uniform irradiation throughout the film along y. The inscription rate levels off as shown by the experimental results. One relationship that is in poor agreement with the model prediction is the plot of grating efficiency as a function of the intersection angle.

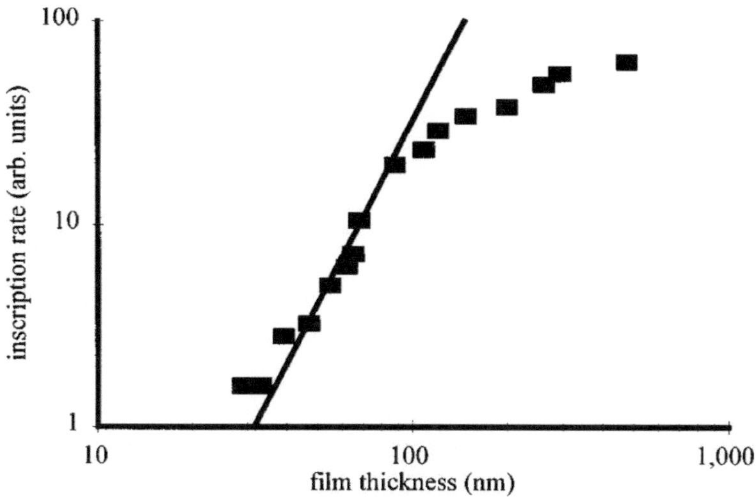

Fig. 5.10 The calculated grating efficiency as a function of film thickness compared with the experimental result (Reprinted from Ref. [50] with the permission of AIP Publishing)

This fluid mechanics model has been improved by considering the depth dependence of photoinduced driving force and velocity distribution in the film surface [58]. In a realistic case, light intensity will exponentially decay beneath the film surface due to the strong absorption of the azo polymer film. Therefore, the body force decaying exponentially in the y direction is adopted to formulate the SRG dynamics, which is derived analytically as a function of film thickness and interference wave number. This model has been tested by comparing film thickness dependence of the diffraction efficiency observed experimentally with the theoretical value in a range of film thickness.

The isomerization-driven free volume expansion model can explain several key experimental results and make reasonable predictions. This viscoelastic flow model seems to be more consistent with the results from time-resolved X-ray and VIS scattering experiments [59]. The photoinduced expansion due to the isomerization of azo chromophore, which is required for the free volume expansion and to induce a local pressure in the material, has been confirmed by investigations with ellipsometry [60] and neutron reflectometry [61]. However, similar to the thermal gradient model, this model has not taken account of the important observation that the SRG formation is strongly affected by the polarization states of the writing beams [10, 12, 42].

5.4.3 Photoinduced Translation Diffusion of Azo Dyes

This model is based on the anisotropic diffusion of azobenzene dyes in polymer matrices, which is induced by their photoinduced *trans–cis* isomerization cycles

[62, 63]. The essential feature of the model is that the dye molecules undergo one-dimensional (1D) random walk along their excitation direction. The following assumptions are adopted to simplify the equation. (i) Rotation–translation coupling is neglected when the molecules are moving, i.e., translation does not imply rotation and vice versa. (ii) When a molecule has an orientation and goes through a *trans–cis–trans* cycle, it moves in a direction parallel to its former orientation. (iii) The time scale for bleaching or orientation redistribution is very short compared to that of diffusion. Consequently, during diffusion, the angular distribution n is independent of time.

Based on the above assumption, at a point M_0 (x_0, y_0, z_0), the photoinduced flux J of dye molecules going in the direction of positive z is given by the following equation:

$$JdS = -\iiint_{r,\ \vartheta,\ \varphi} \exp\left(-\frac{r}{L}\right) dN_S(r, \theta, \varphi) \tag{5.8}$$

where exp(−r/L) is the probability for an excited molecule not to collide during its migration from P to M_0 and (r, θ, φ) are spherical polar coordinates.

Considering a light intensity grating in the x, z plane whose grating wave vector ***k*** (=2π/Λ) is parallel to the z-axis, where Λ is the grating spacing period. Two typical cases of light polarization, parallel to z-axis (*p*-polarized) or parallel to x-axis (*s*-polarized), are considered in the model. The number dN_S, the molecules pointing toward M_0 in volume *dV*, is obtained for these two polarization states. Final diffusion equation is given as

$$\frac{\partial N}{\partial t} = -divJ + D_0 \Delta N \tag{5.9}$$

where D_0 is the spontaneous diffusion coefficient of the dye in a polymer matrix. By solving Eq. 5.9, $N = N(\boldsymbol{r}, t)$, which is the concentration of *trans* molecules at a point ***r*** and a time *t*, can be obtained.

This light-driven random walk model has been developed to consider the molecule dragged by moving chromophores [64]. The model assumes that a diffusion motion of the azo chromophores undergoes a one-dimensional random walk, where each random step follows an isomerizing absorption. When each azo chromophore is put in motion by light, the distance over which an azobenzene-containing polymer can be dragged is calculated. More recently, a more sophisticated version of photoinduced molecular diffusion model (PIMD model) has been developed [65, 66]. This model considers the motion of a dye molecule from a hole in the polymer matrix to another hole, which corresponds to the motion through free space available in the polymer matrix. Monte Carlo calculations based on PIMD model, which incorporates fundamental molecular dipole interactions with an illumination field, can accurately simulate the molecular transport and topography modifications in photoisomerizable azobenzene-containing polymer films.

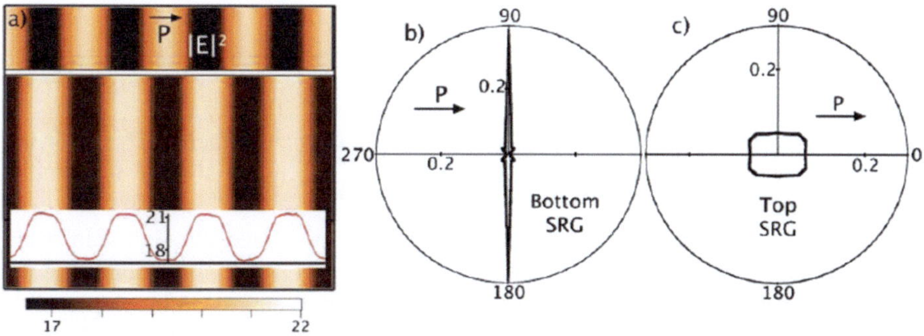

Fig. 5.11 SRG formation under p-type configuration; the incident intensity is shown in the top part of (**a**). The simulated topography in the bottom of (**a**) reveals a well-defined SRG. Along with the topography, molecular orientations were retrieved in (**b**) for the high-intensity regions (*bottom SRG*) and (**c**) for the low-intensity regions (*top SRG*) revealing strong photo-orientation (Reprinted with permission from Ref. [66]. Copyright (2009) American Chemical Society)

Figure 5.11 shows the SRG topography and molecular orientations obtained by the simulation.

5.4.4 *Electromagnetic Gradient Force Model*

This model attributes the SRG formation to the electromagnetic gradient force induced by optical field [67, 68]. The model takes account of most observations available at the time. These observations are summarized as: only azo polymers show large amplitude surface relief structures, laser beams with *s*-polarization cannot produce SRG with substantial surface modulation, the experimentally measurable surface profile is produced by edge diffraction, and the recording process is not a bulk process and needs an unconstrained surface. In this model, forces leading to the migration of polymer chains are generated by dipole interaction of azo chromophores with the gradient of the electric field present in the polymer films.

The polarization **P** induced in the polymer layer near the surface of the polymer film is considered to be a key parameter, which is assumed to be in the same direction as the incident fields. Therefore, the time-averaged optically induced gradient force density is given by

$$\begin{aligned} f &= \langle [\boldsymbol{P}(\boldsymbol{r},t)\cdot\nabla]\boldsymbol{E}(\boldsymbol{r},t)\rangle = \langle [\varepsilon_0\boldsymbol{E}(\boldsymbol{r},t)\cdot\nabla]\boldsymbol{E}(\boldsymbol{r},t)\rangle \\ &= \frac{1}{2}\varepsilon_0\chi'\boldsymbol{E}(\boldsymbol{r})\cdot\nabla\boldsymbol{E}(\boldsymbol{r}) \end{aligned} \tag{5.10}$$

where $< >$ represents the time average and χ and χ' are the polymer susceptibility and its real part.

The transport induced by the force is then given:

$$\boldsymbol{v}(x, y, z) = \mu \boldsymbol{f}(x, y, z) \tag{5.11}$$

where $\boldsymbol{v}(x, y, z)$ is the flow velocity and μ is a proportionality factor. The information about the material responsive behavior, such as polymer viscosity, is included in this parameter.

The surface deformation induced by the linearly polarized 2D Gaussian beam was analyzed by this model [68]. The film surface is taken as the x–y plane and the z-axis is pointing into the medium. Considering the polymer as an incompressible medium, the surface deformation induced by a linearly polarized Gaussian beam is correlated with the light intensity through the following equation:

$$S(x, y, t) = \int_0^t v_z(x, y, 0)dt' = \frac{1}{4} h \mu \varepsilon_0 \chi' \frac{\partial^2 I(x, y)}{\partial x^2} t \tag{5.12}$$

Here, h is the effective thickness of the mobile layer, and Cartesian coordinates x and y are in the film plane. Figure 5.12 depicts the 1D component of the optical intensity I(x), optical field $\boldsymbol{E}$(x), gradient d$\boldsymbol{E}$(x)/dx, and the gradient force $\boldsymbol{f}$(x), where the arrows indicate the force directions. It indicates that an optical field component in the direction of the optical field gradient is essential for the mass transport process. Figure 5.13 shows some typical AFM images of the surface deformation induced by the single laser beam irradiation. The prediction of above model is well consistent with the surface deformation profile induced by a Gaussian laser beam at low intensities [69]. The strong polarization dependence of the SRG formation can be well explained by this model.

The above study also indicates that in the higher-intensity regime, a different mechanism exists to dominate the surface deformation process [68]. Owing to this mechanism, two different types of topographic gratings can be simultaneously formed in the two-beam interference experiment. Based on the results from the single-beam experiments, the phase relationship between these surface-relief-gratings and the interference intensity pattern have been deduced. Several related models have also been developed considering the gradient forces. A single dipole model has been developed to take account of the interaction and dynamics of induced dipole with optical fields [70]. Moreover, the driving force has also been attributed to the Coulomb interaction between an optical electric field and its induced polarization charges [71]. The model includes not only the optically induced gradient force but also the force due to the spatial modulation of the refractive index.

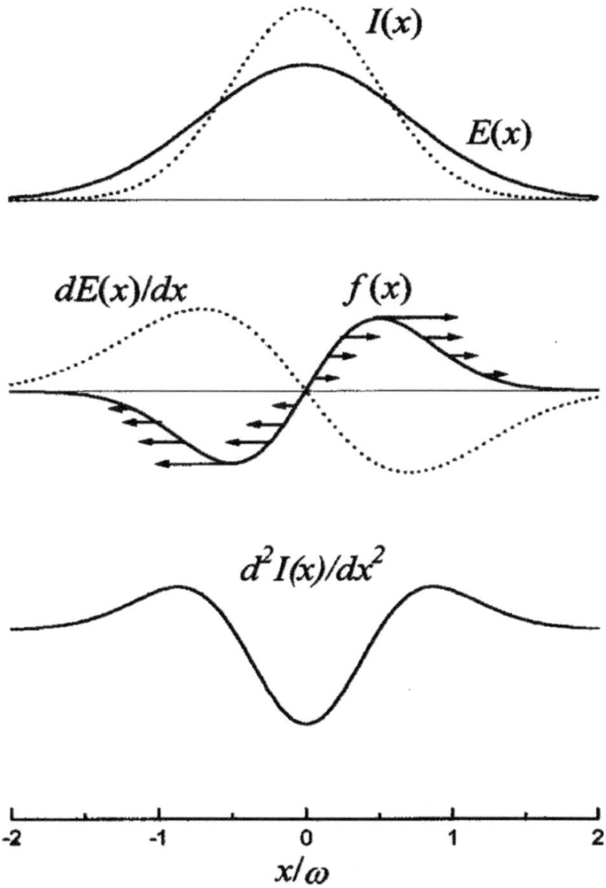

Fig. 5.12 The gradient force distribution and related optical field (Reprinted from Ref. [68] with the permission of AIP Publishing)

5.4.5 *Mean-Field Model*

This model attributes SRG formation to the photoinduced orientation of dye molecules, and mass transport is then induced by the anisotropic intermolecular interactions [72]. Figure 5.14 illustrates the recording setup together with the side-by-side and end-to-end geometries. Due to the side-by-side attractive interaction between the chromophores, the chromophores are supposed to shift to the area with higher orientational order. Therefore, even when the intensity of the incident light is spatially uniform, the mass transport can be induced by the polarization modulation. The model predicts that the chromophores will tend to migrate into regions characterized by chromophore orientation perpendicular to the grating vector induced by the light irradiation.

Considering the photoinduced alignment effect, as discussed in Chap. 4, the orientation order parameter S_{eff} is related to the ellipticity ε $(0 \leq \varepsilon \leq 1)$ of the polarized light through the equation:

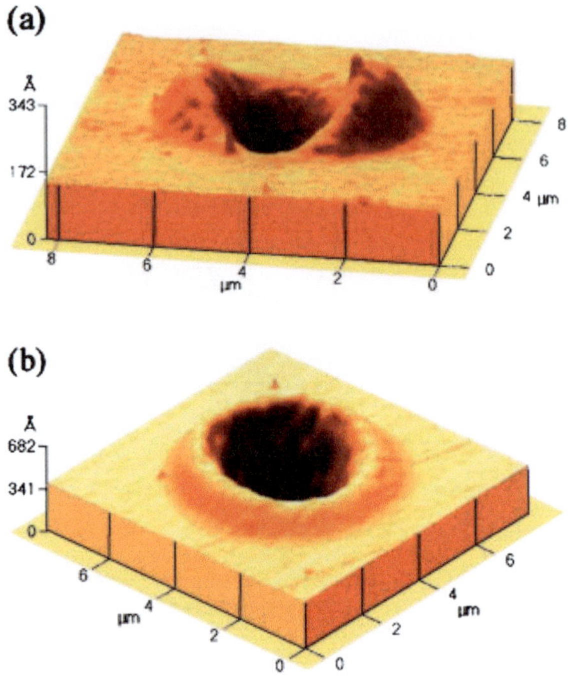

Fig. 5.13 Surface deformation induced by Gaussian beams ($\omega = 2.1\ \mu$m), (**a**) linear polarization: $I_0 = 238$ mW/cm^2, t = 50 min. (**b**) Circular polarization: $I_0 = 230$ mW/cm^2, t = 50 min (Reprinted from Ref. [69] with the permission of AIP Publishing)

$$S_{\text{eff}} = \varepsilon + \frac{1}{2}(1 - \varepsilon)|E_x| \tag{5.13}$$

where the E_x is the x component of the normalized electric field in the grating vector direction.

Due to this effect, the energy functional involving the relief height *h(x)* is given as follows:

$$\begin{aligned} F[h] = & \int_{-\infty}^{\infty} \left\{ \sqrt{1 + \left[\frac{dh(x)}{dx}\right]^2} - 1 \right\} dx \\ & - K \int_{-\infty}^{\infty} S_{\text{eff}}^2(x) h(x) dx + \alpha \int_{-\infty}^{\infty} [h(x) - d] dx \end{aligned} \tag{5.14}$$

In the above equation, the first term is the contribution from the surface tension, which is proportional to the line integral along x. The second term is the first-order contribution to the energy functional due to the mean-field interactions proportional to square of S_{eff}. The last term is included to ensure conservation of volume, α and d represent the Lagrange multiplier and the original film thickness, respectively.

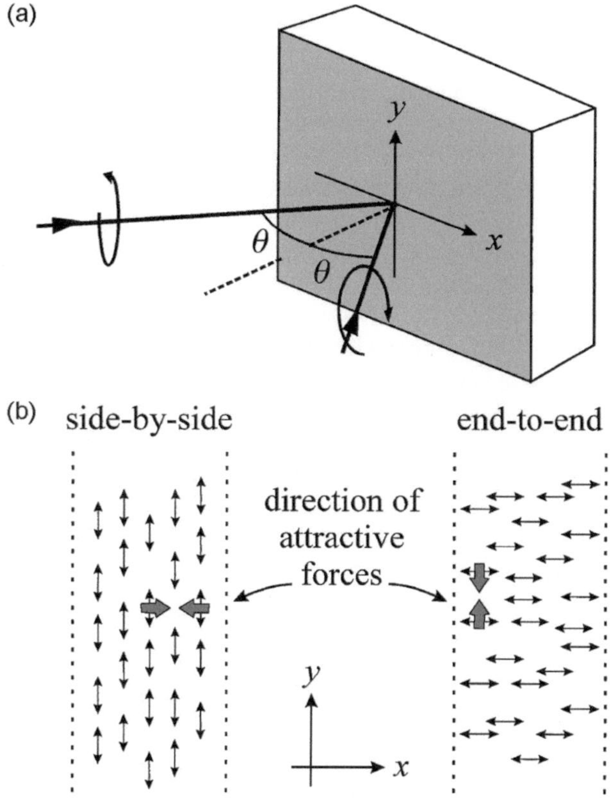

Fig. 5.14 (**a**) The recording setup. (**b**) Illustration of the side-by-side and end-to-end geometries. The attractive interaction is mainly between chromophores aligned side by side, and, hence, the attractive forces are directed along the x- and y-axes in the two cases, respectively (Reprinted from Ref. [72] with the permission of APS Publishing)

The unknown constant K expresses the relative strength between mean-field forces and surface tension.

By using standard variational calculus to minimize the energy functional, the expression of $h(x)$ as a function of S_{eff} can be obtained. The surface profiles of three polarizations, *s*-/*p*-polarized, ±45°, and RCP/LCP, have been calculated by using the model [72]. The results show the agreement with experimental SRGs recorded under these polarization conditions. Although this model has originally been developed for SRGs formed on liquid crystalline polymers, the authors indicate that it should be valid in the case of glassy polymers if photoinduced isomerization could provide the necessary mobility in glassy polymers. This model predicts that the chromophores will tend to migrate into regions characterized by chromophore orientation perpendicular to the grating vector induced by the light irradiation. Unlike the other models discussed above, this model claims the chromophores are transported from the low light intensity regions to high-intensity regions.

5.4.6 *Other Models and Theories*

The models and theories discussed above are well known and often cited by different authors to interpret and elucidate the experimental results. As the observations cannot always be satisfactorily explained, more models and theories considering different mechanisms have been developed. According to the models and theories, the SRG formation has been attributed to electrical field forces caused by the inhomogeneous spatial distribution of the refractive index and the permittivity in the illuminated films [73], statistical reorientation of side chains in the polymer by repeated *trans–cis–trans* isomerization processes [74], elastic deformation of the polymer due to interaction between dipoles ordered through polarized light irradiation [75], stretch along the polarization direction to compensate the entropy decrease produced by the photoinduced reorientation of azobenzene chromophores [76], and mechanical stress caused by the photoinduced orientation anisotropy of azobenzene moieties [77]. The formation and erasure of photoinduced surface-relief-gratings (SRGs) on azobenzene-containing polymers have also been simulated using a calculation model based on the moving particle semi-implicit method [78].

5.5 More Observations and Insights

Above models and theories have been established on the basis of the available experimental results obtained in different stages. On the other hand, predications made by the models and theories have stimulated more extensive investigations on the mechanisms and general understanding. The exact mechanism of the SRG formation is still a controversial issue in many aspects. In recent years, more experimental methods have been established to reveal the different facets of the mechanism. This section briefly discusses some of the results.

5.5.1 *Concomitant Bulk Variations*

SRG formation is accompanied by other light-induced bulk variations occurring during the process. Understanding those processes can supply valuable information to fully elucidate the SRG formation mechanism. As discussed in Sect. 5.3, when exposed to the interference pattern of the polarized light, polarization gratings are formed during the SRG formation. Confocal Raman microscopic study indicates that the variations in diffraction efficiency observed after light irradiation for a short time are not coming from the SRG but from a significant increase in the birefringence [79]. In this initial stage, the azo chromophore distribution functions become highly anisotropic caused by the light irradiation.

X-ray scattering measurements show that buried density gratings are formed during the SRG formation process. The buried density grating is another type of modulated volume structure below polymer film surface formed during the SRG formation process. The formation of a density variation grating was first observed during the thermal erasure treatment of the SRG [80, 81]. After erasure of the SRG, an intrinsic density grating is formed in the azo polymer film with smooth surface. Figure 5.15 illustrates the mass density distribution before and after erasure. The authors indicted that at the density grating (B), the created density difference might be caused by a liquid crystalline arrangement of mesogenic groups within the high-density regions. Later studies with in situ time-resolved coherent X-ray and visible

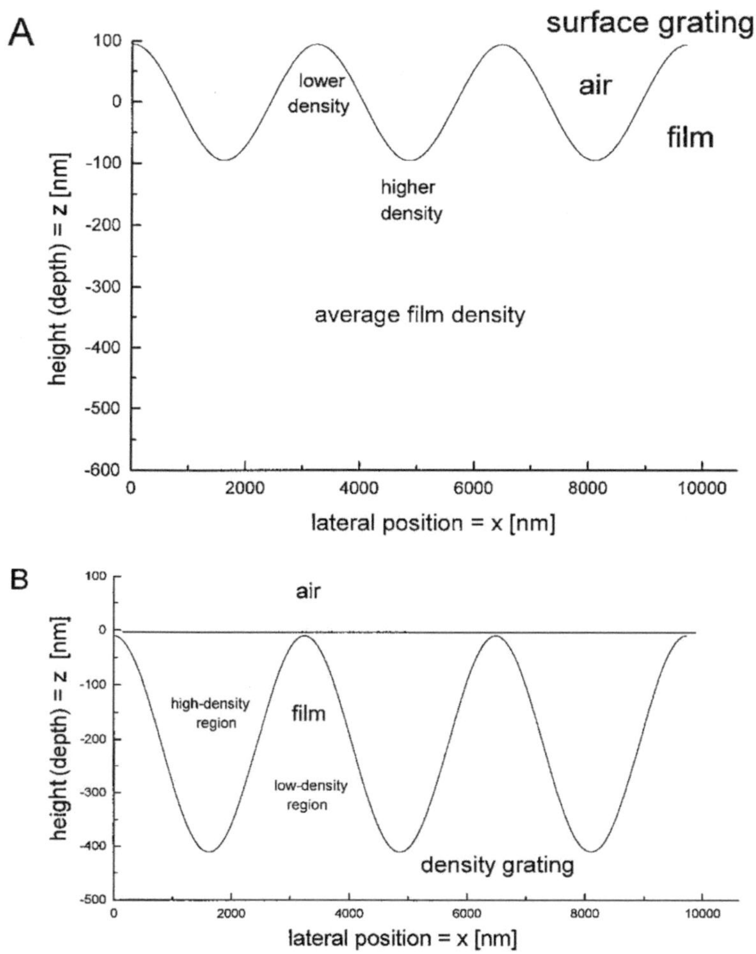

Fig. 5.15 Scheme of the mass density distribution within the surface grating (**a**) and within the density grating after annealing at $T > T_g$ (**b**) (Reprinted with permission from Ref. [80]. Copyright (2000) John Wiley and Sons)

light scattering measurements show that the subsequent formation of a density grating below surface and SRG on top of the film occurs during grating–inscribing process [59, 82]. The SRG formation is accompanied by a density grating growth just below the film surface. The possible correlation between the chromophore orientation and density grating formation has been indicated by the authors [80, 83].

5.5.2 Photoinduced Mechanical Property Variations

Whether photoinduced softening occurs during the SRG inscription is a controversial issue. It has been reported that immediately after shining light into the absorption band of the azo chromophores, the azo polymer films becomes softening [67]. Softening of the polymer films was detected by the indentation of an AFM tip before and after irradiation at constant force. This "light-induced plasticization" of the polymer surface has been attributed to repeated *trans–cis–trans* isomerization cycling, which occurs during the SRG formation. Softening induced by high-power visible light has also been observed by measuring the elastic compliance of thin polymer films doped with an azobenzene dye by quartz crystal resonators [84]. Meanwhile, weak plasticization has been suggested based on the results from electromechanical spectroscopy [85, 86].

On the other hand, different conclusion has been drawn from AFM force distance measurements [87]. The elastic properties were obtained for azo polymer thin films by performing nano-indentation. By fitting the shape of many indentation and retraction curves, a range of modulus values was calculated. The result shows that the average modulus of the film is 219 ± 70 kPa before irradiation, whereas it is 336 ± 62 kPa during irradiation. From both this average and the overall distribution, it is concluded that there is no statistically significant difference between the modulus. It is inferred from the lack of substantial photoinduced softening that the mechanism of the mass transport does not involve a decrease in the bulk elastic properties of the material.

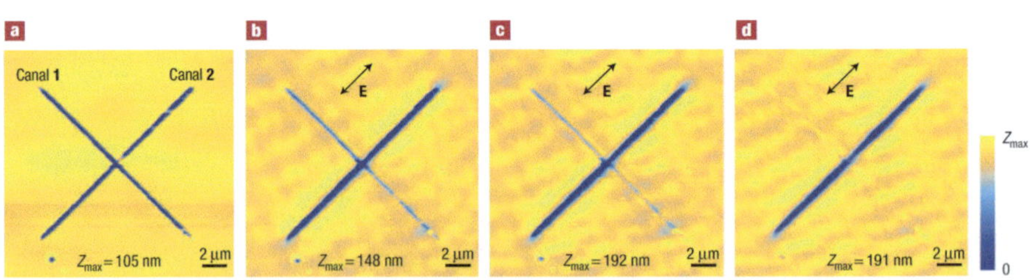

Fig. 5.16 One-direction fluidity of a pDR1M film. (**a**) In the initial state, two perpendicular canals produced by SFM lithography are presented on the non-irradiated film surface (isotropic in plane, spin coating film). (*b*)–(*d*) Images of the surface after 30 min (**b**), 150 min (**c**), and 195 min (**d**) illumination by linearly polarized light show that only canal 1 fills up gradually. The orientation of the E-vector of the light field and the value of Z_{max} are indicated on the images (Reprinted by permission from Macmillan Publishers Ltd: [Nature Materials] Ref. [88], copyright (2005))

Other investigations show that the light-induced softening actually involves more complicated variations. The mechanical properties of a thin film of an azobenzene-containing polymer depend on the polarization direction of a low-intensity far-field illumination [88]. As shown in Fig. 5.16, it is amazing to observe that the film visually behaves as a liquid along one direction and as a solid in others (one-direction liquid state). Moreover, alteration in the local hardness has been observed by AFM for the relief structure on azo polymer films induced by light irradiation [89]. Compared to nonpatterned regions of the same film, local hardness is increased at positions of crests but reduced in troughs of SRG. At room temperature, it shows relative hardness variation up to 50 %, which decreases linearly with grating height and temperature. This observation is believed to be caused by the increased/decreased density of parallel-aligned polymer chains in crests/troughs, which is a result of cooperative motion driven by the *trans–cis–trans* isomerization during grating formation.

5.5.3 Real-Time Imaging of SRG Formation

For a long time, the only way to monitor the SRG grating formation in a real-time manner is through the diffraction efficiency measurement. As discussed above, the diffraction efficiency is contributed from different periodic structures formed during the SRG inscription. Even without this complexity, it is not easy to quantitatively correlate the diffraction efficiency variation with the grating profile development. The method to monitor the grating profile, such as AFM, is usually performed in the ex situ manner and cannot give real-time images. One of the obstacles to the real-time imaging stems from the conventional manner to inscribe the gratings, where the interfering beams come from the front side (free surface) of the films. In recent years, in order to image the SRG formation in real-time manner, the interference pattern is projected from the backside of the film through the substrate. Although it seems to be unclear whether these two irradiation manners could cause SRG-forming mechanism distinction, the backside irradiation manner has provided a valuable way to image SRG formation vividly.

As one of the approaches, the near-field microscopy has been used to image the SRG formation on thin solid films of an azo polymer obtained from the sol–gel method [90]. The interference pattern produced by two laser beams is projected from the backside of the film through the substrate. On the front side, the single-mode optical fiber tip with a combined shear-force and near-field optical microscope is used to simultaneously image the surface topography and the transmitted light intensity. The results show that the material deformation follows two distinct regimes characterized by different kinetics. When using s-/s-polarized beams (intensity interference), SRG builds up on a short time scale and possesses the same spatial phase with the interference pattern. On the other hand, with *p*-/*p*-polarized beams, the irradiation first addresses molecules oriented along the light pattern gradient and photoinduced mass transport occurs from bright to dark areas,

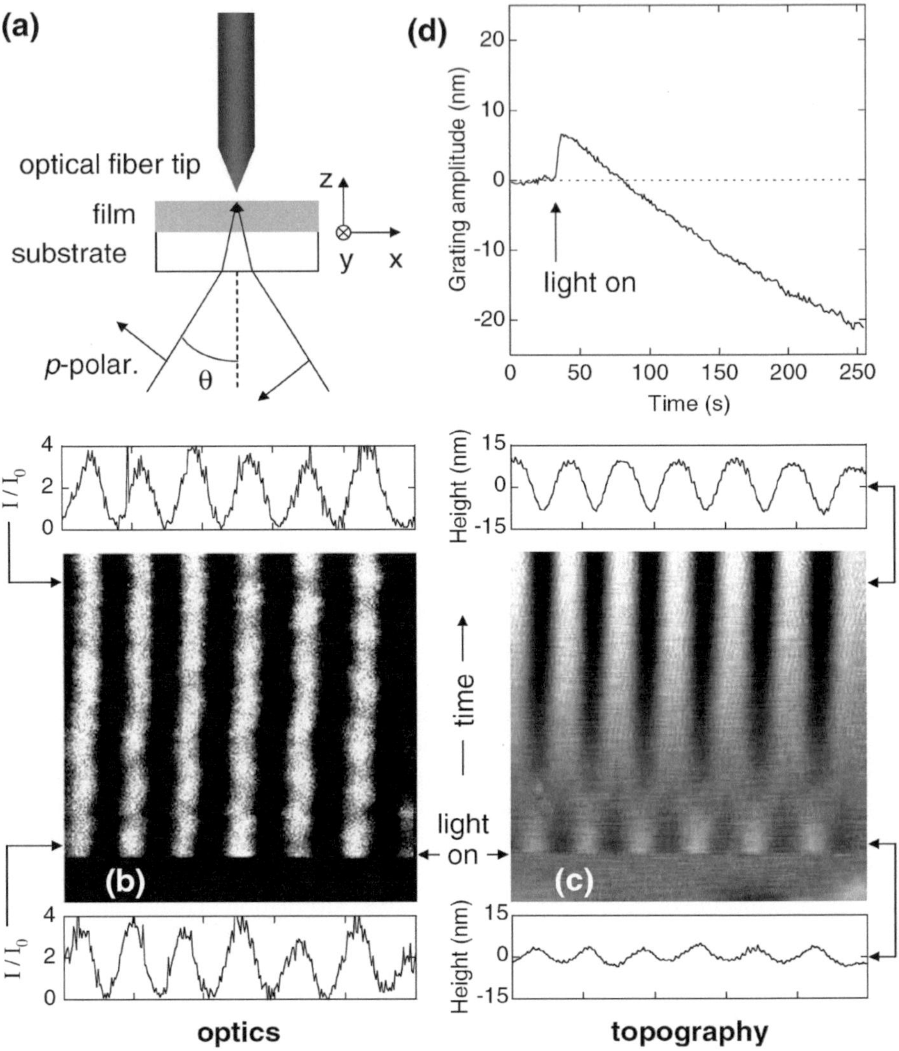

Fig. 5.17 In situ and real-time measurement of the deformation induced by two p-polarized interfering laser beams in a 450-nm-thick sol–gel film. (**a**) Experimental setup. (**b**) 5×5 μm^2 optical image of the interference pattern at the sample surface. Intensity profiles at 10 and 200 s exposure times are plotted. The measured value of the contrast is close to the calculated value $A \approx 0.93$. (**c**)5×5 μm^2 surface topography image acquired during illumination. Topography profiles at 10 and 200 s exposure times are plotted. (**d**) Variation vs exposure time of the amplitude of the photoinduced SRG (Reprinted from Ref. [90] with the permission of AIP Publishing)

which leads to the out-of-phase SRG (Fig. 5.17). Further investigations with the near-field microscopy show that the intensity-driven mechanism critically depends on the polymer matrix, while the polarization-driven mechanism occurs with almost the same efficiency in different materials [91]. The kinetics of the intensity-driven deformation scales with the light power, while the kinetics of the polarization-driven mass transport scales with the amplitude of the electromagnetic field pattern [92].

Another progress has been made by using in situ interferometric AFM setup to monitor the SRG formation [37]. The topography change during the process has been traced by changing polarization and phase of the impinging interference pattern at the same time. The method elucidates the absolute correspondence between the local distribution of electric field vectors and the local topography of the relief grating. For the polarization-modulated interference (±45 combination), the grating amplitude maxima correspond to the region with the vertical $\boldsymbol{E}$-field vector. During the first few minutes of irradiation, the grating profile shows two maxima within one period of the interference pattern. The maxima are situated at the positions of the horizontally and vertically oriented field vector. The topography maxima corresponding to the horizontal orientation disappear after 5 min of irradiation, while the other maxima grow continuously and finally reaching a height of up to 90 % of the total thickness of polymer film.

More recently, both polarized confocal Raman microspectroscopy and scanning near-field optical microscopy with a resolution of 60 nm have been used to characterize photoinduced grating structures of azobenzene-doped polymer films on a glass substrate [93]. During SRG formation by exposing sample to polarization-modulated interference (±45 combination), the anisotropy induced in the film shows distinct orientation distributions in different grating regions. Although the overall alignment of chromophores is barely pronounced, the redistribution of the azo dyes is clearly different in different regions. The orientation distribution of chromophores remains almost isotropic at the bottom of the grating; on the top region of the distribution, it shows an asymmetric maximum at 55° relative to the actinic light. More marked redistribution can be observed on the slopes of the grating. The near-field optical measurement with high spatial resolution shows that during the optical inscription of surface gratings, lateral two-dimensional optical contrast occurs concomitantly, which could be caused by fine intensity modulations of the interfering beams. These fine details have not been observed before by using other methods.

5.6 Photoinduced Mass Transport Beyond SRG

Although above investigations mainly focus on the SRGs formed on azo polymer surfaces upon exposure of the films to interference pattern, this discovery shows far-reaching impact on different research areas far beyond the original scope. The exact nature and scope specified by the term "photoinduced mass transport" and possible applications are still being actively explored at this stage. Other terms, such as "mass migration" and "directional photofluidization" [94], have also been used in the literature. As it seems impossible to comprehensively list all aspects of this rapidly developing research frontier, the section can only highlight some important developments.

SRGs can be inscribed on solid films of different types of materials containing azo chromophores, which is not limited to azo polymers. Amorphous molecular materials (also named molecular glass) are one typical type of the materials, which have recently been developed to contain azo chromophores for optical and photonic applications [95]. The *azo molecular glasses*, which consist of well-designed low-molecular-weight molecules, show glass transition behavior similar to amorphous polymers [96–98]. Due to the well-defined molecular structures, *azo amorphous molecular materials* often show better controllable properties compared with polymers. By using the common spin coating method, thin solid films with a smooth surface can be feasibly prepared by using the amorphous molecular materials. Films of azo molecular glasses have been used for inscribing SRGs [96, 98]. SRGs can be inscribed on azo molecular glass films with a faster rate compared to those typically observed for azo polymers. Owing to its ability to form surface relief structures in a very efficient way, two-dimensional quasicrystal structures with rotation symmetry as high as 60-folds can be inscribed on the molecular glass films [99]. Figure 5.18 shows tenfold quasicrystal produced with dual-beam multiple exposure method. Because of their unique characteristics, amorphous molecular materials possess desirable features for studying the structure–property relationship as well as for device applications.

Although interferometric apparatus is a typical way used to investigate mass transport on solid films, the response to light irradiation has been investigated by using different optical setups. The edge diffraction of laser radiation with polarization perpendicular to the edge and irradiation with a polarized beam with Gaussian intensity distribution are two ways to demonstrate the effects of the gradient force [67, 69]. Optical *near-field effect* is another way to produce the unique light source to study the mass transport. Viscoelasticity change induced by optical near field has been observed on a urethane–urea azo copolymer film covered with monolayers of polystyrene microspheres [100]. Phase imaging from tapping mode atomic force microscopy (TMAFM) shows that the area affected by the optical near field from the polystyrene microsphere becomes relatively softer and the vicinal area becomes harder (Fig. 5.19). On the basis of this principle, the surface relief produced by the optical near field has been used to record images of living specimens, such as a moving *Paramecium*, which is then observed by AFM [101]. As a more typical way to employ the near-field effect, topographic modifications on the azo polymer film surface are created by mass transport driven by the optical near field from noble metal nanostructures [102]. By using the near-field polarization effect, three spatial components of the optical near field of complex metal nanostructures can be mapped [103].

Recently, *spiral-shaped relief structures* on the azo polymer film have been fabricated upon the illumination of focused Laguerre–Gauss beams with helical wave fronts and an optical vortex at their axis [104]. The induced spiral reliefs are sensitive to the vortex topological charge and to the wave front handedness of the beams. This mass transport effect is unexpected as the doughnut-shaped intensity profile of Laguerre–Gauss beams contains no information about the wave front handedness. A phenomenological model has been proposed to explain the

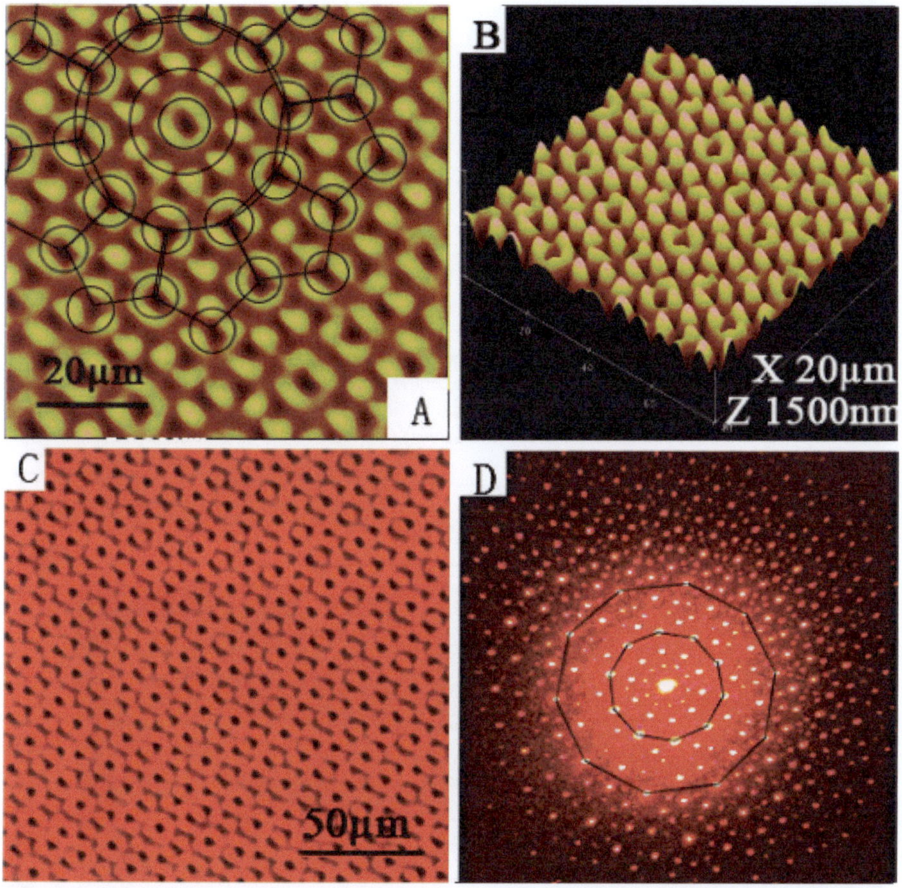

Fig. 5.18 Tenfold quasicrystal produced with the $\Lambda = 6.3$ μm interference pattern: (**a**) AFM 2D view image, (**b**) AFM 3D view image, (**c**) optical micrograph, (**d**) photograph of the He–Ne laser diffraction pattern (Reprinted with permission from Ref. [99]. Copyright (2008) American Chemical Society)

observation. By adding a new surface-related term in the light-driven mass current, which depends on the optical field but is independent of its gradients, the main qualitative features of the relief patterns can be explained by this phenomenological model. Later, a model of the mass migration process is established based on anisotropic light-driven molecular diffusion, which considers an enhanced molecular diffusion in proximity of the free polymer surface [105].

Mass transport has been intensively investigated for azo polymer films, but it is definitely not limited to two-dimensional film surfaces. Actually, this effect can be observed for almost any azo polymer architectures within nano-/microscale to cause the deformations. These types of structures can be obtained through both the self-assembly and microfabrication methods such as soft lithography, photolithography, and others. As a typical case of the self-assembling approaches, the

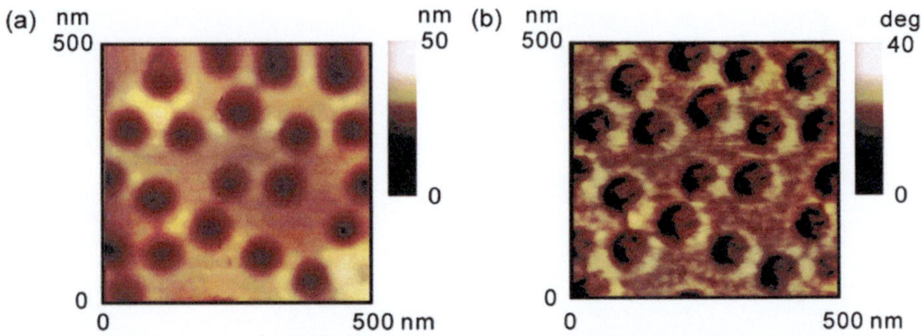

Fig. 5.19 TMAFM images of the sample surface prepared using PS microspheres with a 100 nm diameter. (**a**) Two-dimensional image of surface shape. (**b**) Two-dimensional image of the phase shift. Figures (*a*) and (*b*) were obtained at the same time (Reprinted with permission from Ref. [100]. Copyright (2000) American Chemical Society)

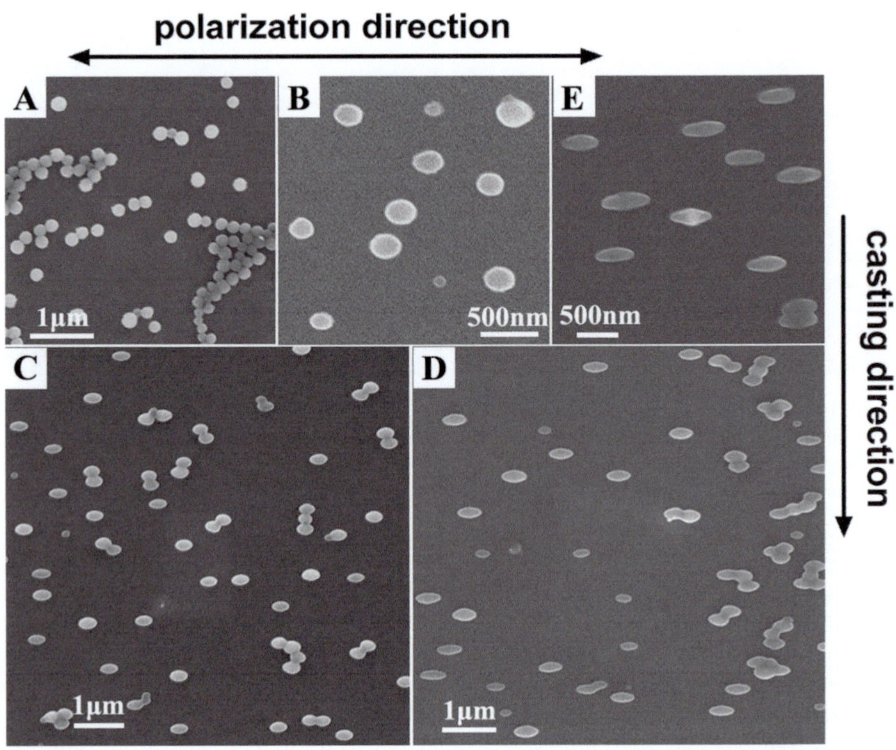

Fig. 5.20 SEM images of colloidal spheres before irradiation (**a**) and after irradiation for different periods of time: (**b**) 5 min, (**c**) 10 min, (**d**) 15 min, (**e**) 20 min (Reprinted with permission from Ref. [107]. Copyright (2006) American Chemical Society)

uniform colloidal spheres can be obtained through the molecular self-assembly in a selective solvent [106, 107]. Figure 5.20 shows the SEM images of microspheres before irradiation and after irradiation for different time periods. Upon the light irradiation, the colloidal spheres show elongated deformation along the polarization direction of the laser beam. The colloidal spheres can be deformed continuously to give ellipsoidal and rodlike particles as the irradiation time increases. The photoinduced mass transport effect has also been used to process different submicron/micron structures fabricated by microfabrication methods such as soft lithography. The photoprocessible and photofabricated azo polymer structures include microwires [108], micrometer hemisphere array [109], and funnel-shaped plasmonic tip arrays [110] as some typical examples. For more contents about the photoinduced mass transport effects at micro-/nanoscale, we refer the interested reader to some recent review articles [8, 111].

5.7 Spontaneous Patterning with a Uniform Light Field

SRGs discussed above are formed by exposure of polymer films to an interference pattern formed by two laser beams, where both polarization and intensity of the light can be varied in space. The surface modulations are considered as a holographic record reflecting the spatial variation of the light field. However, it has been observed with surprise that under irradiation with a uniform laser beam at normal incidence, submicrometer hexagonal relief patterns form on the azo polymer films [112]. In order to make a distinction between this patterning behavior and surface modulations induced by interfering light field, this effect is termed photoinduced spontaneous patterning or self-structured surface patterns, which is a unique type catalogued under the mass transport driven by the light.

The phenomenon was first observed on the film of a poly(methyl methacrylate)-based copolymer with the degree of functionalization of 55 mol% [112]. After irradiation with the Ar^+ laser beam with the wavelength of 514 nm for 80–100 min, the structures appear as regular spaced peaks with similar relief amplitude in 60–70 nm. One difference with the typical condition to inscribe SRGs is that a high light intensity (such as 300 mW/cm^2) is required to induce the surface patterning. A light intensity threshold is observed and no structures can be formed if the intensity is lower than 200 mW/cm^2. Above this threshold, the periodicity of the induced structures shows no correlation with the irradiation time and beam intensity. The structure growth rate $1/\tau$ is proportional to $|P - P_{th}|$, where P and $P_{th} = 250$ mW/cm^2 are the light power and threshold power.

This phenomenon is strongly reminiscent of the laser-induced periodic surface structure (LIPSS) reported for different materials including polymers [113]. However, pulse laser with a short wavelength (193 and 248 nm) is typically required for inducing LIPSS, where the ordinarily induced structures are ripples on the polymer surfaces [114, 115]. Significant differences between LIPSS and spontaneous patterning on azo polymers can be easily recognized by comparing these observations.

Later, several factors including feedback effects have been investigated for this unique mass transport behavior of azo polymers under different experimental conditions [116]. The experiments show that feedback most likely originates from the polymer film itself, which is different from the case for LIPSS.

This surface patterning induced by a single laser beam has been observed for different azo polymers both using a continuous and pulsed laser [117, 118]. These self-structured surface patterns have also been observed on series of azo molecular glasses [119, 120]. Figure 5.21 shows typical patterns formed on the azo molecular glass films. The results show that the surface patterning is closely related to the type of azo chromophores. To understand the structure–property relationship, a series of epoxy-based azo polymers have been synthesized and used to study the spontaneous patterning behavior upon irradiation with laser at different wavelengths [46]. Figures 5.22 and 5.23 show synthetic route of the azo polymers and representative images of the self-structured surface patterns. The results show that the photoinduced surface pattern formation is affected by the structure of azo chromophores, excitation wavelength, and light polarization condition. The efficient excitation wavelength is highly correlated with the absorption band position, which is mainly determined by the electron-withdrawing groups on the azo chromophores.

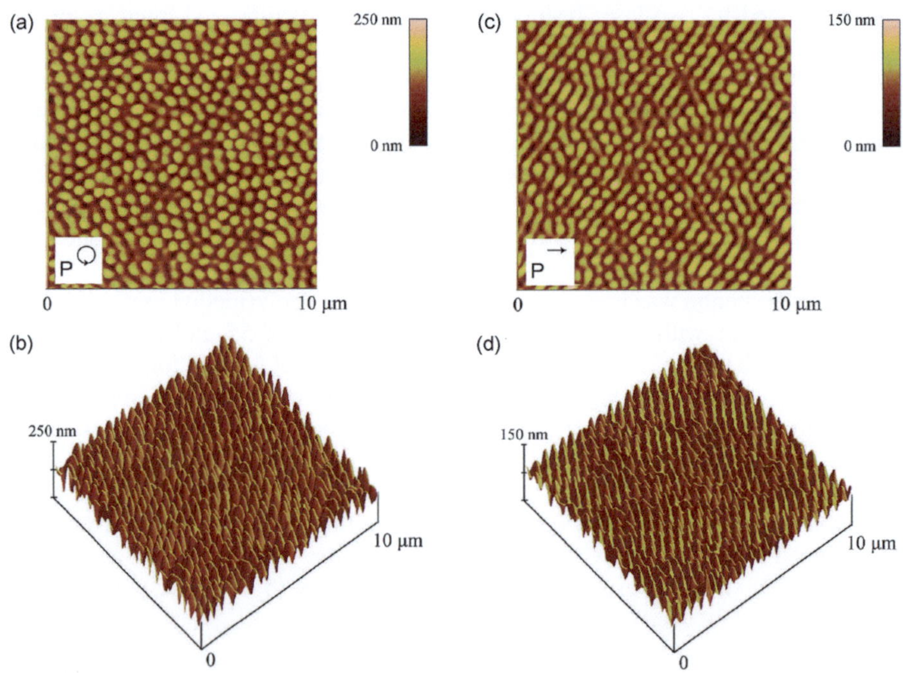

Fig. 5.21 Typical AFM images (10 μm × 10 μm) of photoinduced self-structured patterns formed on azo molecular glass films with different incident laser beam polarizations: (**a**) circularly polarized, 2D view; (**b**) circularly polarized, 3D view; (**c**) linearly polarized, 2D view; (**d**) linearly polarized, 3D view (Reprinted with permission from Ref. [119]. Copyright (2010) American Chemical Society)

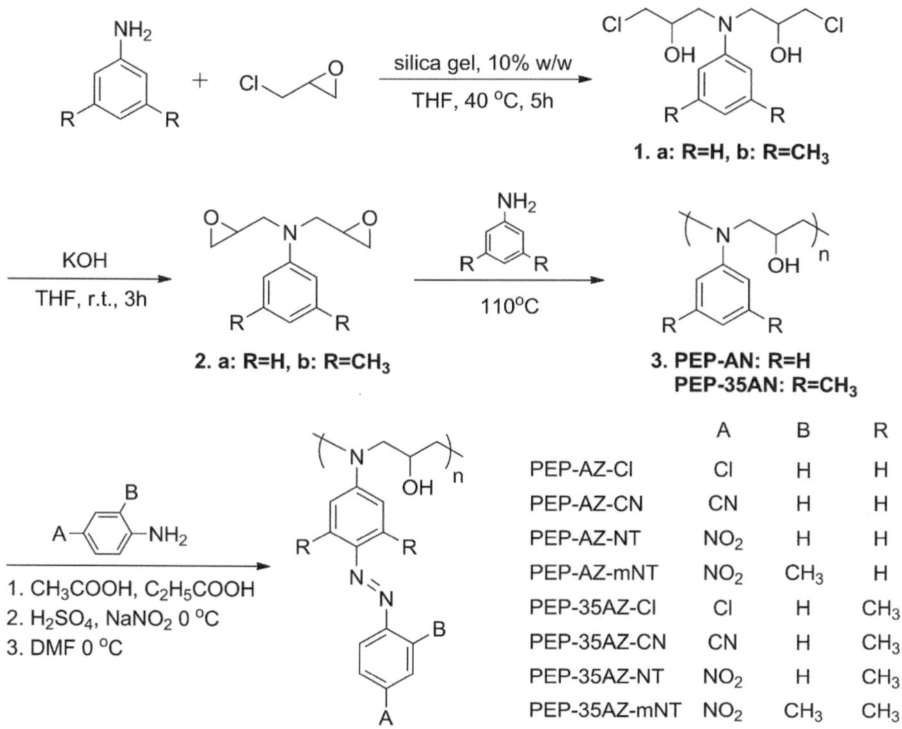

	A	B	R
PEP-AZ-Cl	Cl	H	H
PEP-AZ-CN	CN	H	H
PEP-AZ-NT	NO_2	H	H
PEP-AZ-mNT	NO_2	CH_3	H
PEP-35AZ-Cl	Cl	H	CH_3
PEP-35AZ-CN	CN	H	CH_3
PEP-35AZ-NT	NO_2	H	CH_3
PEP-35AZ-mNT	NO_2	CH_3	CH_3

Fig. 5.22 Synthetic route of the epoxy-based azo polymers with the high density of azo chromophores (Reprinted with permission from Ref. [46]. Copyright (2011) American Chemical Society)

The methyl substituents on the azo chromophores show ability to enhance the photoinduced surface pattern formation in some cases.

Although both SRGs and spontaneous patterning are induced by the light irradiation, the SRG formation is induced by irradiating the polymer films with interfering light, which is distinct with spontaneous patterning using the uniform laser beam irradiation. As discussed above, the SRG formation efficiency is also dependent on the azo chromophore structure and wavelength of the incident laser light. More careful comparison shows that the self-structured surface pattern formation needs a higher energy input and shows stricter wavelength requirement compared with those of the SRG formation [46]. The SRG formation can be observed for almost every type of azo polymers, but some of them cannot form spontaneous patterning under typical light irradiation conditions.

Recently, the mechanism of spontaneous patterning induced by a single-beam irradiation has been investigated. A phenomenological model based on Fick's law of diffusion is proposed to account for the molecular motion [121]. The patterning process is attributed to propagation of waves guided in the film and substrate, which is treated as a planar waveguide with the effective index. The wavelength of the incident pump beam, together with the guiding properties, determines the optical

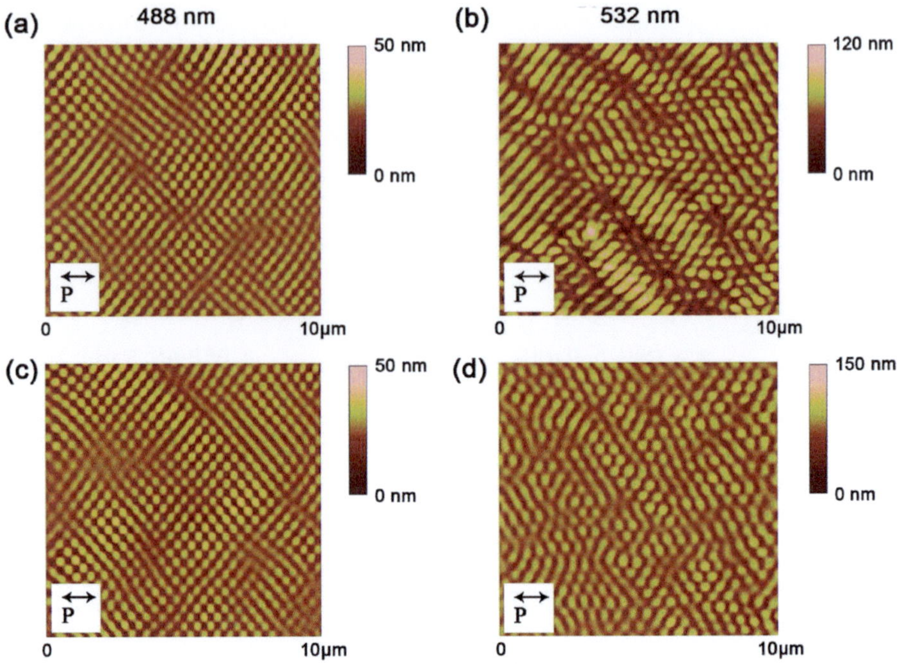

Fig. 5.23 AFM images (10 μm × 10 μm) of the photoinduced surface patterns formed on PEP-AZ-CN and PEP-35AZ-CN films upon irradiation at 488 and 532 nm: (**a**) PEP-AZ-CN, 488 nm; (**b**) PEP-AZ-CN, 532 nm; (**c**) PEP-35AZ-CN, 488 nm; (**d**) PEP-35AZ-CN, 532 nm. The intensity of the laser light was 200 mW/cm^2. The irradiation time with the 488 nm laser was 30 min and the irradiation time with the 532 nm laser was 15 min (Reprinted with permission from Ref. [46]. Copyright (2011) American Chemical Society)

frequency of light propagating inside the sample. As a result, the grating period depends on a given propagation constant and wavelength of the light propagating in the sample. More recently, the photoinduced spontaneous patterning on azo polymer films has been attributed to phase separation, which is caused by an instability created by the photoactivated transitions between two immiscible states of the polymer [122]. In this process, polarized light has a striking effect on pattern formation through enhancing the phase separation. The domain formation can be significantly enhanced along the polarization direction, which is caused by direct coupling between the polarization and the dipole moment of the azo polymer to create an additional energy contribution.

References

1. Rochon, P., Batalla, E., Natansohn, A.: Optically induced surface gratings on azoaromatic polymer films. Appl. Phys. Lett. **66**, 136–138 (1995)

2. Kim, D.Y., Tripathy, S.K., Li, L., Kumar, J.: Laser-induced holographic surface relief gratings on nonlinear optical polymer films. Appl. Phys. Lett. **66**, 1166–1168 (1995)
3. Natansohn, A., Rochon, P.: Photoinduced motions in azobenzene-based amorphous polymers: Possible photonic devices. Adv. Mater. **11**, 1387–1391 (1999)
4. Viswanathan, N.K., Kim, D.Y., Bian, S.P., Williams, J., Liu, W., Li, L., Samuelson, L., Kumar, J., Tripathy, S.K.: Surface relief structures on azo polymer films. J. Mater. Chem. **9**, 1941–1955 (1999)
5. Delaire, J.A., Nakatani, K.: Linear and nonlinear optical properties of photochromic molecules and materials. Chem. Rev. **100**, 1817–1845 (2000)
6. Yager, K.G., Barrett, C.J.: All-optical patterning of azo polymer films. Curr. Opin. Solid State Mater. Sci. **5**, 487–494 (2001)
7. Natansohn, A., Rochon, P.: Photoinduced motions in azo-containing polymers. Chem. Rev. **102**, 4139–4175 (2002)
8. Lee, S.W., Kang, H.S., Park, J.K.: Directional photofluidization lithography: Micro/nanostructural evolution by photofluidic motions of azobenzene materials. Adv. Mater. **24**, 2069–2103 (2012)
9. Wang, X.L., Yin, J.J., Wang, X.G.: Epoxy-based polymers functionalized with bisazo chromophores: Synthesis, characterization and photoresponsive behavior. Polymer **52**, 3344–3356 (2011)
10. Kim, D.Y., Li, L., Jiang, X.L., Shivshankar, V., Kumar, J., Tripathy, S.K.: Polarized laser induced holographic surface relief gratings on polymer films. Macromolecules **28**, 8835–8839 (1995)
11. Ramanujam, P.S., Holme, N.C.R., Hvilsted, S.: Atomic force and optical near-field microscopic investigations of polarization holographic gratings in a liquid crystalline azobenzene side-chain polyester. Appl. Phys. Lett. **68**, 1329 (1996)
12. Holme, N.C.R., Nikolova, L., Ramanujam, P.S., Hvilsted, S.: An analysis of the anisotropic and topographic gratings in a side-chain liquid crystalline azobenzene polyester. Appl. Phys. Lett. **70**, 1518–1520 (1997)
13. Lee, T.S., Kim, D.Y., Jiang, X.L., Li, L., Kumar, J., Tripathy, S.: Synthesis and optical properties of polyureas with azoaromatic groups in the main chain. Macromol. Chem. Phys. **198**, 2279–2289 (1997)
14. Lee, T.S., Kim, D.Y., Jiang, X.L., Li, L., Kumar, J., Tripathy, S.: Photoinduced surface relief gratings in high-T_g main-chain azoaromatic polymer Films. J. Polym. Sci. A **36**, 283–289 (1998)
15. Wu, Y.L., Natansohn, A., Rochon, P.: Photoinduced birefringence and surface relief gratings in novel polyurethanes with azobenzene groups in the main chain. Macromolecules **34**, 7822–7828 (2001)
16. Wang, X.G., Balasubramanian, S., Kumar, J., Tripathy, S.K., Li, L.: Azo chromophore-functionalized polyelectrolytes. 1. Synthesis, characterization, and photoprocessing. Chem. Mater. **10**, 1546–1553 (1998)
17. He, J.A., Bian, S.P., Li, L., Kumar, J., Tripathy, S.K., Samuelson, L.A.: Surface relief gratings from electrostatically layered azo dye films. Appl. Phys. Lett. **76**, 3233–3235 (2000)
18. Lee, S.H., Balasubramanian, S., Kim, D.Y., Viswanathan, N.K., Bian, S., Kumar, J., Tripathy, S.K.: Azo polymer multilayer films by electrostatic self-assembly and layer-by-layer post azo functionalization. Macromolecules **33**, 6534–6540 (2000)
19. Camilo, C.S., dos Santos Jr., D.S., Rodrigues Jr., J.J., Vega, M.L., Campana, F.S.P., Oliveira Jr., O.N., Mendonca, C.R.: Surface-relief gratings and photoinduced birefringence in layer-by-layer films of chitosan and an azopolymer. Biomacromolecules **4**, 1583–1588 (2003)
20. Sukwattanasinitt, M., Wang, X.G., Li, L., Jiang, X.L., Kumar, J., Tripathy, S.K., Sandman, D. J.: Functionalizable self-assembling polydiacetylenes and their optical properties. Chem. Mater. **10**, 27–29 (1998)

21. Sukwattanasinitt, M., Lee, D.C., Kim, M., Wang, X.G., Li, L., Yang, K., Kumar, J., Tripathy, S.K., Sandman, D.J.: New processable, functionalizable polydiacetylenes. Macromolecules **32**, 7361–7369 (1999)
22. Yang, S.Z., Li, L., Cholli, A.L., Kumar, J., Tripathy, S.K.: Azobenzene-modified poly (L-glutamic acid) (AZOPLGA): Its conformational and photodynamic properties. Biomacromolecules **4**, 366–371 (2003)
23. Che, P.C., He, Y.N., Wang, X.G.: Hyperbranched azo-polymers synthesized by azo-coupling reaction of an AB_2 monomer and postpolymerization modification. Macromolecules **38**, 8657–8663 (2005)
24. Li, W.H., Dohi, T., Hara, M., Nagano, S., Haba, O., Yonetake, K., Seki, T.: Phototriggered mass migration consorted with surface dewetting in thin films of a liquid crystalline azobenzene-containing dendrimer. Macromolecules **45**, 6618–6627 (2012)
25. Zettsu, N., Seki, T.: Highly efficient photogeneration of surface relief structure and its immobilization in cross-linkable liquid crystalline azobenzene polymers. Macromolecules **37**, 8692–8698 (2004)
26. Zettsu, N., Ogasawara, T., Arakawa, R., Nagano, S., Ubukata, T., Seki, T.: Highly photosensitive surface relief gratings formation in a liquid crystalline azobenzene polymer: new implications for the migration process. Macromolecules **40**, 4607–4613 (2007)
27. Gao, J., He, Y.N., Xu, H.P., Song, B., Zhang, X., Wang, Z.Q., Wang, X.G.: Azobenzene-containing supramolecular polymer films for laser-induced surface relief gratings. Chem. Mater. **19**, 14–17 (2007)
28. Priimagi, A., Lindfors, K., Kaivola, M., Rochon, P.: Efficient surface-relief gratings in hydrogen-bonded polymer-azobenzene complexes. ACS Appl. Mater. & Inter. **1**, 1183–1189 (2009)
29. Labarthet, F.L., Buffeteau, T., Sourisseau, C.: Inscription of holographic gratings using circularly polarized light: Influence of the optical set-up on the birefringence and surface relief grating properties. Appl. Phys. B **74**, 129–137 (2002)
30. Ramanujam, P.S., Pedersen, M., Hvilsted, S.: Instant holography. Appl. Phys. Lett. **74**, 3227–3229 (1999)
31. Leopold, A., Wolff, J., Baldus, O., Huber, M.R., Bieringer, T., Zilker, S.J.: Thermally induced surface relief gratings in azobenzene polymers. J. Chem. Phys. **113**, 833–837 (2000)
32. Baldus, O., Leopold, A., Hagen, R., Bieringer, T., Zilker, S.J.: Surface relief gratings generated by pulsed holography: A simple way to polymer nanostructures without isomerizing side-chains. J. Chem. Phys. **114**, 1344–1349 (2001)
33. Rodríguez, F.J., Sánchez, C., Villacampa, B., Alcalá, R., Cases, R., Millaruelo, M., Oriol, L.: Surface relief gratings induced by a nanosecond pulse in a liquid-crystalline azopolymethacrylate. Appl. Phys. Lett. **87**, 201914 (2005) (3 pages)
34. Rodríguez, F.J., Sánchez, C., Villacampa, B., Alcalá, R., Cases, R., Millaruelo, M., Oriol, L.: Fast and stable recording of birefringence and holographic gratings in an azopolymethacrylate using a single nanosecond light pulse. J. Chem. Phys. **123**, 204706 (2005) (7 pages)
35. Sánchez, C., Alcalá, R., Hvilsted, S., Ramanujam, P.S.: Biphotonic holographic gratings in azobenzene polyesters: Surface relief phenomena and polarization effects. Appl. Phys. Lett. **77**, 1440–1442 (2000)
36. Sánchez, C., Cases, R., Alcalá, R., López, A., Quintanilla, M., Oriol, L., Millaruelo, M.: Biphotonic holographic recording in a liquid crystalline cyanoazobenzene side-chain polymethacrylate. Polarization, intensity, and relief gratings. J. Appl. Phys. **89**, 5299–5306 (2001)
37. Yadavalli, N.S., Santer, S.: In-situ atomic force microscopy study of the mechanism of surface relief grating formation in photosensitive polymer films. J. Appl. Phys. **113**, 224304 (2013) (12 pages)
38. Eichler, H.J., Günter, P., Pohl, D.W.: Laser-induced dynamic gratings. Springer-Verlag, Berlin (1986)

39. Gaylord, T.K., Moharam, M.G.: Thin and thick gratings: Terminology clarification. Appl. Opt. **20**, 3271–3273 (1981)
40. Petit, R. (ed.): Electromagnetic theory of gratings. Springer-Verlag, Berlin (1980)
41. Nikolova, L., Ramanujam, P.S.: Polarization holography. Cambridge University Press, New York (2009)
42. Barrett, C.J., Natansohn, A.L., Rochon, P.L.: Mechanism of optically inscribed high-efficiency diffraction gratings in azo polymer films. J. Phys. Chem. **100**, 8836–8842 (1996)
43. Fukuda, T., Matsuda, H., Shiraga, T., Kimura, T., Kato, M., Viswanathan, N.K., Kumar, J., Tripathy, S.K.: Photofabrication of surface relief grating on films of azobenzene polymer with different dye functionalization. Macromolecules **33**, 4220–4225 (2000)
44. He, Y.N., Wang, X.G., Zhou, Q.X.: Epoxy-based azo polymers: synthesis, characterization and photoinduced surface-relief-gratings. Polymer **43**, 7325–7333 (2002)
45. Andruzzi, L., Altomare, A., Ciardelli, F., Solaro, R., Hvilsted, S., Ramanujam, P.S.: Holographic gratings in azobenzene side-chain polymethacrylates. Macromolecules **32**, 448–454 (1999)
46. Wang, X.L., Yin, J.J., Wang, X.G.: Self-structured surface patterns on epoxy-based azo polymer films induced by laser light irradiation. Macromolecules **44**, 6856–6867 (2011)
47. Zhou, Y.Q., Tang, B., Wang, X.G.: Photoinduced deformation behavior of a series of newly synthesized epoxy-based polymers bearing *push-pull* azo chromophores. Polymer **60**, 292–301 (2015)
48. Wang, D.R., Ye, G., Zhu, Y., Wang, X.G.: Photoinduced mass-migration behavior of two amphiphilic side-chain azo diblock copolymers with different length flexible spacers. Macromolecules **42**, 2651–2657 (2009)
49. Kim, M.J., Lee, J.D., Chun, C.M., Kim, D.Y., Higuchi, S.J., Nakayama, T.: Control of photodynamic motions of azobenzene-derivative polymers by laser excitation wavelength. Macromol. Chem. Phys. **208**, 1753–1763 (2007)
50. Barrett, C.J., Rochon, P.L., Natansohn, A.L.: Model of laser-driven mass transport in thin films of dye-functionalized polymers. J. Phys. Chem. **109**, 1505–1516 (1998)
51. Jiang, X.L., Li, L., Kumar, J., Kim, D.Y., Shivshankar, V., Tripathy, S.K.: Polarization dependent recordings of surface relief gratings on azobenzene containing polymer films. Appl. Phys. Lett. **68**, 2618–2620 (1996)
52. Viswanathan, N.K., Balasubramnian, S., Li, L., Tripathy, S.K., Kumar, J.: A detailed investigation of the polarization-dependent surface-relief-grating formation process on azo polymer films. Jpn. J. Appl. Phys. **38**, 5928–5937 (1999)
53. Labarthet, F.L., Buffeteau, T., Sourisseau, C.: Analyses of the diffraction efficiencies, birefringence, and surface relief gratings on azobenzene-containing polymer films. J. Phys. Chem. B **102**, 2654–2662 (1998)
54. Helgert, M., Fleck, B., Wenke, L., Hvilsted, S., Ramanujam, P.S.: An improved method for separating the kinetics of anisotropic and topographic gratings in side-chain azobenzene polyesters. Appl. Phys. B **70**, 803–807 (2000)
55. Sobolewska, A., Miniewicz, A.: Analysis of the kinetics of diffraction efficiency during the holographic grating recording in azobenzene functionalized polymers. J. Phys. Chem. B **111**, 1536–1544 (2007)
56. Sobolewska, A., Bartkiewicz, S.: Three gratings coupling during the holographic grating recording process in azobenzene-functionalized polymer. Appl. Phys. Lett. **92**, 253305 (2008) (3 pages)
57. Yager, K.G., Barrett, C.J.: Temperature modeling of laser-irradiated azo-polymer thin films. J. Chem. Phys. **120**, 1089–1096 (2004)
58. Sumaru, K., Yamanaka, T., Fukuda, T., Matsuda, H.: Photoinduced surface relief gratings on azopolymer films: Analysis by a fluid mechanics model. Appl. Phys. Lett. **75**, 1878–1880 (1999)

59. Geue, T.M., Saphiannikova, M.G., Henneberg, O., Pietsch, U., Rochon, P.L., Natansohn, A. L.: Formation mechanism and dynamics in polymer surface gratings. Phys. Rev. E **65**, 052801 (2002) (4 pages)
60. Tanchak, O.M., Barrett, C.J.: Light-induced reversible volume changes in thin films of azo polymers: The photomechanical effect. Macromolecules **38**, 10566–10570 (2005)
61. Yager, K.G., Tanchak, O.M., Godbout, C., Fritzsche, H., Barrett, C.J.: Photomechanical effects in azo-polymers studied by neutron reflectometry. Macromolecules **39**, 9311–9319 (2006)
62. Lefin, P., Fiorini, C., Nunzi, J.M.: Anisotropy of the photoinduced translation diffusion of azo-dyes. Opt. Mater. **9**, 323–328 (1998)
63. Lefin, P., Fiorini, C., Nunzi, J.M.: Anisotropy of the photo-induced translation diffusion of azobenzene dyes in polymer matrices. Pure Appl. Opt. **7**, 71–82 (1998)
64. Bellini, B., Ackermann, J., Klein, H., Grave, C., Dumas, P., Safarov, V.: Light-induced molecular motion of azobenzene containing molecules: A random-walk model. J. Phys.: Condens. Matter **18**, S1817–S1835 (2006)
65. Juan, M.L., Plain, J., Bachelot, R., Royer, P., Gray, S.K., Wiederrecht, G.P.: Stochastic model for photoinduced surface relief grating formation through molecular transport in polymer films. Appl. Phys. Lett. **93**, 153304 (2008) (3 pages)
66. Juan, M.L., Plain, J., Bachelot, R., Royer, P., Gray, S.K., Wiederrecht, G.P.: Multiscale model for photoinduced molecular motion in azo polymers. ACS Nano **3**, 1573–1579 (2009)
67. Kumar, J., Li, L., Jiang, X.L., Kim, D.Y., Lee, T.S., Tripathy, S.: Gradient force: The mechanism for surface relief grating formation in azobenzene functionalized polymers. Appl. Phys. Lett. **72**, 2096–2098 (1998)
68. Bian, S., Williams, J.M., Kim, D.Y., Li, L., Balasubramanian, S., Kumar, J., Tripathy, S.: Photoinduced surface deformations on azobenzene polymer films. J. Appl. Phys. **86**, 4498–4508 (1999)
69. Bian, S., Li, L., Kumar, J., Kim, D.Y., Williams, J., Tripathy, S.K.: Single laser beam-induced surface deformation on azobenzene polymer films. Appl. Phys. Lett. **73**, 1817–1819 (1998)
70. Lee, J.D., Kim, M.J., Nakayama, T.: A single-dipole model of surface relief grating formation on azobenzene polymer films. Langmuir **24**, 4260–4264 (2008)
71. Inoue, N., Nozue, M., Yamane, O., Umegaki, S.: Driving force for formation of a surface relief grating on an azobenzene-containing polymer. J. Appl. Phys. **104**, 023106 (2008) (6 pages)
72. Pedersen, T.G., Johansen, P.M., Holme, N.C.R., Ramanujam, P.S.: Mean-field theory of photoinduced formation of surface reliefs in side-chain azobenzene polymers. Phys. Rev. Lett. **80**, 89–92 (1998)
73. Baldus, O., Zilker, S.J.: Surface relief gratings in photoaddressable polymers generated by cw holography. Appl. Phys. B **72**, 425–427 (2001)
74. Bublitz, D., Fleck, B., Wenke, L.: A model for surface-relief formation in azobenzene polymers. Appl. Phys. B **72**, 931–936 (2001)
75. Gaididei, Y.B., Christiansen, P.L., Ramanujam, P.S.: Theory of photoinduced deformation of molecular films. Appl. Phys. B **74**, 139–146 (2002)
76. Saphiannikova, M., Neher, D.: Thermodynamic theory of light-induced material transport in amorphous azobenzene polymer films. J. Phys. Chem. B **109**, 19428–19436 (2005)
77. Toshchevikov, V., Saphiannikova, M., Heinrich, G.: Microscopic theory of light-induced deformation in amorphous side-chain azobenzene polymers. J. Phys. Chem. B **113**, 5032–5045 (2009)
78. Barada, D., Itoh, M., Yatagai, T.: Computer simulation of photoinduced mass transport on azobenzene polymer films by particle method. J. Appl. Phys. **96**, 4204–4210 (2004)
79. Labarthet, F.L., Bruneel, J.L., Buffeteau, T., Sourisseau, C.: Chromophore orientations upon irradiation in gratings inscribed on azo-dye polymer films: A combined AFM and confocal Raman microscopic study. J. Phys. Chem. B **108**, 6949–6960 (2004)

80. Pietsch, U., Rochon, P., Natansohn, A.: Formation of a buried lateral density grating in azobenzene polymer films. Adv. Mater. **12**, 1129–1132 (2000)
81. Geue, T., Schultz, M., Grenzer, J., Pietsch, U., Natansohn, A., Rochon, P.: X-ray investigations of the molecular mobility within polymer surface gratings. J. Appl. Phys. **87**, 7712–7719 (2000)
82. Henneberg, O., Geue, T., Saphiannikova, M., Pietsch, U., Natansohn, A., Rochon, P., Finkelstein, K.: Investigation of material flow on inscribing a polymer surface grating probing X-ray and VIS light scattering. Colloids Surf. A Physic Eng. Asp. **198–200**, 107–111 (2002)
83. Geue, T.M., Saphiannikova, M.G., Henneberg, O., Pietsch, U., Rochon, P.L., Natansohn, A. L.: X-ray investigations of formation efficiency of buried azobenzene polymer density gratings. J. Appl. Phys. **93**, 3161–3166 (2003)
84. Srikhirin, T., Laschitsch, A., Neher, D., Johannsmann, D.: Light-induced softening of azobenzene dye-doped polymer films probed with quartz crystal resonators. Appl. Phys. Lett. **77**, 963–965 (2000)
85. Mechau, N., Neher, D., Börger, V., Menzel, H., Urayama, K.: Optically driven diffusion and mechanical softening in azobenzene polymer layers. Appl. Phys. Lett. **81**, 4715–4717 (2002)
86. Mechau, N., Saphiannikova, M., Neher, D.: Dielectric and mechanical properties of azobenzene polymer layers under visible and ultraviolet irradiation. Macromolecules **38**, 3894–3902 (2005)
87. Yager, K.G., Barrett, C.J.: Photomechanical surface patterning in azo-polymer materials. Macromolecules **39**, 9320–9326 (2006)
88. Karageorgiev, P., Neher, D., Schulz, B., Stiller, B., Pietsch, U., Giersig, M., Brehmer, L.: From anisotropic photo-fluidity towards nanomanipulation in the optical near-field. Nat. Mater. **4**, 699–703 (2005)
89. Veer, P.U., Pietsch, U., Mueller, A.D.: Alteration of the mechanical properties of azopolymer film in the process of surface relief grating formation. Appl. Phys. Lett. **94**, 231911 (2009) (3 pages)
90. Garrot, D., Lassailly, Y., Lahlil, K., Boilot, J.P., Peretti, J.: Real-time near-field imaging of photoinduced matter motion in thin solid films containing azobenzene derivatives. Appl. Phys. Lett. **94**, 033303 (2009) (3 pages)
91. Fabbri, F., Garrot, D., Lahlil, K., Boilot, J.P., Lassailly, Y., Peretti, J.: Evidence of two distinct mechanisms driving photoinduced matter motion in thin films containing azobenzene derivatives. J. Phys. Chem. B **115**, 1363–1367 (2011)
92. Fabbri, F., Lassailly, Y., Monaco, S., Lahlil, K., Boilot, J.P., Peretti, J.: Kinetics of photoinduced matter transport driven by intensity and polarization in thin films containing azobenzene. Phys. Rev. E **86**, 115440 (2012) (6 pages)
93. Florio, G.D., Bründermann, E., Yadavalli, N.S., Santer, S., Havenith, M.: Polarized 3D Raman and nanoscale near-field optical microscopy of optically inscribed surface relief gratings: Chromophore orientation in azo-doped polymer films. Soft Matter **10**, 1544–1554 (2014)
94. Lee, S.W., Shin, J.H., Lee, Y.H., Fan, S.H., Park, J.K.: Directional photofluidization lithography for nanoarchitectures with controlled shapes and sizes. Nano Lett. **10**, 296–304 (2010)
95. Shirota, Y.: Photo- and electroactive amorphous molecular materials: Molecular design, syntheses, reactions, properties, and applications. J. Mater. Chem. **15**, 75–93 (2005)
96. Nakano, H., Takahashi, T., Kadota, T., Shirota, Y.: Formation of a surface relief grating using a novel azobenzene-based photochromic amorphous molecular materials. Adv. Mater. **14**, 1157–1160 (2002)
97. Ishow, E., Bellaïche, C., Bouteiller, L., Nakatani, K., Delaire, J.A.: Versatile synthesis of small NLO-active molecules forming amorphous materials with spontaneous second-order NLO response. J. Am. Chem. Soc. **125**, 15744–15745 (2003)
98. Kim, M.J., Seo, E.M., Vak, D.J., Kim, D.Y.: Photodynamic properties of azobenzene molecular films with triphenylamines. Chem. Mater. **15**, 4021–4027 (2003)

99. Guo, M.C., Xu, Z.D., Wang, X.G.: Photofabrication of two-dimensional quasi-crystal patterns on UV-curable molecular azo glass films. Langmuir **24**, 2740–2745 (2008)
100. Ikawa, T., Mitsuoka, T., Hasegawa, M., Tsuchimori, M., Watanabe, O., Kawata, Y., Egami, C., Sugihara, O., Okamoto, N.: Optical near field induced change in viscoelasticity on an azobenzene-containing polymer surface. J. Phys. Chem. B **104**, 9055–9058 (2000)
101. Kawata, Y., Murakami, M., Egami, C., Sugihara, O., Okamoto, N., Tsuchimori, M., Watanabe, O., Nakamura, O.: Nonoptically probing near-field microscopy for the observation of biological living specimens. Appl. Phys. Lett. **78**, 2247–2249 (2001)
102. Hubert, C., Rumyantseva, A., Lerondel, G., Grand, J., Kostcheev, S., Billot, L., Vial, A., Bachelot, R., Royer, P., Chang, S.H., Gray, S.K., Wiederrecht, G.P., Schatz, G.C.: Near-field photochemical imaging of noble metal nanostructures. Nano Lett. **5**, 615–619 (2005)
103. Hubert, C., Bachelot, R., Plain, J., Kostcheev, S., Lerondel, G., Juan, M., Royer, P., Zou, S.L., Schatz, G.C., Wiederrecht, G.P., Gray, S.K.: Near-field polarization effects in molecular-motion-induced photochemical imaging. J. Phys. Chem. C **112**, 4111–4116 (2008)
104. Ambrosio, A., Marrucci, L., Borbone, F., Roviello, A., Maddalena, P.: Light-induced spiral mass transport in azo-polymer films under vortex-beam illumination. Nat.Commun. **3**, 989 (2012) (9 pages)
105. Ambrosio, A., Maddalena, P., Marrucci, L.: Molecular model for light-driven spiral mass transport in azopolymer films. Phys. Rev. Lett. **110**, 146102 (2013) (5 pages)
106. Li, Y.B., He, Y.N., Tong, X.L., Wang, X.G.: Photoinduced deformation of amphiphilic azo polymer colloidal spheres. J. Am. Chem. Soc. **127**, 2402–2403 (2005)
107. Li, Y.B., He, Y.N., Tong, X.L., Wang, X.G.: Stretching effect of linearly polarized Ar^+ laser single-beam on azo polymer colloidal spheres. Langmuir **22**, 2288–2291 (2006)
108. Liu, B., He, Y.N., Wang, X.G.: Fabrication of photoprocessible azo polymer microwires through a soft lithographic approach. Langmuir **22**, 10233–10237 (2006)
109. Liu, B., He, Y.N., Fan, P.W., Wang, X.G.: Azo polymer microspherical cap array: Soft-lithographic fabrication and photoinduced shape deformation behavior. Langmuir **23**, 11266–11272 (2007)
110. Lee, S.W., Shin, J.H., Lee, Y.H., Park, J.K.: Fabrication of the funnel-shaped three-dimensional plasmonic tip arrays by directional photofluidization lithography. ACS Nano **4**, 7175–7184 (2010)
111. Priimagi, A., Shevchenko, A.: Azopolymer-based micro- and nanopatterning for photonic applications. J. Polym. Sci. B **52**, 163–182 (2014)
112. Hubert, C., Fiorini-Debuisschert, C., Maurin, I., Nunzi, J.M., Raimond, P.: Spontaneous patterning of hexagonal structures in an azo-polymer using light-controlled mass transport. Adv. Mater. **14**, 729–732 (2002)
113. Sipe, J.E., Young, J.F., Preston, J.S., van Driel, H.M.: Laser-induced periodic surface structure. I theory. Phys. Rev. B **27**, 1141–1154 (1983)
114. Bolle, M., Lazare, S., Le Blanc, M., Wilmes, A.: Submicron periodic structures produced on polymer surfaces with polarized excimer laser ultraviolet radiation. Appl. Phys. Lett. **60**, 674–676 (1992)
115. Bolle, M., Lazare, S.: Characterization of submicrometer periodic structures produced on polymer surfaces with low-fluence ultraviolet laser radiation. J. Appl. Phys. **73**, 3516–3524 (1993)
116. Hubert, C., Fiorini-Debuisschert, C., Rocha, L., Raimond, P., Nunzi, J.M.: Spontaneous photoinduced patterning of azo-dye polymer films: The facts. J. Opt. Soc. Am. B **24**, 1839–1846 (2007)
117. Tsutsumi, N., Fujihara, A.: Self-assembled spontaneous structures induced by a pulsed laser on a surface of azobenzene polymer film. J. Appl. Phys. **101**, 033110 (2007) (5 pages)
118. Ambrosio, A., Girardo, S., Camposeo, A., Pisignano, D., Maddalena, P.: Controlling spontaneous surface structuring of azobenzene-containing polymers for large-scale nano-lithography of functional substrates. Appl. Phys. Lett. **102**, 093102 (2013) (4 pages)

119. Yin, J.J., Ye, G., Wang, X.G.: Self-structured surface patterns on molecular azo glass films induced by laser light irradiation. Langmuir **26**, 6755–6761 (2010)
120. Wang, X.L., Yin, J.J., Wang, X.G.: Photoinduced self-structured surface pattern on a molecular azo glass film: Structure-property relationship and wavelength correlation. Langmuir **27**, 12666–12676 (2011)
121. Leblond, H., Barille, R., Ahmadi-Kandjani, S., Nunzi, J.M., Ortyl, E., Kucharski, S.: Spontaneous formation of optically induced surface relief gratings. J. Phys. B: Mol. Opt. Phys. **42**, 205401 (2009) (7 pages)
122. Galinski, H., Ambrosio, A., Maddalena, P., Schenker, I., Spolenak, R., Capasso, F.: Instability-induced pattern formation of photoactivated functional polymers. PNAS **111**, 17017–17022 (2014)

Chapter 6
Photoresponsive Liquid Crystal Elastomers

Abstract In this chapter, one very interesting function of azo polymers is discussed, which is unique for the azobenzene-containing liquid crystal polymers with light cross-linking. These azo polymer networks, which are classified as azo liquid crystal elastomers, show many interesting properties and potential applications in actuators, shape memory, and others. Similar to the above two chapters, the discussion starts from the basic concepts and background as the introduction. Then, material design and preparation, structure–property relationship, mechanism, and models are sequentially discussed. Meanwhile, several selected cases are presented to show the characteristics and potential applications of the functions.

Keywords Liquid crystal elastomer • Polysiloxane • Polyacrylate • Side-on • Theory • Photomechanical response • Mechanism • Deformation mode • Possible application

Liquid crystalline elastomers (LCEs) have been intensively investigated in recent years owing to their fascinating properties and potential applications [1–4]. LCEs are lightly cross-linked polymer networks containing mesogenic groups through covalent bond linkages. This unique molecular architecture, which combines anisotropic orientation of liquid crystals (LCs) with the rubbery elasticity of polymer networks, endows the materials with many specific properties and effects, such as stimuli-induced reversible deformation and anisotropic shape changes. On the basis of the sensitivity of mesophases to various physical stimuli, LCEs can be classified into thermoresponsive, electroresponsive, and photoresponsive among others. LCEs containing azobenzene moieties are one of the most important types of photoresponsive LCEs, which are discussed in this section.

6.1 Liquid Crystal Elastomers: A Brief Introduction

LCEs can be considered as a special category of liquid crystal polymers (LCPs), which have intrinsic liquid crystalline order under proper conditions. The classification of LCPs can follow those of liquid crystals (LCs), which have been classified into nematics, smectics, and cholesterics based on the symmetry of the mesophases

X. Wang, *Azo Polymers*, Soft and Biological Matter,
DOI 10.1007/978-3-662-53424-3_6

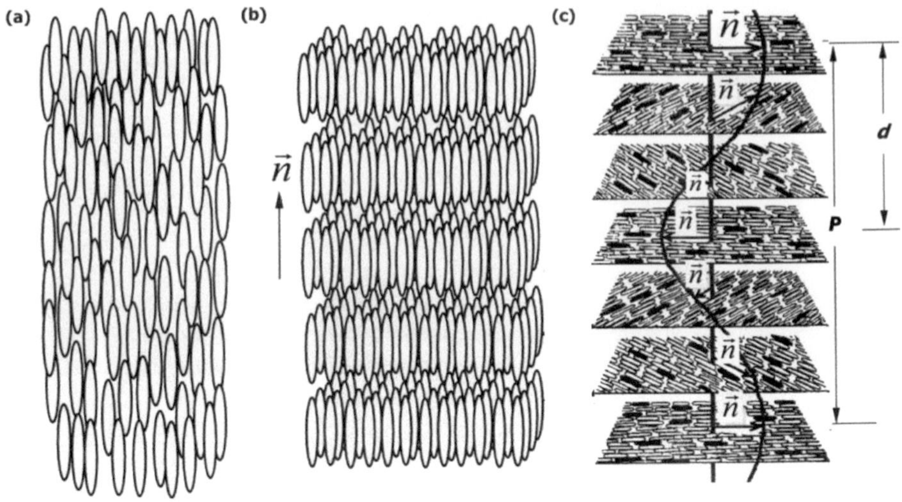

Fig. 6.1 The main types of LC phases: (**a**) nematic, (**b**) smectic, and (**c**) cholesteric. **n** is the director of liquid crystal; P is the pitch of the helix; chiral molecules are shaded (Reprinted with permission from Ref. [6]. Copyright (2003) Elsevier)

[5, 6]. Figure 6.1 illustrates three types of the LC structures. In nematic LCs, the rigid rodlike groups show an orientational order but lack long-range order for their positions (Fig. 6.1a), which has the $D_{\infty h}$ symmetry in Schoenflies notation. The long-range orientation order is characterized by the director **n**, which is a unit vector representing the principal axis of the local alignment. All smectics possess orientation order and layered structures with a well-defined interlayer spacing. Main types of smectics include smectic A (Fig. 6.1b), smectic C, and hexatic smectics. Cholesterics are LCs with molecular orientation showing regular spatial period, where some of the molecules need to be chiral to induce the alignment (Fig. 6.1c). Similarly, LCEs can also be classified into nematic, cholesteric, smectic elastomers [4]. The nematic LCE with uniform spatial orientation is the most widely studied type among them.

One unique characteristic of LCEs, which is in common with LCPs, is related to the characteristics of polymeric architecture. When a polymer with incorporated mesogenic groups enters an LC phase under proper conditions, the polymer conformation is strongly affected by the LC alignment [7]. Therefore, the spatial orientation of mesogenic groups and the average shape of polymer backbone are highly correlated with each other. It was indicated by de Gennes [8] that when LCPs form cross-linked networks, the anisotropic average shape of polymer chains is fixed by the space restriction. For cross-linked networks, the correlation between the average shape and the orientational order can introduce many new physical properties related to these structural characteristics [1–4]. The polymer chains take elongated shape when the mesogens orient in the order phase, while in the isotropic phase, they recover a random coil form driven by entropy. For the simplest nematic LCEs, the nematic–isotropic phase transition can cause a reversible change in the average shape of the polymer backbone from elongated to coiled shapes. Therefore, at macroscopic level, a reversible shape deformation of an elastomer object is achieved through the conformation variation.

LCEs were synthesized in a pioneering work based on side-chain LC polysiloxane by introducing the lightly cross-linked networks [9]. Years later, LCEs with mesogenic groups in main chain (polymalonates), side chain (polyacrylates and polymethacrylates), and combined main chain/side chain were prepared in isotropic solutions by reactions of linear LC polymers [10]. These cross-linked polymers show the liquid crystalline phase structures as the linear polymers possess. Phase transition temperatures remain nearly unchanged when the degree of cross-linking is low ($\leq$2 mol%). The obtained LCEs are very soft above T_g and can be easily stretched for more than 100 %. For higher degrees of cross-linking, the phase transition temperatures and properties of the cross-linked polymers are altered. On the other hand, when the degree of cross-linking is high enough, the densely cross-linked systems belong to another interesting category named LC thermosets [11].

Without special treatment during the fabrication, LCEs always exist in a polydomain form, where $\mathbf{n}(\mathbf{r})$ is not a continuous function of position $\mathbf{r}$. For LCs, *disclinations* and *walls* are typical defects, which can be observed under polarized optical microscope. If these defects are fixed by cross-linking, the LCEs obtained will possess polydomain textures. As a milestone in the development, a two-step cross-linking strategy has been established to obtain the *monodomain* elastomers [12]. In the first step of the procedure, the LC polymer melt is weakly cross-linked to form a gel. It is then stretched in a uniaxial manner and the second-step cross-linking is initiated to fix the uniaxial alignment. The term, *liquid single-crystal elastomers* (LSCEs), has been coined to define the monodomain LCEs as well.

As the general well-developed strategy, monodomain LCEs can be prepared by first aligning monomers or polymers by a mechanical field and then fixing the monodomain structure by polymerization/cross-linking reactions [13]. Specifically, the orientation of nematic monomers and prepolymers can be achieved by strong magnetic field and then fixed by polymerization/cross-linking reactions. The techniques for fabricating LC display cell can also be applied to induce the monomer orientation before cross-linking. The orientation of the monomers is induced by anchoring forces of alignment layers in LC cells, for instance, where the monomer melt is filled in rubbed poly(vinyl alcohol) (PVA)-coated glass cell under vacuum [14]. Recently, it has been reported that monodomain orientation can be achieved by capillary flow to fill poly(dimethylsiloxane) (PDMS) channel with a nematic monomer mixture followed by the cross-linking [15]. However, the methods based on the surface forces are usually too weak to induce polymer orientation in bulk samples, which are suitable for monomer thin films before polymerization.

A proper way to introduce the cross-linkages is a necessary step for forming network structures. The *cross-linking* can be introduced through different ways, which is critically important to obtain the final structures of LCEs. Typically, cross-linking can be realized by using copolymers with small amount of reactive groups and adding bi- or trifunctional cross-linking agents [12], be achieved through

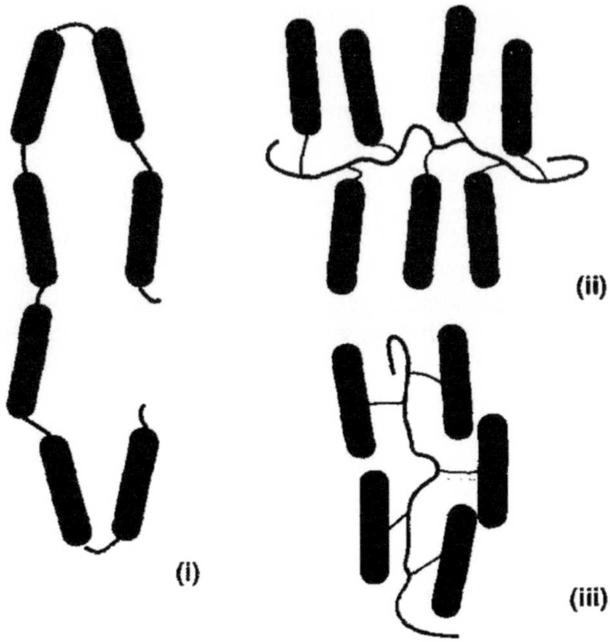

Fig. 6.2 Three types of polymer liquid crystals. (*i*) Main chain (MC) with the nematic element part of the backbone; the backbone has a prolate (elongated) shape. (*ii*) Side chain (SC) with nematic element pendant to the backbone; the backbone has an oblate (flattened) shape because of the character of the coupling of the rods to the backbone. (*iii*) Side chain with the pendant nematic elements naturally parallel to the backbone, thus creating a prolate backbone (Reprinted with permission from Ref. [19]. Copyright (1996) Elsevier)

chemical reactions initiated by lights with photoinitiators [16], or be accomplished under irradiation with gamma ray [17, 18]. These methods have been well established and efficiently used to prepare monodomain preorientated LCEs.

LCEs can be more specifically classified according to the same classification of LCPs [19], which can be classified into main chain and side chain according to the positions of mesogenic groups relative to polymer backbones (Fig. 6.2). In recent years, various LCEs have been synthesized and intensively studied, which include both side-chain and main-chain types as documented in the initial reports [9, 10]. Side-chain LCE, which is an overwhelming majority, is further classified into end-on and side-on LCEs depending on the manner of mesogenic groups linking to the backbone. On the other hand, according to mesophase structures, LCEs can be categorized as nematic elastomers [12], cholesteric elastomers [20], and smectic elastomers [21]. The LCEs have been actively explored for applications as artificial muscles, micropumps and microvalves for microfluidic devices, opto-mechanical shutters, and others [22, 23]. For more details, we refer the reader to the monograph and review articles as well as references there [4, 22, 23].

6.2 LCEs Containing Azo Functional Groups

As discussed in the previous chapters, some types of azo polymers are LCPs, for which the liquid crystallinity is induced either by the intrinsic mesogenic nature of azobenzene moieties and/or through copolymerization with monomers with mesogenic groups. One interesting property of the LC azo polymers is the photochemically induced isothermal phase transition. This phase transition behavior was first demonstrated for a series of side-chain LC polymers with different spacer lengths and molecular weights, which were doped with a small amount of 4-butyl-4′-methoxyazobenzene (BMAB) [24]. The phase transition from nematic (N) to isotropic (I) states is triggered by exposure to 366 nm UV light. This phase transition caused by the *trans*-to-*cis* isomerization is reversible when irradiated with visible light at 525 nm, which causes the transition from I to N states induced by the *cis*-to-*trans* isomerization of BMAB. More importantly, similar isothermal phase transition behavior has also been observed for LC copolymers containing both mesogenic groups and azo chromophores [24] and LC homopolymers containing mesogenic donor–acceptor azobenzene moieties [25].

Although azobenzene-containing LCEs are cross-linked LC azo polymers, which should show similarity, it was unclear at that time whether the azo polymer networks could show similar phase transition triggered by the *trans*–*cis* isomerization. Azo LCEs have been widely known and then intensively investigated until the first report published by Finkelmann et al. in 2001 [26]. Some pioneering works and major types of LCEs containing azo functional groups will be discussed below. Typically, the materials are named as azo LCEs or azobenzene-containing LCEs [4]. Other names, such as azobenzene-containing LCNs or azo LCNs, have also been frequently used by authors from the term *liquid crystal polymer networks* (LCNs) [27]. The name LCN can be more properly used to include glassy materials with the network structure in addition to the elastomers. In the following discussion, both LCEs and LCNs are used by following the original literature without further distinction.

6.2.1 Polysiloxane-Based Azo LCEs

A well-designed polysiloxane-based azo LCE was first prepared to demonstrate large and reversible shape changes induced by light irradiation [26]. The LCE has a poly(oxy(methylsilylene)) (PHMS) backbone with pendant groups mainly composed of mesogenic $M4OCH_3$ (62 %) and non-liquid crystalline $MOCH_3$ (7 %) (Fig. 6.3). A small amount of M4CN (1 %) is introduced to exploit the IR dichroism of CN groups to measure the nematic order parameter. The percentages given in bracket were calculated based on HMS units. The elastomer was synthesized by the hydrosilylation addition reaction and through the two-stage cross-linking approach. V_1 (10 %) and photoisomerizable linker AV_2 (20 %) as the first-stage and second-stage linkers were used in the reaction. The monodomain structure was obtained by applying a uniaxial mechanical extension in a swollen state just after the addition reaction.

As shown both by experiments and theoretical analysis, large contraction can be photoinduced for the polysiloxane-based azo LCE. Similar to a typical LCE, the contraction of the LCE can also be induced by temperature variation. The relative contraction versus temperature T is defined as

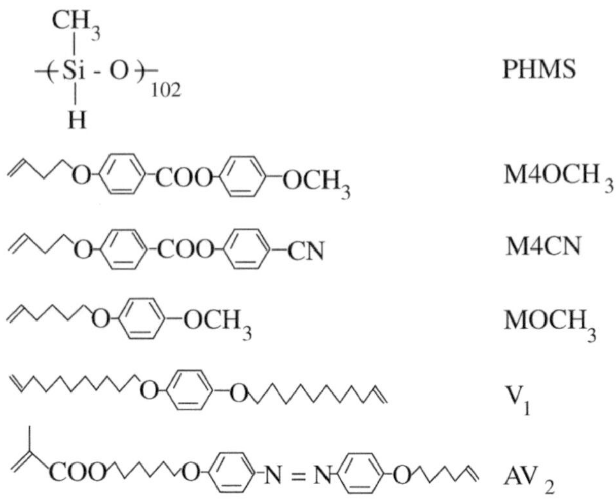

Fig. 6.3 Chemical structure of the compounds used to prepare elastomer AE4 (Reprinted from Ref. [26] with the permission of APS Publishing)

$$e_{T0}(T) = L(T)/L_0 \tag{6.1}$$

where L and L_0 are the sample lengths at the temperature T and $T_0 = 298$ K.

Due to the *trans–cis* isomerization, the shape change from rodlike to bent is induced for the azobenzene moieties in the LCE upon the light irradiation. When irradiated with UV light, the azo LCE show gradual contraction with time t increase. Figure 6.4 shows the contraction fraction of the elastomers, $1-e_{T0}(t)$, plotted against the time when exposed to UV irradiation at different temperatures. The fractional contraction shows the dependence on the temperature. The elastomer at 313 K can achieve the same contraction 22 % as it experiences on heating from 313 K to the isotropic state. Upon the light irradiation, the elastomer at 298 K exhibits a 17 % contraction, which is smaller relative to the 24 % change found on heating to isotropy. The inset in the figure shows the recovery of the elastomer at 298 K, after the illumination is switched off. These optical effects are large and comparable to thermomechanical effects.

The results given in Fig. 6.4 have been explained by considering the order parameter change caused by the light irradiation when azobenzene moieties change from a mesogenic group to non-mesogenic one [26]. As a simple version of this explanation, the molecular interaction parameter in Maier–Saupe theory is reduced as the concentration of *cis* isomer ($\Delta\rho$) increases, which is defined as the concentration difference of the *trans* isomer of the azobenzene moieties before and after light irradiation, i.e., $\Delta\rho = \rho_0 - \rho$. Therefore, the relative contraction at the time t can be correlated with T through the following equation:

$$e_{T0}(t) = e_{T0}[T(t)] = e_{T0}\left[T_0 + \frac{2T_{\mathrm{ni}}\Delta\rho(t)}{\rho_0}\right] \tag{6.2}$$

where T_{ni} is the nematic–isotropic transition temperature.

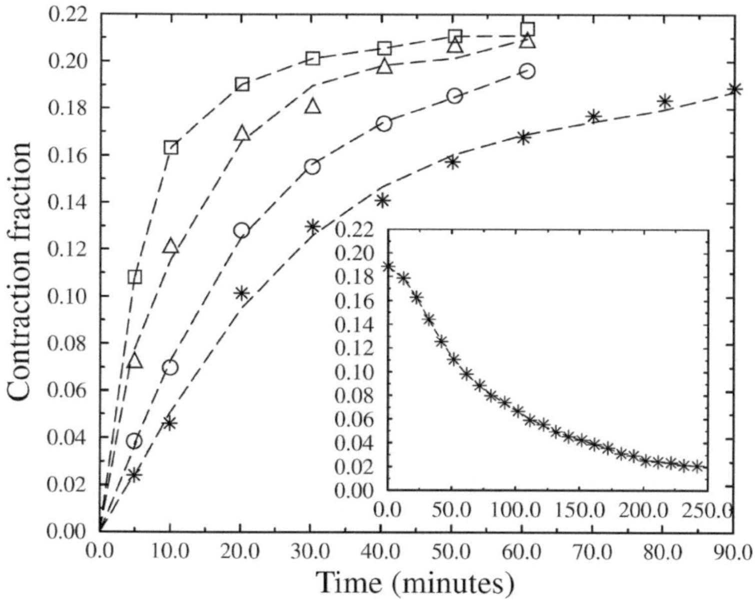

Fig. 6.4 Contraction fraction, $1-e_{T0}(t)$, versus time exposed to UV radiation for $T_0 = 298$ K (*), 303 K (o), 308 K (Δ), and 313 K (□). *Dashed lines* are guides for the eye. Inset: Recovery of the contraction for the 298 K elastomer after the 90 min of illumination (Reprinted from Ref. [26] with the permission of APS Publishing)

A thorough investigation has been performed by using a series of monodomain nematic liquid crystal elastomers containing different proportions of photoisomerizable mesogenic moieties [28]. The LCE samples were obtained by the hydrosilylation of the terminal vinyl group in the mesogenic moiety with the Si–H bond of the polysiloxane chain. The two-step cross-linking technique was used to obtain the monodomain samples, where a uniaxial stress was applied after the first stage of cross-linking and then the monodomain nematic alignment was fixed by the second-step reaction. The study shows that when *trans–cis* isomerization is triggered by UV light irradiation, the azo chromophores turn from a rodlike to a kinked shape. Depending on the proportion and position of the photosensitive groups in the cross-linked polymer network, the LCE samples exhibit different types and magnitudes of the responses. The contraction caused by the photoisomerization is attributed to the destabilization of the nematic phase and reduced order parameter. The reduction in the orientational order parameter can result in a significant change in the macromolecular shape to produce a large-scale uniaxial contraction. As a result, the effect of UV light irradiation is dramatically enhanced near the critical temperature for the nematic to isotropic transition.

6.2.2 Polyacrylate-Based Azo LCEs

This type of LCEs has been initially developed by Ikeda's group to demonstrate the photoinduced bending [29]. The LCE films were prepared by thermal polymerization of azobenzene-containing monomer in the existence of cross-linker with a ratio of 9:1 mol/mol (Fig. 6.5). The monomer 1 (A6AB2) possesses the monotropic nematic phase, which only shows orientational order in temperature range from 95 to 91 °C when cooling from an isotropic phase. The cross-linker (monomer 2, DA6AB) shows no mesomorphic phase behavior. The polymerization was performed at a temperature 5 °C higher than the melting temperature of the mixture. During the polymerization process, an isotropic-nematic phase transition occurred to give the LCE films.

As presented in Fig. 6.5, upon irradiation with linearly polarized UV light (366 nm), the LCE film shows the precisely controlled bending behavior. The film always bends toward the light-coming side with the bending parallel to the direction of the light polarization. When exposed to the visible light, the bent film can completely recover the original flat form. The mechanism of the deformation has been given by considering the domain structure by the authors. Distinct from the monodomain LCEs, the films are polydomain with a random macroscopic orientation. The polarized light can be most efficiently absorbed by the azobenzene moieties in the light polarization direction, which induces the reorientation of the azo groups. As the UV light can only penetrate the surface to a limited depth, the photoinduced orientation and volume contraction at the surface layer cause the film bending.

The polymerization of A6AB2 and DA6AB initiated by (1,1′-azobis(cyclohexane-1-carbonitrile) has also been carried out in a glass cell coated with polyimide alignment layers [30]. The directionally rubbed polyimide layers were used to induce the orientation of the LC monomers, which was then stabilized by the polymerization/cross-linking reaction. This process can produce the films with the orientational order parameters in the range from 0.062 to 0.120, which increases with the cross-linking density. As an alternative, the LCE films have been prepared from similar azo monomers and cross-linker by photopolymerization initiated with a photoinitiator [31]. The polymerization typically takes 2 h under the irradiation with a high-pressure mercury lamp (500 W) through glass filters. The photoinduced bending can be demonstrated in solvent and in air. For the former, the film is placed horizontally in toluene; for the latter, the film is heated to 90 °C. Upon exposure to unpolarized UV light at 366 nm, the films bend toward the irradiated side along the rubbing direction. In the case of the films placed in toluene (i.e., in LC gel state), the bending can be completed in 20 s at the light intensity of 3 mW/cm^2. When irradiated with the unpolarized visible light, the bent films can completely recover the original form. The bending of the films along the rubbing direction of the alignment layers has been attributed to preferentially aligned azobenzene moieties along the rubbing direction. The decrease in alignment order of azobenzene moieties produced by light irradiation results in the anisotropic volume contraction in the surface region, which causes the anisotropic bending along the rubbing direction [30].

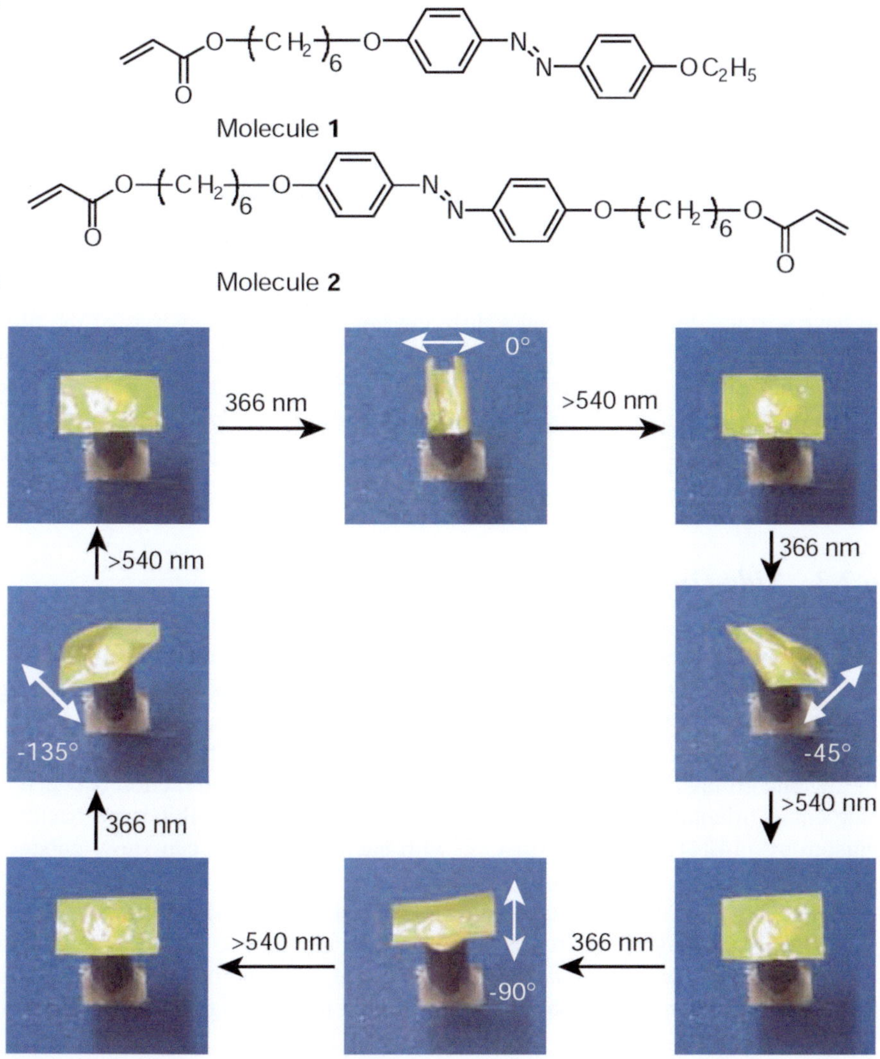

Fig. 6.5 (**a**) Chemical structures of the liquid crystal monomer (molecule 1) and cross-linker (molecule 2) used for preparation of the film. (**b**) Photographic frames of the film bending in different directions in response to irradiation by linearly polarized light of different angles of polarization (*white arrows*) at 366 nm and being flattened again by visible light longer than 540 nm. The flat film (4.5 mm × 3 mm × 7 μm) lays on a copper stick fixed to a copper plate; a stage under the plate was set at 85 °C to control the temperature of the film, which was covered by a piece of *blue paper*. The bending time for the four different bending directions was within 10 s; when the light intensity of 366 nm linearly polarized light was 3.5 mW cm^{-2} after exposure to visible light longer than 540 nm (547 nm, 24.2 mW cm^{-2}; 577 nm, 26.8 mW cm^{-2}), the bent film reverted to the flat state in about 10 s (Reprinted by permission from Macmillan Publishers Ltd: [Nature] Ref. [29], copyright (2003))

Another distinctive approach has been developed to prepare LCN showing large amplitude light-induced motion [32]. An azobenzene-containing diacrylate monomer (A6MA) was blended with an LC diacrylate monomer (C6M), where the A6MA had proportions between 2 wt % and 20 wt % in the mixture. Structures of the similar monomers can be seen in the following section (Sect. 6.5.4 and Fig. 6.16 there). A trace amount of LC756 chiral dopant was also added to ensure the formation of exclusively right-handed twisted samples. Then, the LC mixture with the twisted alignment mode was photopolymerized by using Irgacure 819 (2 wt %) as the initiator. This type of the twisted networks is heavily cross-linked and shows elastic moduli on the order of 1 GPa. When such samples are cut to have surface directors making a different angle with the long axis of the films, uniaxial bending or helical coiling deformation modes can be induced by UV light irradiation. Such a heavily cross-linked system should be more properly termed liquid crystal polymer networks (LCNs). To simplify the presentation, no strict distinction will be made in the following sections.

6.2.3 Side-On Nematic Elastomer

The abovementioned LCEs all contain mesogenic groups in an end-on manner, i.e., the rodlike groups are attached to the polymer backbone at one end through covalent bonds. On the other hand, side-on LCE is characterized by laterally tethered mesogenic groups to the polymer main chain. By this way to introduce mesogenic groups, the side-on LCEs should have enhanced interaction between the side-chain orientation and main-chain conformation and exhibit large backbone anisotropy [14]. Therefore, the LCEs with such a structure can show enhanced shape deformation upon the light irradiation.

Side-on nematic elastomer containing azobenzene moieties has been developed to exploit this advantage [33]. In this approach, two monomers, one of them contains azo group, were prepared and used to fabricate the side-on nematic elastomer (Fig. 6.6). As the monomer has strong absorption in UV range, the polymerization

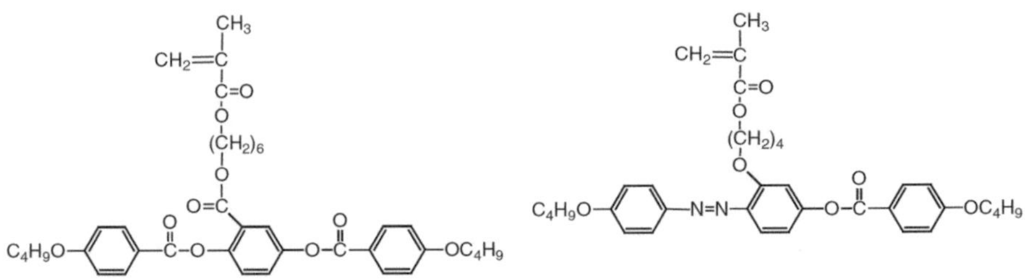

Fig. 6.6 Structures of the side-on LC monomers (Reprinted with permission from Ref. [33]. Copyright (2003) John Wiley and Sons)

was initiated by a near-infrared photoinitiator, 1,3,3,1′,3′,3′-hexamethyl-11-chloro-10,12-propylene-tricarbocyanine triphenylbutyl borate (CBC). A series of LCEs, with different ratios of the monomer 1 and monomer 2, was synthesized by the photopolymerization with 0.2–1 mol% of CBC and 10 mol% of 1,6-hexanediol dimethacrylate as the cross-linker. For the preparation, the melted mixture of the monomer, initiator, and cross-linker was filled in the cells covered with the alignment layers. After slowly cooling to the nematic phase temperature 70 °C, the polymerization was carried out upon the irradiation with the red light ($\lambda > 600$ nm). The solid thin films with a thickness about 20 μm can be obtained by this method.

The side-on nematic LCE films also show contraction upon the UV light irradiation. When irradiated with UV light (365 nm), these films contract quickly with the relative contraction up to 12–18 % (Fig. 6.7). For the azobenzene-containing LCE films, the extents of the contractions are in the same range as those induced by the temperature to cause the *N-I* transition. The contractions of the films depend on the molar percentage of the azo monomer unit and the light intensity. The LCE films with different percentage of the azo monomer unit show different contraction fraction. The reversal expansion can be observed after switching off the UV light in the time period of 30–60 min, which is attributed to the *cis*-to-*trans* back isomerization of the azo moieties. On the other hand, when irradiated with the visible light ($\lambda = 450$–470 nm) in order to induce the *cis–trans* isomerization, no such expansion can be observed. This feature is distinct from the reversible bent and unbent deformation induced by UV and visible lights for polydomain LCE films.

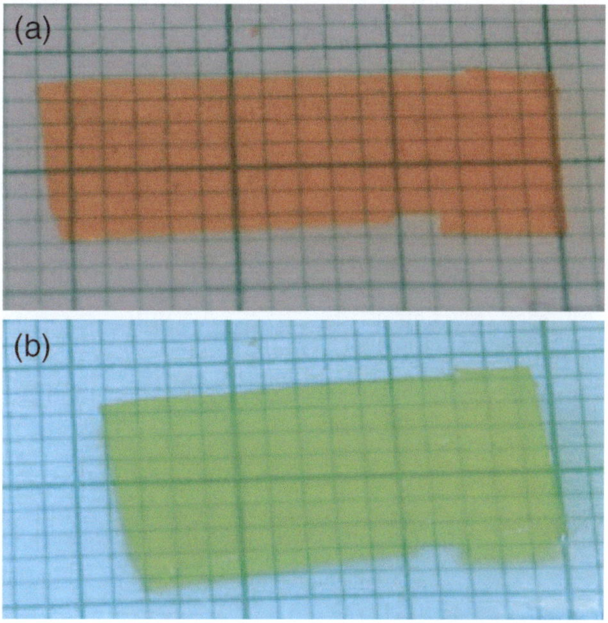

Fig. 6.7 LCE films, (**a**) before UV irradiation and (**b**) under UV irradiation for 130 s (Background is a graduated paper) (Reprinted with permission from Ref. [33]. Copyright (2003) John Wiley and Sons)

6.3 Molecular Theory of Light-Induced Deformation

The photoinduced deformation discussed above is triggered by the *trans–cis* photoisomerization of the azo chromophores. Subsequent responses, such as order–disorder transition, and chromophore orientation, will be activated by the photochemical reactions. When LCPs form cross-linked networks, the anisotropic confirmations of polymer chains are fixed by the space restriction. The anisotropic conformations can also undergo a reversible variation. The mechanical stress related to these variations then causes the light-induced deformation of LCEs. As discussed above, when irradiated with unpolarized UV light, the solid monodomain LCE films show the contraction along the orientation direction and the spontaneous recovery after switching off the light. This observation on the fundamental deformation behavior is used here to discuss molecular origin of the light-induced deformation.

6.3.1 Deformation Caused by Order–Disorder Transition

The photoinduced contraction and expansion of monodomain LCEs are typically observed for azobenzene-containing LCE thin films. Finkelmann et al. have elucidated the mechanism in their original article [26]. The rodlike *trans*-azobenzene moieties are mesogenic groups that are favorable for forming the LC phase. When the *trans–cis* isomerization occurs, the bent *cis* isomers can no longer act as the mesogenic groups, which will cause the decrease of the orientation order or even the nematic to isotropic phase transition. In this case, the photoinduced orientation order decrease is equivalent to the corresponding *order–disorder* variation caused by raising the temperature. Therefore, the effect of the illumination corresponds to an apparent elevation from the real temperature T_0 to an effective temperature T_{eff}. In the photostationary state, the current volume fraction of *trans* isomers (φ) can be given as a function of the volume fraction of the azobenzene moieties (φ_0), the illumination intensity (I), the absorption efficiency (η), and the lifetime of the *cis* isomer (τ_c).

As a simplified model [4], the rate of *trans*-to-*cis* isomerization is given as $-\eta I\varphi$ and back reaction is $(\varphi_0 - \varphi)/\tau_c$. Assuming these two rates are equal in the photostationary state, the current volume fraction of *trans* isomers can be obtained as

$$\varphi = \varphi_0 \frac{1}{1 + \tau_c \eta I} \tag{6.3}$$

and the volume fraction of *cis* isomers φ_c equal to $\varphi_0 - \varphi$ is obtained as

$$\varphi_c = \varphi_0 \frac{\tau_c \eta I}{1 + \tau_c \eta I} \approx \varphi_0 \tau_c \eta I \tag{6.4}$$

This *trans*-to-*cis* isomerization will dilute the density of the mesogenic groups in the system. As the first-order approximation, the nematic to isotropic transition temperature decreases from T_{ni} to $T_{\text{ni}}(1-\varphi_c)$, which is equivalent to raising the effective temperature to

$$T_{\text{eff}}(T_0, I) = T_0 + T_{\text{ni}} \varphi_0 \frac{\tau_c \eta I}{1 + \tau_c \eta I} \approx T_0 + T_{\text{ni}} \varphi_0 \tau_c \eta I \tag{6.5}$$

When the T_{eff} is raised to a temperature higher than T_{ni} upon the light irradiation, the nematic to isotropic phase transition occurs in the system. This type phase transition behavior can be theoretically analyzed with Maier–Saupe mean field theory or Landau–de Gennes theory. Considering the effect of the phase transition on the polymer conformation, the photoresponsive effect causes the deformation equivalent to the thermal contraction of LCE at higher temperature. Consequently, the UV illumination induces the contraction of the films along the LC orientation direction through the *trans–cis* isomerization of azobenzene moieties. For a thorough discussion about the mechanism, we refer the interested reader to the monograph by Warner and Terentjev [4].

6.3.2 Deformation Caused by Photoinduced Orientation

A quantitative model has been developed to elucidate the microscopic mechanism for the light-induced orientation to cause deformation of elastomer films [34]. The deformation is attributed to the light-induced mechanical stress originating from *reorientation* of azo chromophores with respect to the electric vector $\boldsymbol{E}$ of the light. Such light-induced reorientation of azo chromophores and its effect on the adjacent segments have been discussed in Chap. 4. This model assumes that a regular cubic network consists of Kuhn segments, and N_{ch} rodlike azo chromophores are covalently attached to the main chain of a segment. The orientation anisotropy of azo chromophores with respect to $\boldsymbol{E}$ is given by means of an effective orientation potential acting on each chromophore:

$$V(\Theta) = C I_{\text{p}} \cos^2 \Theta \tag{6.6}$$

where C is proportional constant, I_{p} is the intensity of the light, and Θ is the angle between the long axis of the azobenzene unit and $\boldsymbol{E}$. Owing to the covalent bonding between azo chromophores and chain segments, the reorientation of chromophores under light irradiation results in the reorientation of chain segments. The orientation of a Kuhn segment with respect to $\boldsymbol{E}$ can be described by the three Euler angles (Ω).

Under this condition, the effective potential of a Kuhn segment in the field of the light wave can be given:

$$U(\Omega) = CI_{\mathrm{p}}N_{\mathrm{ch}}\left\langle \cos^2\Theta(\Omega)\right\rangle \tag{6.7}$$

where $\cos^2\Theta$ is averaged over all chromophores in the Kuhn segment whose orientation is defined by Ω.

The free energy expression can then be given to include the orientation interaction of chromophores with the linearly polarized light and the entropic elasticity of network strands. The equilibrium elongation is determined by minimizing the free energy. This study shows that the photoelastic behavior of azobenzene-containing elastomers is very sensitive to their chemical structure. The elastomers can demonstrate uniaxial contraction or expansion along the polarization direction of the light, if the azo chromophores preferably orientate parallel or perpendicular to the main chains. The concave or convex bending deformations of the films can be induced depending on the contraction or expansion of the surfaces. The magnitude of the light-induced deformation decreases as the degree of cross-linking increase. Further theoretical investigation shows that photomechanical properties of azobenzene-containing elastomers with random spatial distribution of network strands can be described in a very good approximation by this regular cubic network model, if the length of the strands is appropriately chosen [35].

6.4 Photomechanical Responses and Their Mechanisms

Although the light-induced variations of the orientational order and reorientation direction have been proved to be the major causes to trigger the deformation by above models and theories, the exact responsive behavior of azo LEC objects to the light irradiation could be very complicated. The deformation modes depend on whether the film is monodomain or polydomain, whether the light is polarized or unpolarized, the preorientation manners of the samples, and light irradiation conditions, among others. Understanding these diversified modes of the photoresponsive behavior can bridge the gap between the fundamental principles and real applications.

6.4.1 Deformation Modes

As discussed above, two basic photoresponsive modes observed for azo LCEs are contraction and bending. When irradiated with unpolarized UV light, the solid monodomain LCE films show the contraction along the orientation direction, which can spontaneously recover the original shape after switching off the light.

This deformation mode is caused by the uniform decrease of the orientational order or the change of the orientation direction upon the light irradiation. On the other hand, both monodomain and polydomain LCE films can show reversible bending deformation upon light irradiation. Upon the irradiation with unpolarized UV light, the bending of the monodomain LCE films occurs along the alignment direction of the mesogens, while the bending of polydomain films can be any direction controlled by the polarization direction of the actinic light.

The uniform contraction is a basic deformation mode related to the polymer network, which is mainly observed for the LCE thin films responding to uniform photomechanical stress. Bending and other more complicated deformation modes are attributed to nonuniform elastic responses of the LCEs, which are related to photoinduced nonuniform stress and strain in films, beams, plates, and others [36]. The elastic deformation can be attributed to inhomogeneous illumination owing to the light absorption when penetrating through the solid objects. It results in the nonuniform variation in orientation order or alignment direction of the azobenzene moieties in the LCE objects. Reversible bidirectional bending of LCE films can also be induced by photoinduced orientation of the azobenzene moieties. For systems with densely loading azo chromophores, the photoinduced strain will show a nonmonotonic dependence on the intensity of the incident light. By adjusting the initial orientational states of LCEs and the light irradiation conditions, more complicated deformation modes can be observed, which can be potentially exploited for various applications.

6.4.2 Reversible Bending Deformation

Several typical reversible bending modes have been observed and attributed to the change in LC orientation order and the orientation direction under the irradiation with the light [37]. For the monodomain LCE films, the bending behavior depends on the initial alignment of photoactive azo mesogens in the films. When the alignment is parallel (homogeneous) and normal (homeotropic) to the film surface, the homogeneous films always bent toward the actinic UV light source, while the homeotropic films bent away from the light source [38]. A qualitative explanation is given by the authors as illustrated in Fig. 6.8. As the absorption of the azobenzene moieties for UV light is large, which has the extinction coefficient about $2.0 \times 10^4\,\mathrm{L\,mol^{-1}\,cm^{-1}}$, more than 99 % of the incident photons will be absorbed by the surface layer within a thickness of 1 mm. Upon exposure to actinic UV light, the orientation order change for LCE films with a high concentration of azobenzene moieties occurs more significantly in the film surface layer. Therefore, the two different types of the photoinduced bending behavior are actually equivalent to the relative surface contraction and expansion with respect to the unaffected bulk volume.

A quantitative model has been established to describe the mechanical responses to the inhomogeneous illumination [36]. It considers an LCE beam or plate of thickness w, width W, and length L (Fig. 6.9). If light incident from one side is linearly

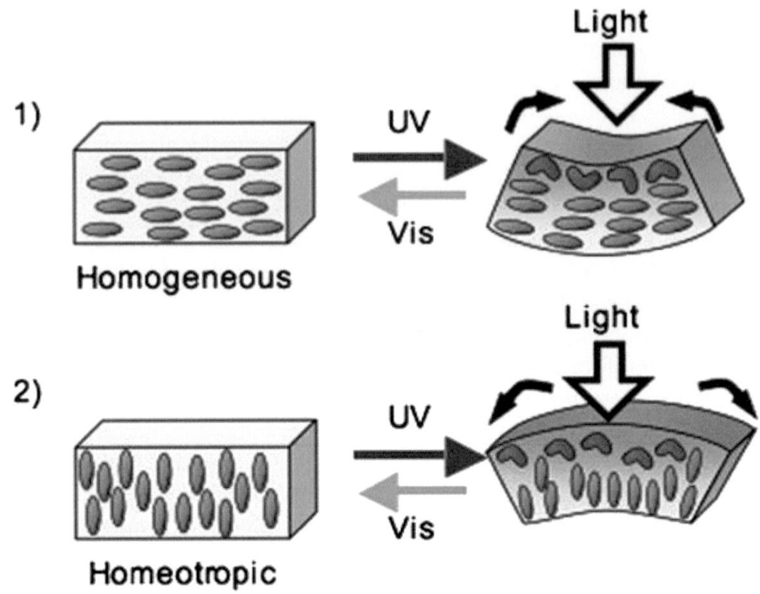

Fig. 6.8 Schematic illustration of the bending mechanism in the homogeneous and the homeotropic films (Reprinted with permission from Ref. [38]. Copyright (2006) John Wiley and Sons)

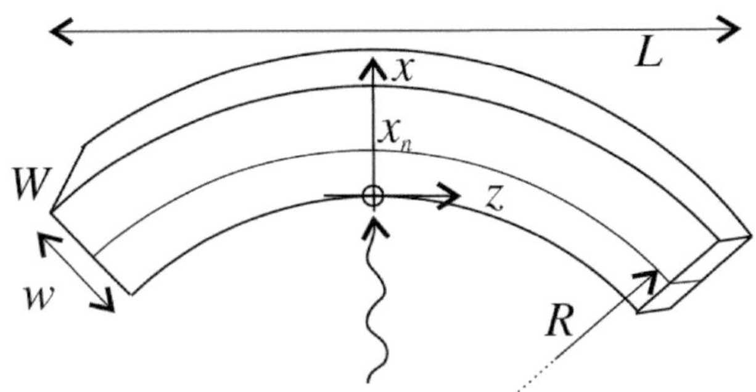

Fig. 6.9 A photobeam, initially straight and illuminated from below, contracts more where the light is less attenuated. The curvature induced is R. A neutral surface is at x_n (Reprinted from Ref. [36] with the permission of APS Publishing)

absorbed, the light intensity $I(x)$ is exponentially attenuated with the distance penetrated x so that $I(x) = I(0)\exp(-x/d)$. The attenuation length d characterizes the density of azo chromophores in the *trans*-form when coupling to the optical field. When ignoring the distinction between the current and reference configurations for considering the small strain, the axial elastic strain is then approximately proportional to the light intensity. Under this condition, the relations are established to explicitly correlate the curvature and the position of neutral planes with the intensity and penetration of the radiation. The result shows that the curvature can be tuned by altering the attenuation length, which shows the maximum relative to w. Quantitative models have also been established to describe the deformations of polydomain LCEs caused by both polarized and unpolarized lights [39, 40].

In addition to the uniformly aligned monodomain and polydomain samples, the LCN actuators can be developed by varying the orientation of the director and density of cross-linking through the thickness of the films. Submillimeter bending radii are achieved using a splayed molecular orientation [41]. Figure 6.10 is the schematic illustration of three typical molecular alignments and reaction of the actuator. Typically, films with a splayed or twisted director alignments show greater amplitude and faster bending compared with uniaxial planar systems with the same chemical composition. Films with an internal gradient in cross-linking density are prepared using a mix of mesogenic mono- and diacrylates, an optical absorber and a photoinitiator that are active at the same wavelengths [42]. A two-way bending, reversal of the bending direction, can be achieved for the films by exposure to a single continuous radiation.

Exposure of azo LCE or LCN films to blue-green light (442–514 nm) can induce reorientation of the azobenzene moieties, which has been discussed in Chap. 4. Owing to the dichroic absorbance of *trans* isomer, irradiation with polarized blue-green light results in a statistical buildup of azobenzene moieties in the direction perpendicular to the light polarization direction. Upon the irradiation with a linearly polarized laser beam in different polarization directions, the reorientation orthogonal to the beam polarization results in expansion (a) or contraction (b) of the incident surface of the film and forces its bending away or toward the laser beam, respectively [43]. Blue-green laser exposure has been used to generate forward and reverse bending of polydomain LCN cantilevers under different exposure

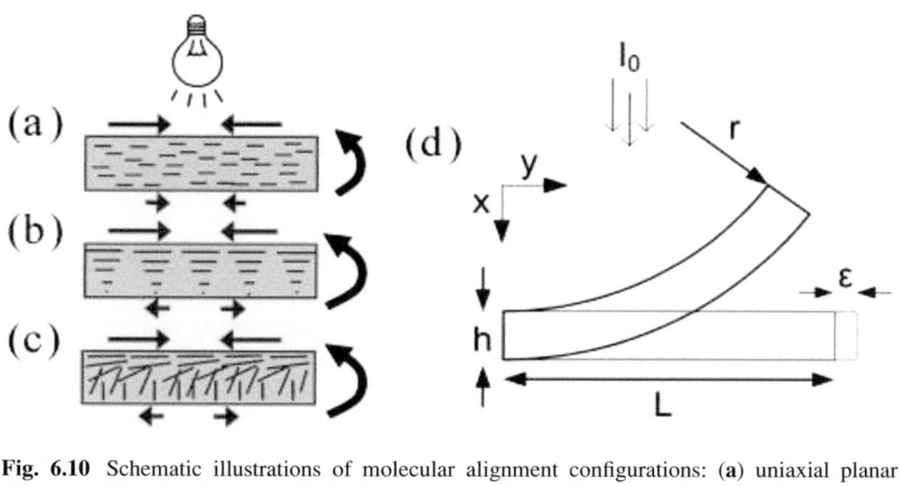

Fig. 6.10 Schematic illustrations of molecular alignment configurations: (**a**) uniaxial planar alignment, (**b**) twisted nematic (TN) alignment, (**c**) splayed alignment. (**d**) Model of a photomechanical actuator with bending radius r under incident light with intensity I_0 (Reprinted with permission from Ref. [41]. Copyright (2007) Springer)

conditions [44]. Relationship between the photomechanical response and the thermomechanical properties of the LCN material has been investigated [45]. It shows that increasing the cross-link density of the polydomain LCN will reduce the magnitude of the bending angle of the cantilevers. The equilibrium photodriven bending angle is dependent on the temperature, which is strongly related to the temperature dependence of the storage modulus of the sample. The angle-controlled bending has also been demonstrated for monodomain azo LCN cantilevers [46]. Monodomain cantilever shows a faster and larger-angle bending compared with its polydomain counterpart. The differences between the photomechanical responses of azo LCNs upon exposure to either UV and blue-green irradiation have been investigated and elucidated by spectroscopic methods [47].

6.4.3 Nonlinear and Complex Deformation Modes

Nonlinear response of polydomain elastomers has been theoretically studied by considering the rotation of the domain directors and the reduction of orientational order [48]. The model predicts that the photoinduced strain will show a nonmonotonic dependence on the intensity of the incident light. At high intensity, most domains become isotropic irrespective of their initial orientation and the sample will recover its original shape. This nonlinear response in the high dye-loaded limit is theoretically explored for a cantilever motion [49]. This investigation considers the bend of a heavily dye-doped system, where Beer's law suggests only a tiny penetration of the radiation. This opto-elastic process is attributed to the photo-bleaching and consequential deeper penetration of the light. Understanding such nonlinear absorptive effects is very important as heavily doped systems are commonly used in practical application. Theoretical analysis shows the evolution in time from initial Beer's law to a finally deeply penetrating optical profile in the heavily absorbing media [50]. The spatiotemporal decay of the Poynting flux and the nonlinear population dynamics of the photoactive molecules under intense irradiation are described by the solutions of a group of the coupled, nonlinear, partial differential equations. Study of the deep optical penetration dynamics has indicated the important role of the moving fronts of the light penetration [51].

In addition to nonlinear light absorption behavior, the bend and developed curvature depend on director field distribution. The general principles have been studied for the weak response, where bend in each direction is treated independently of that in others [52], and the strong spontaneous distortion [53]. For the weak response, a cantilever bend is caused by the asymmetry in strains about the midplane, which arises either from director distributions or from gradient in the radiation. For strong spontaneous distortion, the bend in one direction causes stretch in another direction if that is also bending and can cause suppression of one of the bends. Under this condition, more complicated deformation modes can be induced by the radiation.

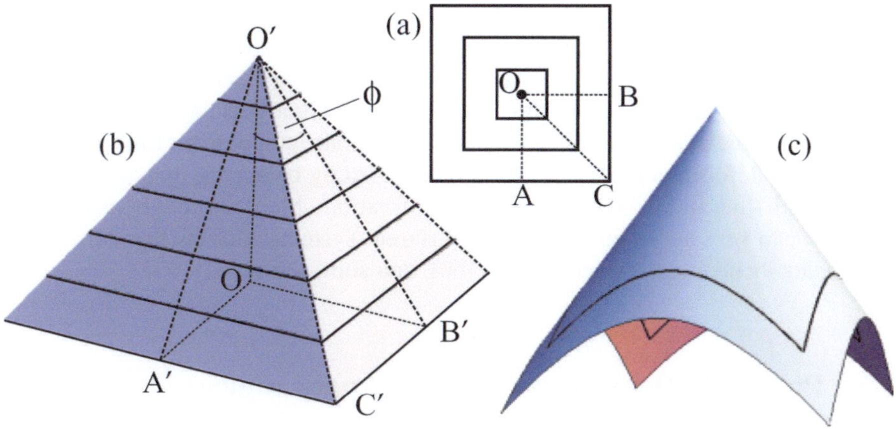

Fig. 6.11 (**a**) A square representation of a $m = 1$ defect. (**b**) Ignoring bend energy, on cooling the defect rises to being a pyramidal cone. (**c**) Relieving the bend energy of the creased edges of the pyramid yields a circular cone, where the integral lines of n are the cusped trajectories shown (Reprinted from Ref. [54] with the permission of APS Publishing)

More complicated deformation modes can be induced by exploiting the director spanning around the defects in LCE sheets, where the director $\boldsymbol{n}(\boldsymbol{r})$ is not a smooth function of the position $\boldsymbol{r}$. Disclination is a typical type of defects in nematics to specify the discontinuity in the director field [5]. A disclination is typically classified according to its strength m ($\pm$ integer or half integer). Theoretical investigation shows that a subset of texture domains in nematic glasses can be compatibly combined to yield analogs of the traditional disclination textures seen in standard liquid crystals [54]. A complex distribution of strain can be produced and concentrated at the defects. Subject to the compatibility constraints of the construction, flat sheets with topological defect can work as shape blueprinting of three-dimensional structures (Fig. 6.11). The disclination-mediated thermo-optical response in nematic glass sheets will reversibly induce three-dimensional shapes from flat sheets of the material, which has been predicted theoretically [55] and realized by using LCNs containing light-sensitive dyes [56]. Based on the above understanding, rich and diverse topographical transformations of initially flat films have been demonstrated by using photoresponsive azo LCN [57]. Photomechanical responses of films with defect strength from $-5/2$ to $+5/2$ are proved to show the various programmed patterns when irradiated with unpolarized 445 nm light.

6.4.4 Mechanics of Photoinduced Deformation

As discussed above, different deformation modes can be induced by light irradiation. In all these cases, the deformation is triggered by the *trans–cis* photoisomerization of the azobenzene moieties, which are closely dependent on the polymer and network structures. If the stress and strain induced in the LCEs can be reasonably correlated with the local light intensity, the problem can be treated as a classical mechanical problem to analyze the stress–strain relationship for elastic materials. The original study on the photomechanical analyses has been carried out on films, beams, plates, and others [36] and on mono-/polydomain films [58]. The equilibrium deformation of the objects is obtained by balancing the photoinduced strains with the strains related to the deformation. More thorough mechanical analyses can be seen in references for different systems such as thin-film plates [59], splay–bend and twist cantilevers [60], and solid shells [61].

6.5 Prototype Devices and Possible Applications

By using azobenzene-containing LCEs, different prototype devices have been developed on the basis of the abovementioned deformation and motion modes. Using light as the driving force shows some obvious advantages such as the noncontact actuating manner, precise position control, and faster responsive rate compared with thermal actuation. Due to the tailor-ability and processibility, devices with different functions can be fabricated. Several selected examples are presented below to demonstrate the possible applications.

6.5.1 Miniature Swimming Devices

Miniature swimming devices were developed by dissolving azo dyes into an LCE sample [62]. The device with large (more than 60° bending) and fast response characteristics has been fabricated to show the "swimming" motion in response to nonuniform illumination of visible light. The monodomain LCE consists of a poly (methylsiloxane) backbone functionalized with side-chain mesogenic groups (Fig. 6.12). The transparent birefringent monodomain samples were obtained by the two-step cross-linking scheme as discussed above in Sect. 6.2. An azo dye (Disperse Orange I) was dissolved into the nematic LCE samples by swelling the network with a solution of dye in toluene. The solvent was subsequently removed through vacuum evaporation.

The photoinduced deformation of the dye-doped LCE samples was demonstrated by irradiation with green light ($\lambda = 514$ nm) from a continuous-wave Ar^+ laser (Fig. 6.13a). For a piece of the samples (5 mm × 5 mm × 0.32 mm) supported at one end, the dominant motion mode is a large and rapid light-induced bend (>45° bend angle after illumination for 60 ms at 600 mW). The axis of the cylindrical bend is parallel to the sample surface and perpendicular to the nematic director. The force developed by the dye-doped samples is remarkably fast, which has a rise time of 20 ms at 600 mW and a relaxation time of 75 ms. As a result of momentum transfer, the dye-doped LCE sample floating on water shifts away from the light

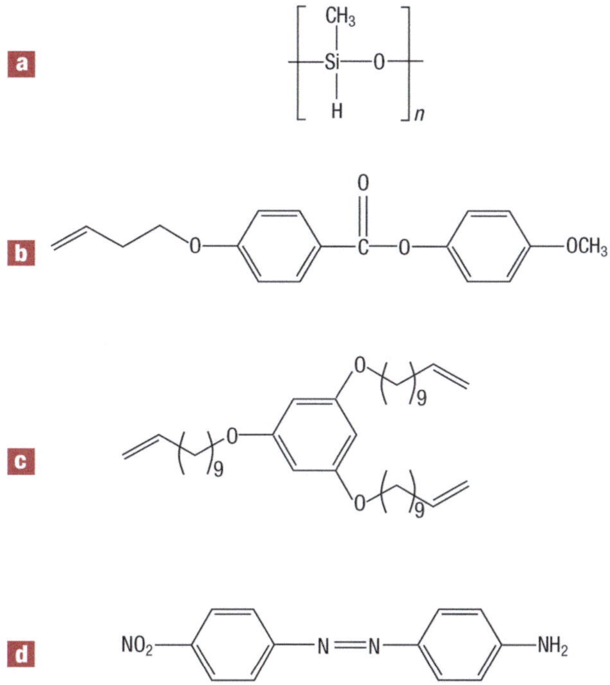

Fig. 6.12 The chemical structures of the nematic LCE and dye. (**a**) The methylsiloxane monomer backbone, (**b**) the mesogenic side chain, (**c**) the trifunctional cross-linker, (**d**) the azo dye Disperse Orange I (Reprinted by permission from Macmillan Publishers Ltd: [Nature Materials] Ref. [62], copyright (2004))

spot when illuminated with light from above. Fig. 6.13b illustrates the proposed mechanism to understand how the sample shape change and interaction with the fluid below it can cause the swimming action. Such an action can mimic that of flatfish such as skates or rays in the miniature scale.

6.5.2 Light-Driven Plastic Motors

Beyond the typical contraction/expansion and bending motions, rotational motions of azobenzene-containing LCEs and their composite materials have been demonstrated [63]. The light-driven plastic motor was fabricated by using laminated films composed of an LCE film and a flexible polyethylene (PE) sheet. The LCE films were prepared by photopolymerization of a mixture of an LC monomer and an LC diacrylate both containing an azobenzene moiety. The ratio of the monomer to cross-linker was 20/80 mol/mol, containing 2 mol% of a photoinitiator. The photopolymerization was conducted at a temperature at which the mixture exhibited a smectic phase in a glass cell coated with rubbed polyimide alignment layers. The LCE films show a glass transition temperature around room temperature and enough flexibility at this temperature, which allows the LCE films to work at room temperature in air.

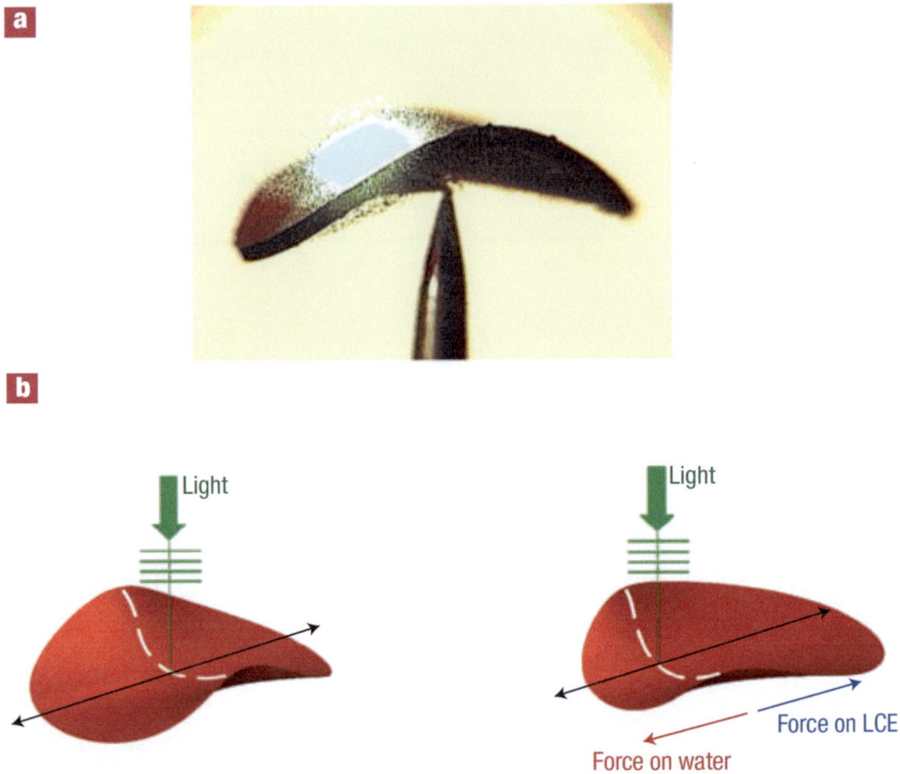

Fig. 6.13 Illustration of the mechanism underlying the locomotion of the dye-doped LCE sample. (**a**) A single video frame showing the shape deformation of an LCE sample immediately following illumination. Here the sample is supported by a pin in air and the image is recorded approximately along the nematic director. The illumination is centered on the pin. (**b**) Illustration of how the sample shape changes and hence interacts with the fluid below it. The *left* figure shows the initial deformation of the sample on illumination. The *black arrow* shows the axis of the bend, which is orthogonal to the nematic director whose direction of alignment across the sample is shown as a dashed white curve. The *right* figure shows the motion of the sample as the center of deformation moves back along the axis of the bend toward the light, and momentum is exchanged between the sample and the fluid (Reprinted by permission from Macmillan Publishers Ltd: [Nature Materials] Ref. [62], copyright (2004))

The plastic motors were fabricated by connecting a plastic belt of the LCE laminated film at the ends to work as the belt on a pulley system (Fig. 6.14). By irradiating the belt with UV light from top right and visible light from top left simultaneously, the two pulleys can be driven by the rotation of the belt in a counterclockwise direction at room temperature. The authors have proposed the mechanism of the rotation motion as follows. Upon the light irradiation, the UV light generates a local contraction at the irradiated part of the belt near the right pulley along the alignment direction of the azobenzene mesogens, which is parallel to the long axis of the belt. This contraction acting on the right pulley drives it to

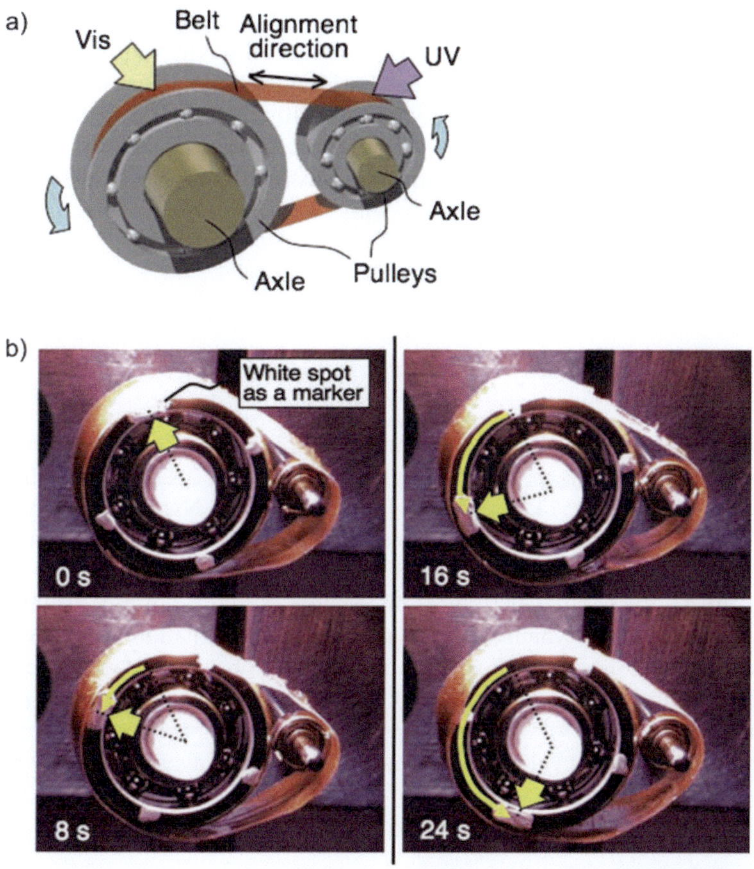

Fig. 6.14 A light-driven plastic motor with the LCE laminated film. (**a**) Schematic illustration of a light-driven plastic motor system, showing the relationship between light irradiation positions and a rotation direction. (**b**) Series of photographs showing time profiles of the rotation of the light-driven plastic motor with the LCE laminated film induced by simultaneous irradiation with UV (366 nm, 240 mWcm^{-2}) and visible light (>500 nm, 120 mWcm^{-2}) at room temperature. Diameter of pulleys: 10 mm (*left*), 3 mm (*right*). Size of the belt: 36 mm × 5.5 mm. Thickness of the layers of the belt: PE, 50 μm; LCE, 18 μm (Reprinted with permission from Ref. [63]. Copyright (2008) John Wiley and Sons)

rotate in the counterclockwise direction. At the same time, the visible light produces a local expansion at the irradiated part of the belt near the left pulley, which causes a counterclockwise rotation of the left pulley. These contraction and expansion, produced simultaneously at the different parts along the long axis of the belt, drive the pulleys to rotate around the same direction. Owing to the rotation, new parts of the belt are exposed to UV and to visible light, which enables the motor system to rotate continuously. Reverse rotation of this belt can also be induced just by altering the irradiation positions of the UV and visible light.

In addition to the rotational motion, the similar laminated LCE films can also be used for the photoinduced movements (such as inchworm walk) on the surface [64]. The inchworm walk can be induced for laminated films bent with asymmetric end shapes, irradiating with UV (366 nm, 240 mW cm^{-2}) and visible light (>540 nm, 120 mW cm^{-2}) at room temperature. Upon exposure to UV light, the film extends forward because the sharp edge acts as a stationary point, and the film retracts from the rear side by irradiation with visible light because the flat edge acts as a stationary point.

6.5.3 High-Frequency Photodriven Oscillator

The high-frequency oscillation has been demonstrated by using azobenzene-containing LCN cantilever photoactuated with an Ar^+ laser with the multi-wavelength (457, 488, 514 nm) output [27]. The azo LCNs were prepared by the thermal copolymerization of two azobenzene-containing liquid crystal monomers in the aligned nematic phase. Upon exposure to the laser beam at low intensity (0.3 W cm^{-2}), the cantilever (typical dimensions 5 mm × 1 mm × 50 μm) shows both direction bending, which depends on the relative orientation of the polarization with the director as discussed above. It is interesting to observe that when the intensity of the laser beam is increased to 0.8 W cm^{-2}, the cantilever shows an oscillation at a frequency of nearly 30 Hz (Fig. 6.15). The distance traveled by the tip of the cantilever during a half cycle is approximately 7 mm at an average velocity of nearly

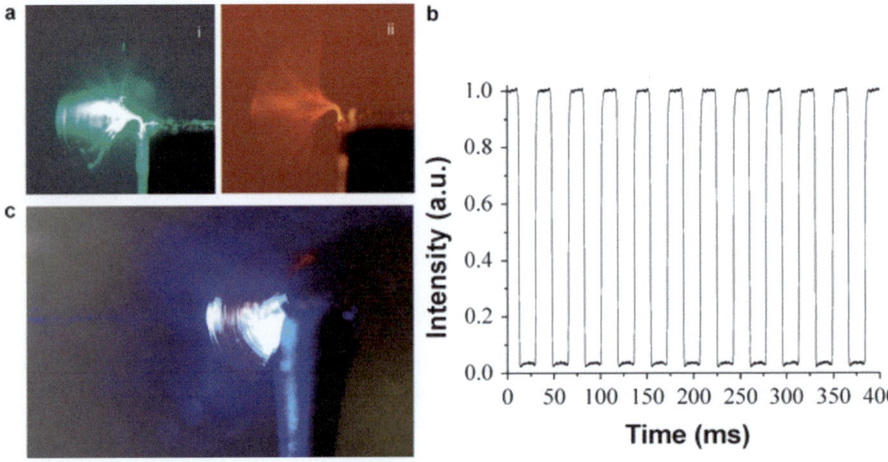

Fig. 6.15 (**a**) Oscillation of an azo LCN cantilever upon exposure to a 0.8 W cm^{-2} Ar^+ laser beam (i, color; ii, laser light filtered). The amplitude of the oscillation is greater than 170°. (**b**) The frequency of the oscillating azo LCN cantilever was measured optically and determined to be 28.4 Hz. (**c**) The oscillation can also be driven with similar intensity from a He–Cd (442 nm) laser beam (Reprinted with permission from Ref. [27]. Copyright (2008) Royal Society of Chemistry)

200 mm s^{-1} (4.8°ms^{-1}). Above the threshold intensity requirement, increasing laser intensity increases the amplitude of the oscillation from very small (<10°) to very large (170°). The cantilever thickness shows significant influences on the intensity threshold to drive the cantilever oscillation and the corresponding frequency.

By reducing the cantilever length from 5 mm to 2.7 mm, the oscillation frequency can increase from 32 to 120 Hz upon the irradiation of a focused 100 mW laser of multiple wavelengths [65]. Such increase of the experimental frequency is fully consistent with the calculated natural resonant frequency (f) for a non-damped cantilever, which depends on the cantilever length (l) following the relationship $f \sim l^{-2}$. Meanwhile, the frequency shows a linear correlation with the thickness (t), i.e., $f \sim t$. When reducing the chamber pressure from 1 atm to 0.03 atm, the amplitude of the oscillatory increases from 110° to 252°. The conversion efficiency of light energy to mechanical work is approximately calculated to be higher than 0.1 % for the system. The large increase in amplitude at low pressure evidences strong hydrodynamic loss, which can potentially be applied in a flying device with hummingbird-like beating wings. The oscillatory behavior can also be demonstrated by using a focused sunlight. By adjusting the orientation of the nematic director to the long axis of the cantilever, in-plane motion can be further extended to an out-of-plane twisting (flexural–torsional oscillation) oscillations [66]. Similar to the in-plane vibrations, the frequency is strongly correlated to the length of the cantilever, while the amplitude and threshold laser intensity for oscillation is strongly correlated to temperature.

6.5.4 Printed Artificial Cilia

Artificial cilia are nature-inspired structures, which can mimic the behavior of cilia for microbes to move through fluids. LCN cilia are fabricated using ink-jet printing technology in combination with self-organizing liquid crystal network actuators [67]. The monomers used to prepare the cilia are shown in Fig. 6.16, where two monomers containing azobenzene moieties, A3MA and DR1A, were used. A3MA and DR1A have the absorption bands with λ_{max} at 358 nm and 490 nm, respectively. The monomeric LC mixture was then deposited with the ink-jet printer, using separate cartridges for the mixtures with the two different dyes. In order to achieve the splay–bend molecular organization through the thickness of the film, the four basic processing steps were adopted to produce the modular cilia: (a) structured deposition of the PVA release layer; (b) spin-coating, curing, and buffing of the polyimide alignment layer; (c) ink-jet deposition of the monomer mixes containing DR1A and A3MA and curing; and (d) dissolving the PVA release layer. The function of the polyimide alignment layer is to warrant an upward bend–splay alignment, where the molecular director points perpendicular to the substrate at the bottom and parallel to the substrate at the top of the cilia. For practical reasons with the available printing setup, the reactive monomers were cured after printing every separate color, which produced a distinct separation of the colors. Polymerization

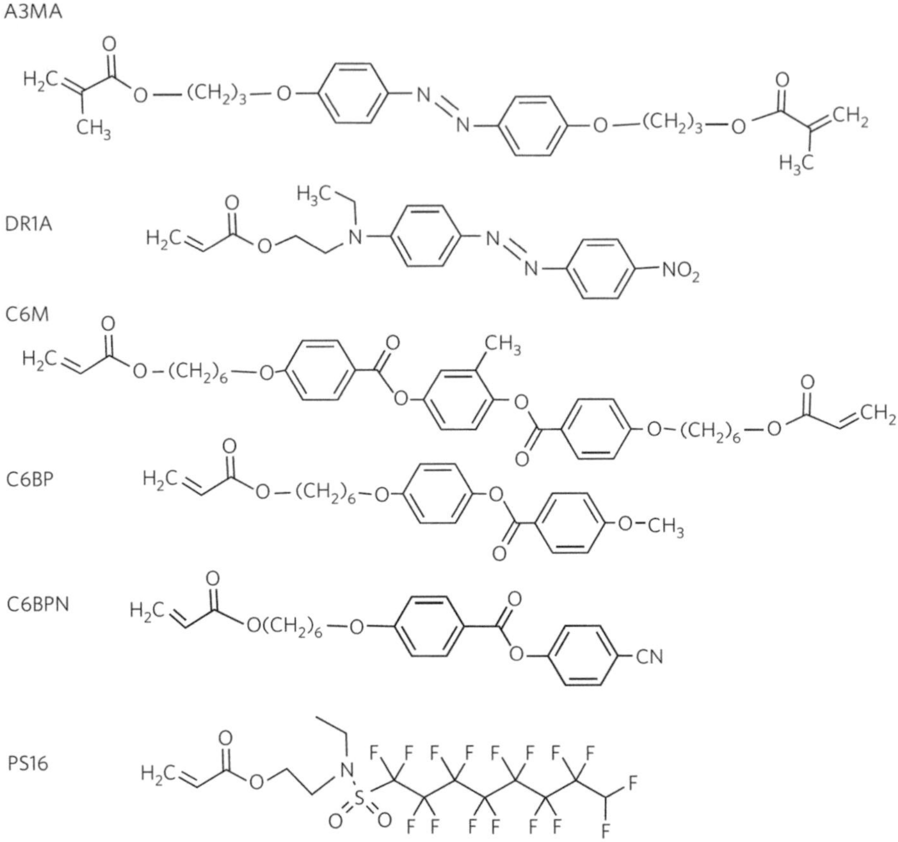

Fig. 6.16 The monomers used to form the actuators. The two photoactive monomers are A3MA and DR1A. The host matrix is built from C6M, C6BP, and C6BPN. The surfactant PS16 is used to obtain the desired alignment at the air interface (Reprinted by permission from Macmillan Publishers Ltd: [Nature Materials] Ref. [67], copyright (2009))

directly after printing helped to preserve the morphology during the exchange of a printer cartridge. As an alternative approach, the curing can also be performed after printing the full structure, which can produce cilia with a strong adhesion between the two parts.

Figure 6.17 shows the images of the microstructured cilia and their response to light. The optimal dye concentration is determined by the molar extinction of the dyes and the thickness of the film. Typically, a concentration of 4 wt% for the A3MA dye and 1 wt% for the DR1A dye can be chosen, as proved by the investigation, which is high enough to give a significant response, but low enough to allow light to penetrate a thickness of 10 μm. The cilia produced by this way can be light driven owing to the large strain gradients induced by the radiation. One obvious advantage of this approach is that by using multiple inks, microactuators with different subunits can be fabricated and selectively actuated by changing the

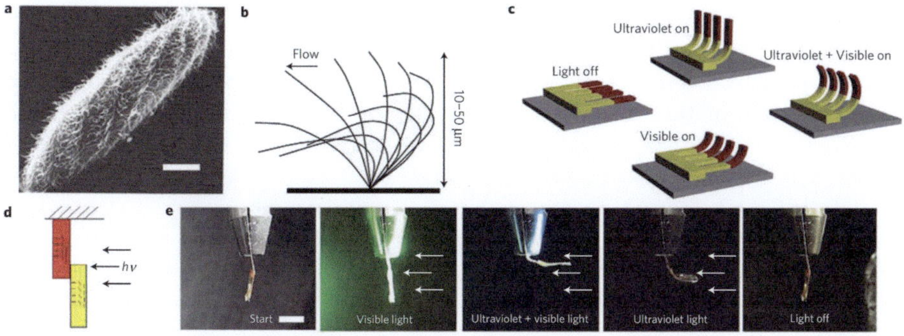

Fig. 6.17 Natural and artificial cilia and their motions. (**a**) Cilia can be found on several microorganisms, such as paramecia (scale bar 20 μm). (**b**) A paramecium uses the beating motion of the cilia, characterized by different forward and backward strokes, for self-propulsion. (**c**) Artificial, light-driven cilia produce an asymmetric motion controlled by the spectral composition of the light. (**d**) Schematic representation of the macroscopic setup, showing the orientation of the molecules. (**e**) Steady-state responses of a 10 μm thick, 3 mm wide, and 10 mm long modular liquid crystal network actuator to different colors of light (scale bar 5 mm) (Reprinted by permission from Macmillan Publishers Ltd: [Nature Materials] Ref. [67], copyright (2009))

wavelength of the light. These artificial cilia can be potentially applied for flowing and mixing in wet environments. Through the roll-to-roll fabrication, the large-scale and low-cost productions of miniaturized active polymer systems are promising for lab-on-a-chip and other applications.

6.5.5 Shape-Memory Applications

Shape-memory materials and devices have many applications in different areas [68]. Thermoresponsive shape-memory polymers are well known for their unique molecular structures to fix the temporary shape by forming physical cross-links and restore the permanent shape after heating them above a transition temperature. Although light has been exploited for various photoresponsive applications, there are few reports concerning the light-induced shape-memory polymers. One of the successful cases has been reported to use an interpenetrating network containing cinnamate groups as functional groups [69]. The function of the material is based on the efficient photoreversible [2 + 2] cycloaddition reactions of cinnamates when exposed to UV light with the wavelengths $\lambda > 260$ nm or < 260 nm. In this case, a mechanically deformed shape is partially fixed through photo-cross-linking reaction upon irradiation with UV light at $\lambda > 260$ nm, and it can restore the original shape after irradiation with UV light at $\lambda < 260$ nm.

Distinct with the above case using cinnamate as photoresponsive groups, LCN containing azo chromophores has been used to demonstrate the capability for undergoing rapid photo-fixing after a short exposure (<5 min) with 442 nm light and thermally or optically restoring the permanent shape [70]. The temporary states in both cantilever and freestanding geometries are photo-fixed by radiation with a linearly polarized 442 nm light, which possesses a mechanism similar to the cantilever deformation discussed above (Fig. 6.18). Restoring the cantilever to

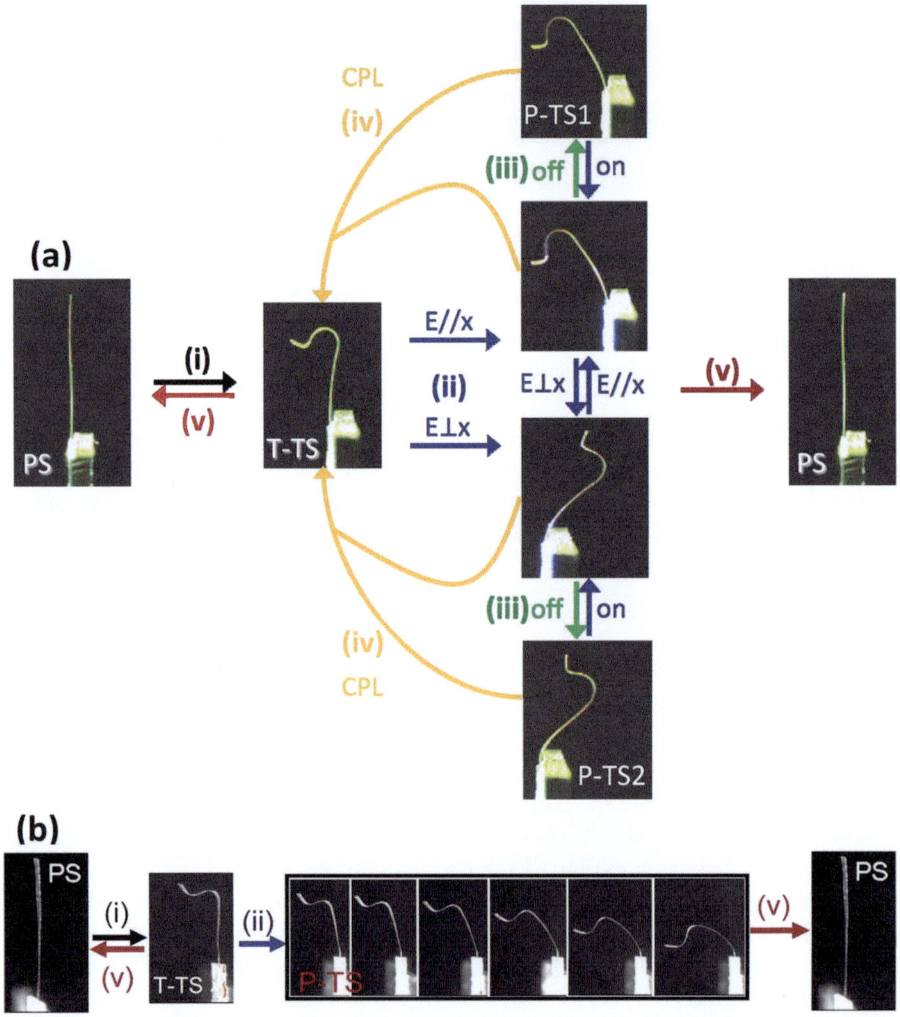

Fig. 6.18 Photoreconfiguration of a thermally fixed temporary shape in (**a**) PD-20CL (80 mW cm^{-2}, 442 nm) and (**b**) MD-20CL (20 mW cm^{-2}, 442 nm): (*i*) thermal fixing, (*ii*) photomechanical deformation with 442 nm light, (*iii*) photo-fixing after removal of 442 nm light, (*iv*) optical recovery of the thermally fixed temporary state with exposure to circularly polarized light, and (*v*) thermal recovery of permanent shape through heating through glass transition temperature. Permanent shape (PS), thermally fixed temporary shape (T-TS), and photo-fixed temporary shapes (P-TS1 and P-TS2) (Reprinted with permission from Ref. [70]. Copyright (2011) Royal Society of Chemistry)

the permanent shape (vertical position) can be achieved by exposing it to circularly polarized light (CPL), which recovers the permanent shape completely. As demonstrated as an interesting instance for possible application, a thermally fixed catapult made of the photoresponsive glassy LCN can transduce light energy into mechanical work to launch an object with a rate of 0.3 m s^{-1} by the "photo-fuel."

References

1. Finkelmann, H., Rehage, G.: Liquid-crystal side-chain polymers. Adv. Polym. Sci. **60–1**, 97–172 (1984). Springer-Verlag, Berlin
2. Zentel, R.: Liquid crystalline elastomers. Angew. Chemie. Int. Ed **28**, 1407–1415 (1989)
3. Terentjev, E.M.: Liquid-crystalline elastomers. J. Phys. Condens. Matter **11**, R239–R257 (1999)
4. Warner, M., Terentjev, E.M.: Liquid crystal elastomers. Oxford University Press, Oxford (2003)
5. de Gennes, P.G., Prost, J.: The physics of liquid crystals. Clarendon, Oxford (1993)
6. Shibaev, V., Bobrovsky, A., Boiko, N.: Photoactive liquid crystalline polymer systems with light-controllable structure and optical properties. Prog. Polym. Sci. **28**, 729–836 (2003)
7. Cotton, J.P., Hardouin, F.: Chain conformation of liquid-crystalline polymer studied by small-angle neutron scattering. Prog. Polym. Sci. **22**, 795–828 (1997)
8. de Gennes, P. G.: Possibilites offertes par la reticulation de polymers en presence d'un critical liquide (French). Phys. Lett. **28A**, 725–726
9. Finkelmann, H., Kock, H.J., Rehage, G.: Investigations on liquid-crystalline polysiloxanes 3. Liquid crystalline elastomers–a new type of liquid-crystalline materials. Makromol. Chemie. Rapid Commun. **2**, 317–322 (1981)
10. Zentel, R., Rechert, G.: Liquid crystalline elastomers based on liquid crystalline side group, main chain and combined polymers. Makromol. Chemie. **187**, 1915–1926 (1986)
11. Shiota, A., Ober, C.K.: Rigid rod and liquid crystalline thermosets. Prog. Polym. Sci. **22**, 975–1000 (1997)
12. Küpfer, J., Finkelmann, H.: Nematic liquid single-crystal elastomers. Makromol. Chemie. rapid Commun. **12**, 717–726 (1991)
13. Lacey, D., Beattie, H.N., Mitchell, G.R., Pople, J.A.: Orientation effects in monodomain nematic liquid crystalline polysiloxane elastomers. J. Mater. Chem. **8**, 53–60 (1998)
14. Thomsen III, D.L., Keller, P., Naciri, J., Pink, R., Jeon, H., Shenoy, D., Ratna, B.R.: Liquid crystal elastomers with mechanical properties of a muscle. Macromolecules **34**, 5868–5875 (2001)
15. Wei, R.B., He, Y.N., Wang, X.G., Keller, P.: Nematic liquid crystalline elastomer grating and microwire fabricated by micro-molding in capillaries. Macromol. Rapid Commun. **34**, 330–334 (2013)
16. Brehmer, M., Zentel, R., Wagenblast, G., Siemensmeyer, K.: Ferroelectric liquid-crystalline elastomers. Macromol. Chem. Phys. **195**, 1891–1904 (1994)
17. Zubarev, E.R., Talroze, R.V., Yuranova, T.I., Vasilets, V.N., Plate, N.A.: Influence of crosslinking conditions on the phase behavior of a polyacrylate-based liquid-crystalline elastomer. Macromol. Rapid Commun. **17**, 43–49 (1996)
18. Zubarev, E.R., Kuptsov, S.A., Yuranova, T.I., Talroze, R.V., Finkelmann, H.: Monodomain liquid crystalline networks: Reorientation mechanism from uniform to stripe domains. Liq. Cryst. **26**, 1531–1540 (1999)
19. Warner, M., Terentjev, E.M.: Nematic elastomers-a new state of matters? Prog. Polym. Sci. **21**, 853–891 (1996)

20. Kim, S.T., Finkelmann, H.: Cholesteric liquid single-crystal elastomers (LSCE) obtained by the anisotropic deswelling method. Macromol. Rapid Commun. **22**, 429–433 (2001)
21. Nishikawa, E., Finkelmann, H., Brand, H.R.: Smectic A liquid single crystal elastomers showing macroscopic in-plane fluidity. Macromol. Rapid Commun. **18**, 65–71 (1997)
22. Ohm, C., Brehmer, M., Zentel, R.: Liquid crystalline elastomers as actuators and sensors. Adv. Mater. **22**, 3366–3387 (2010)
23. Yang, H., Ye, G., Wang, X.G., Keller, P.: Micron-sized liquid crystalline elastomer actuators. Soft Matter **7**, 815–823 (2011)
24. Ikeda, T., Horiuchi, S., Karanjit, D.B., Kurihara, S., Tazuke, S.: Photochemically induced isothermal phase transition in polymer liquid crystals with mesogenic phenyl benzoate side chains. 2. Photochemically induced isothermal phase transition behaviors. Macromolecules **23**, 42–48 (1990)
25. Tsutsumi, O., Kitsunai, T., Kanazawa, A., Shiono, T., Ikeda, T.: Photochemical phase transition behavior of polymer azobenzene liquid crystals with electron-donating and -accepting substituents at the 4,4′-positions. Macromolecules **31**, 355–359 (1998)
26. Finkelmann, H., Nishikawa, E., Pereira, G.G., Warner, M.: A new opto-mechanical effect in solids. Phys. Rev. Lett. **87**, 015501 (2001) (4 pages)
27. White, T.J., Tabiryan, N.V., Serak, S.V., Hrozhyk, U.A., Tondiglia, V.P., Koerner, H., Vaia, R. A., Bunning, T.J.: A high frequency photodriven polymer oscillator. Soft Matter **4**, 1796–1798 (2008)
28. Hogan, P.H., Tajbakhsh, A.R., Terentjev, E.M.: UV manipulation of order and macroscopic shape in nematic elastomers. Phys. Rev. E **65**, 041720 (2001) (10 pages)
29. Yu, Y.L., Nakano, M., Ikeda, T.: Directed bending of a polymer film by light. Nature **425**, 145 (2003)
30. Yu, Y.L., Nakano, M., Shishido, A., Shiono, T., Ikeda, T.: Effect of cross-linking density on photoinduced bending behavior of oriented liquid-crystalline network films containing azobenzene. Chem. Mater. **16**, 1637–1643 (2004)
31. Ikeda, T., Nakano, M., Yu, Y.L., Tsutsumi, O., Kanazawa, A.: Anisotropic bending and unbending behavior of azobenzene liquid-crystalline gels by light exposure. Adv. Mater. **15**, 201–205 (2003)
32. Harris, K.D., Cuypers, R., Scheibe, P., van Oosten, C.L., Bastiaansen, C.W.M., Lub, J., Broer, D.J.: Large amplitude light-induced motion in high elastic modulus polymer actuators. J. Mater. Chem. **15**, 5043–5048 (2005)
33. Li, M.H., Keller, P., Li, B., Wang, X.G., Brunet, M.: Light-driven side-on nematic elastomer actuators. Adv. Mater. **15**, 569–572 (2003)
34. Toshchevikov, V., Saphiannikova, M., Heinrich, G.: Light-induced deformation of azobenzene elastomers: A regular cubic network model. J. Phys. Chem. B **116**, 913–924 (2012)
35. Toshchevikov, V.P., Saphiannikova, M., Heinrich, G.: Theory of light-induced deformation of azobenzene elastomers: Influence of network structure. J. Chem. Phys. **137**, 024903 (2012) (13 pages)
36. Warner, M., Mahadevan, L.: Photoinduced deformations of beams, plates, and films. Phys. Rev. Lett. **92**, 134302 (2004) (4 pages)
37. Ikeda, T.: Photomodulation of liquid crystal orientations for photonic applications. J. Mater. Chem. **13**, 2037–2057 (2003)
38. Kondo, M., Yu, Y.L., Ikeda, T.: How does the initial alignment of mesogens affect the photoinduced bending behavior of liquid-crystalline elastomers? Angew. Chem. Int. Ed. **45**, 1378–1382 (2006)
39. Corbett, D., Warner, M.: Polarization dependence of optically driven polydomain elastomer mechanics. Phys. Rev. E **78**, 061701 (2008)
40. Corbett, D., Warner, M.: Changing liquid crystal elastomer ordering with light – a route to opto-mechanically responsive materials. Liq. Cryst. **36**, 1263–1280 (2009)

41. van Oosten, C.L., Harris, K.D., Bastiaansen, C.W.M., Broer, D.J.: Glassy photomechanical liquid-crystal network actuators for microscale devices. Eur. Phys. J. E **23**, 329–336 (2007)
42. van Oosten, C.L., Corbett, D., Davies, D., Warner, M., Bastiaansen, C.W.M., Broer, D.J.: Bending dynamics and directionality reversal in liquid crystal network photoactuators. Macromolecules **41**, 8592–8596 (2008)
43. Tabiryan, N., Serak, S., Dai, X.M., Bunning, T.: Polymer film with optically controlled form and actuation. Opt. Express **13**, 7442–7448 (2005)
44. Serak, S.V., Tabiryan, N.V., White, T.J., Bunning, T.J.: Azobenzene liquid crystal polymer-based membrane and cantilever optical systems. Opt. Express **17**, 15736–15746 (2009)
45. Lee, K.M., Koerner, H., Vaia, R.A., Bunning, T.J., White, T.J.: Relationship between the photomechanical response and the thermomechanical properties of azobenzene liquid crystalline polymer networks. Macromolecules **43**, 8185–8190 (2010)
46. White, T.J., Serak, S.V., Tabiryan, N.V., Vaiaa, R.A., Bunning, T.J.: Polarization-controlled, photodriven bending in monodomain liquid crystal elastomer cantilevers. J. Mater. Chem. **19**, 1080–1085 (2009)
47. Lee, K.M., Tabiryan, N.V., Bunning, T.J., White, T.J.: Photomechanical mechanism and structure-property considerations in the generation of photomechanical work in glassy, azobenzene liquid crystal polymer networks. J. Mater. Chem. **22**, 691–698 (2012)
48. Corbett, D., Warner, M.: Nonlinear photoresponse of disordered elastomers. Phys. Rev. Lett. **96**, 237802 (2006) (4 pages)
49. Corbett, D., Warner, M.: Linear and nonlinear photoinduced deformations of cantilevers. Phys. Rev. Lett. **99**, 174302 (2007) (4 pages)
50. Corbett, D., van Oosten, C.L., Warner, M.: Nonlinear dynamics of optical absorption of intense beams. Phys. Rev. A **78**, 013823 (2008) (4 pages)
51. Corbett, D., Xuan, C., Warner, M.: Deep optical penetration dynamics in photobending. Phys. Rev. E **92**, 013206 (2015) (5 pages)
52. Warner, M., Modes, C.D., Corbett, D.: Curvature in nematic elastica responding to light and heat. Proc. R. Soc. A **466**, 2975–2989 (2010)
53. Warner, M., Modes, C.D., Corbett, D.: Suppression of curvature in nematic elastic. Proc. R. Soc. A **466**, 3561–3578 (2010)
54. Modes, C.D., Warner, M.: Blueprinting nematic glass: Systematically constructing and combining active points of curvature for emergent morphology. Phys. Rev. E **84** (2011) (8 pages)
55. Modes, C.D., Bhattacharya, K., Warner, M.: Disclination-mediated thermo-optical response in nematic glass sheets. Phys. Rev. E **81**, 060701 (2010) (4 pages)
56. de Haan, L.T., Sánchez-Somolinos, C., Bastiaansen, C.M.W., Schenning, A.P.H.J., Broer, D. J.: Engineering of complex order and the macroscopic deformation of liquid crystal polymer networks. Angew. Chem. Int. Ed. **51**, 12469–12472 (2012)
57. McConney, M.E., Martinez, A., Tondiglia, V.P., Lee, K.M., Langley, D., Smalyukh, I.I., White, T.J.: Topography from topology: Photoinduced surface features generated in liquid crystal polymer networks. Adv. Mater. **25**, 5880–5885 (2013)
58. Dunn, M.L.: Photomechanics of mono- and polydomain liquid crystal elastomer films. J. Appl. Phys. **102**, 013506 (2007) (7 pages)
59. Dunn, M.L., Maute, K.: Photomechanics of blanket and patterned liquid crystal elastomer films. Mech. Mater. **41**, 1083–1089 (2009)
60. Modes, C.D., Warner, M., van Oosten, C.L., Corbett, D.: Anisotropic response of glassy splay-bend and twist nematic cantilevers to light and heat. Phys. Rev. E **82**, 041111 (2010) (8 pages)
61. Modes, C.D., Warner, M.: Responsive nematic solid shells: Topology, compatibility, and shape. Eur. Phys. Lett. **97**, 36007 (2012) (4 pages)
62. Camacho-Lopez, M., Finkelmann, H., Palffy-Muhoray, P., Shelley, M.: Fast liquid-crystal elastomer swims into the dark. Nat. Mater **3**, 307–310 (2004)
63. Yamada, M., Kondo, M., Mamiya, J.I., Yu, Y.L., Kinoshita, M., Barrett, C.J., Ikeda, T.: Photomobile polymer materials: Towards light-driven plastic motors. Angew. Chem. Int. Ed. **47**, 4986–4988 (2008)

64. Yamada, M., Kondo, M., Miyasato, R., Naka, Y., Mamiya, J.I., Kinoshita, M., Shishido, A., Yu, Y.L., Barrett, C.J., Ikeda, T.: Photomobile polymer materials – various three-dimensional movements. J. Mater. Chem. **19**, 60–62 (2009)
65. Serak, S., Tabiryan, N., Vergara, R., White, T.J., Vaia, R.A., Bunning, T.J.: Liquid crystalline polymer cantilever oscillators fueled by light. Soft Matter **6**, 779–783 (2010)
66. Lee, K.M., Smith, M.L., Koerner, H., Tabiryan, N., Vaia, N.A., Bunning, T.J., White, T.J.: Photodriven, flexural-torsional oscillation of glassy azobenzene liquid crystal polymer networks. Adv. Funct. Mater. **21**, 2913–2918 (2011)
67. van Oosten, C.L., Bastiaansen, C.W.M., Broer, D.J.: Printed artificial cilia from liquid-crystal network actuators modularly driven by light. Nat. Mater. **8**, 677–682 (2009)
68. Osada, Y., Matsuda, A.: Shape memory in hydrogels. Nature **376**, 219 (1995)
69. Lendlein, A., Jiang, H.Y., Jünger, O., Langer, R.: Light-induced shape-memory polymers. Nature **434**, 879–882 (2005)
70. Lee, K.M., Koerner, H., Vaia, R.A., Bunning, T.J., White, T.J.: Light-activated shape memory of glassy, azobenzene liquid crystalline polymer networks. Soft Matter **7**, 4318–4324 (2011)

Index

X. Wang, *Azo Polymers*, Soft and Biological Matter,
DOI 10.1007/978-3-662-53424-3

O

P

S

T

W

MIX
Papier aus verantwortungsvollen Quellen
Paper from responsible sources
FSC® C105338

If you have any concerns about our products,
you can contact us on
ProductSafety@springernature.com

In case Publisher is established outside the EU,
the EU authorized representative is:
Springer Nature Customer Service Center GmbH
Europaplatz 3, 69115 Heidelberg, Germany

Printed by Libri Plureos GmbH
in Hamburg, Germany